Cognitive Pluralism

Cognitive Pluralism

Steven Horst

The MIT Press
Cambridge, Massachusetts
London, England

This book was set in Stone Serif Std by Toppan Best-set Premedia Limited.

Library of Congress Cataloging-in-Publication Data is available.

ISBN: 978-0-262-03423-4 (hc), 978-0-262-55261-5 (pb)

For Doretta

Contents

Preface

To the best of my recollection, the idea of cognitive pluralism first occurred to me during an NEH Summer Institute on Meaning held at Rutgers University in 1993. Some initial explorations of the theory found their way into two previous books, *Beyond Reduction* (2007) and *Laws, Mind, and Free Will* (2011), where I presented cognitive pluralism as a way of explaining disunities of scientific understanding. Although the main arguments of those books did not depend on cognitive pluralist speculations, many readers made it clear that they found cognitive pluralism to be the books' most interesting and engaging thesis; and I had intended, even before publishing them, to write a stand-alone work on cognitive pluralism once they were complete. Between the time it took to produce those books and the demands of academic life, it has taken quite a while to bring this book to completion. Looking back, I see that much of the early drafting was done during a sabbatical supported by an NEH Fellowship in 2006–2007. It was not until a subsequent sabbatical in the fall of 2014 that I was able to pull together an enormous amount of draft material written piecemeal over the years and compose the remaining chapters. The anonymous referees who read the prospectus confirmed my suspicion that what I had pulled together was too large for one book, and the volume you now have before you is the result of separating out the material on cognitive architecture, epistemology, and semantics found here from further discussions of metaphysics, which will be reframed into a separate volume.

During the writing process, the book went through a number of different visions of its scope and intended audience. Several chapters, particularly ones aimed at introducing material such as the history of globalist and localist views in neuroscience and the attempts of some historically important philosophers to deal with questions about unities of mind and understanding, did not make the final cut. These are not, for the most part, things that could be published as professional articles, but might be of interest to the educated public, as well as fellow academicians. I hope, as time permits, to make some of these "orphans" presentable and make them available to interested readers at my website at https://wesleyan.academia.edu/StevenHorst and eventually at an online noetic orphanage of their own.

Over the years, I have had many fruitful interchanges at talks and conferences, as well as in print, concerning some of the ideas that have found their way into this book. It is with no small fear of leaving out some who deserve greater credit that I mention any in particular, but I am especially mindful of the occasions afforded by visits to the Society for Mind-Matter Research, the University of Stirling, Universität Münster, the Shalem Center, the Society for Philosophy and Psychology, the Ian Ramsey Centre for Science and Religion at Oxford, Mount Holyoke College, Elizabethtown College, the Boston Colloquium for Philosophy of Science, and the University of California campuses at Riverside, Santa Cruz, and San Diego. I would also like to thank Rob Cummins, Jay Garfield, Yoram Hazony, Nancy Cartwright, and Joseph Rouse, who have in various ways been supportive of this project over the years. Particular thanks go to David Danks, who read and commented on much of the manuscript; Michael Silberstein, who has on several occasions engaged issues about cognitive pluralism in print; and William Seager and Harald Atmanspacher, who commented, along with Silberstein, on a target article (Horst 2014) on cognitive pluralism while I was pulling the book together. Philip Laughlin at MIT Press has been the very model of an acquisitions editor throughout the publication process. Thanks also to a group of Wesleyan undergraduates who partook in a seminar on cognitive pluralism that included draft material for this book.

I should also like to extend my deep personal gratitude to Doretta Wildes, not only for being understanding on all the occasions when I absented myself when in the grip of an idea that had to be written down while the muse still sang, but also for all the other fine things she has brought into my life.

May 14, 2015 (Feast of the Ascension)
Middletown, Connecticut

I From the Standard View to Cognitive Pluralism

1 Introduction: Beliefs, Concepts, and Mental Models

When I was an undergraduate, one of the books that had the greatest impact on my thinking was J. L. Austin's (1962) *How to Do Things with Words*. That book, based on his 1955 William James Lectures at Harvard, was the seminal work in speech act theory, which became my first philosophical love as a student. I am not going to talk about speech act theory in this book. Rather, I mention Austin and his book as a kind of model for what I would like to accomplish here, which is not so much to engage one of the "big contemporary debates" in philosophy—or even those of greater longevity—nor to say that some such debate is all balderdash, but rather to draw attention to some things that have not received much philosophical notice, even though they are right under our noses.

Here is the opening text of Austin's lectures:

> What I shall have to say here is neither difficult nor contentious; the only merit I should like to claim for it is that of being true, at least in parts. The phenomenon to be discussed is very widespread and obvious, and it cannot fail to have been already noticed, at least here and there, by others. Yet I have not found attention paid to it specifically.
>
> It was for too long the assumption of philosophers that the business of a "statement" can only be to "describe" some state of affairs, or to "state some fact," which it must do either truly or falsely. Grammarians, indeed, have regularly pointed out that not all "sentences" are (used in making) statements: there are, traditionally, besides (grammarians') statements, also questions and exclamations, and sentences expressing commands or wishes or concessions. And doubtless philosophers have not intended to deny this, despite some loose use of "sentence" for "statement." (Austin 1962, 1)

Austin's main point here was that, in giving so much exclusive attention to statements, and even more narrowly to one fact about them—the fact that they can be true or false—philosophers of his era had ended up largely ignoring all the other things that might potentially be of interest about language, some of which, in the end, might have important consequences even for our understanding of statements and their truth values.

My point is in a similar vein. Briefly, it might be put something like this:

Philosophers have placed particular emphasis on a number of topics about thoughts and language, each of which is in itself quite respectable: the semantic values of concepts and words, the truth conditions of judgments and sentences, the justification of beliefs, and the forms of inference that are truth preserving. Comparatively little attention has been given to a topic of equal importance, one that might in the end have important implications for theories of semantics, truth, epistemology, and reasoning as well. One name we might give this topic is *understanding*, though what I mean by that will have to emerge from the rest of the book.

The ways we frame our questions in different areas of philosophy—semantics, truth theory, epistemology, logic—tend to force the theories designed to answer them into particular molds. Questions about semantics focus our attention on word-sized units: actual words in a natural language, predicates in a logically regimented language, and individual concepts. Questions about truth focus our attention on sentence-sized units: sentences, propositions, judgments, and beliefs. Questions about logic and justification focus our attention on argument-sized units. Moreover, these three differently sized units are related compositionally: Arguments are formed out of sentence-sized units inserted into the slots of logical argument forms. Sentence-sized units, in turn, are composed of word-sized units inserted into slots of grammatical or logical forms. Units of each of these sizes occur in language, thought, and logic. They are clearly there to be found in both natural languages and logically regimented systems, and I think it not psychologically unrealistic to say that they are important units of thought as well.

I do not object to treating word-sized, sentence-sized, and argument-sized units as being fundamentally important to semantics, truth, epistemology, and logical reasoning; indeed, I think some of the questions of those disciplines can be satisfied *only* by theories framed in terms of such units, even if those theories present an *idealized* picture of thought (and even of language). My concern, rather, is twofold. First, *collectively*, our philosophical theories form, perhaps not quite a comprehensive *theory* of thought and language, but at least a kind of *attractor* toward a particular type of theory, or perhaps a set of constraints on what such a theory would have to look like to accommodate and unify the insights of accounts of semantics, truth, justification, and reasoning. The basic assumption, which we might call the "standard view," is that accounts of thought, understanding, and reasoning need to posit just these three types of units: word-sized concepts, sentence-sized intentional states, and argument-sized inferences—a three-tiered picture of the elements of our cognitive architecture whose predominant features are conspicuously modeled on those of a language.

Second, there are other important mental and linguistic phenomena that seem to require different types of fundamental units—at least if we are to give an account of them that is psychologically realistic—and the entrenchment of the traditional theories about the aforementioned philosophical topics may tend to blind us to these other units. The particular type of unit I am concerned with is what I shall refer to as a *mental model of a content domain*. I shall argue that mental models are fundamental units of understanding, in the sense that understanding comes in the form of models and in domain-sized chunks. We *believe* propositions, but we *understand* things such as Newtonian mechanics, the game of chess, and the etiquette for dining in a restaurant, and we do so through having mental models of those domains. A great deal of what we think of as the semantic properties of concepts is derived from properties of models in which they play a role; many inferences (particularly those involved in intuitive judgment and in expert performance—what contemporary psychologists call System 1 thinking) are produced by model-based processes; and many of our beliefs are warranted only indirectly by being among the commitments of good models. This is a familiar claim in psychology and in AI—indeed, in many ways, the view I am presenting is very similar to Marvin Minsky's (1974, 1985) frame-based account of the mind and stands in a complementary relationship to Philip Johnson-Laird's (1983) work on a slightly different notion of a mental model—but it has found surprisingly little uptake even in philosophy of mind, much less in philosophical accounts of epistemology, semantics, or reasoning.

There are thus far the rudiments of two types of claims here. First, any viable account of cognitive architecture has to posit mental models as fundamental features of human cognition. Second, a model-based account of cognitive architecture ultimately has some revisionary implications for how we should understand meaning, knowledge, and truth as well. Any realistic model-based account of cognition, however, must also deal with the fact that we possess mental models of *many different* content domains, which we think and reason about in different ways. Indeed, if we understand the world through models of particular domains, there *have to be* a very large number of such models, because the list of things we think about is very long, and indeed open-ended. Perhaps we *could* (someday, at Peirce's ideal end of inquiry) come up with a single "super model" that would accommodate everything we understand; but that is not how we *really do* think about the world, either in ordinary life or in the sciences. It is because a psychologically plausible model-based cognitive architecture must encompass *many* models that I have dubbed my theory cognitive *pluralism*.

There are a number of important questions to ask in connection with this thesis: Why would human (and animal) minds be built so as to understand the world through many distinct models? How do different models relate to one another, and how can we use several models in conjunction with one another to triangulate the world? How does the thesis that understanding *in general* is model based differ from the thesis that *some* aspects of the mind are "modular" in a sense that is more narrowly confined to cognitive systems that are fully encapsulated, automatic, nativistic, and perhaps products of natural selection? And how is model-based understanding related to the kinds of thinking that philosophers more commonly study—thinking that is (at least putatively) language-like and can involve explicit arguments?

One feature of models that I emphasize in this book is that models are *idealized*. One way in which they are idealized consists simply in the fact that each model is a model *of* some specific domain and brackets everything else about the world. But most models are also idealized in the deeper sense that they provide *good-enough* ways of thinking and reasoning about their domains *for some particular set of purposes* without these being exact, unidealized, or context-neutral ways of representing their targets. This is perfectly understandable when viewed from the standpoint of asking how an organism that is a product of natural selection and possesses only limited cognitive resources can nonetheless successfully negotiate its environment. Nature builds organisms with cognitive abilities that are relevant to their needs, using whatever types of neural resources are available for a given species, and generally does so with remarkable efficiency. What is potentially more surprising is that even our most sophisticated modes of understanding, such as scientific models, are also domain centered and idealized in the same basic ways, even though they are far more exacting and track features of the world that are more "objective" in the sense of being less tied to idiosyncrasies of the individual or the species. A model is always a model of some particular phenomenon (and not of everything at once), represented in some particular way (and not in some other way). As a result, understanding the world through models of its different features also requires further abilities to recognize when each model can appropriately be applied and how to use multiple models to triangulate the world.

The fact that different models represent the world in different ways also opens up the possibility that two models may not always play well together: they may be formally incommensurable or lead to conflicting anticipations about how things in the world will behave. Two models that are each quite good individually for particular purposes may conflict with each other, and this raises important questions about whether the kinds of understanding they supply *jointly* can be accommodated by a single "super model" that is intended to "unify" them. This is the kind of problem that drives both the reductionist agenda in philosophy of science and the search for a grand unified theory in physics, and many people have assumed that it is only a matter of time, discovery, and hard work before such projects yield their desired results. What cognitive pluralism suggests is that the question of whether we can achieve such explanatory unification depends not only on how the world is but also on empirical psychological facts about our cognitive architecture.

1.1 Overview of the Book

The book is divided into three parts.

Part 1, "From the Standard View to Cognitive Pluralism," presents as a foil for cognitive pluralism a "standard view" that is assumed in mainstream epistemology, semantics, truth theory, and theory of reasoning, which does not accord mental models any special role. Instead the standard view treats the basic types of units found in thinking as analogous to the three sizes of units found in language and logic: units that are word sized, sentence sized, and argument sized. It is sentence-sized judgments or beliefs (or their propositional contents) that are candidates for truth. Sentence-sized units have semantic values that are largely functions of the meanings of their constituent concepts plus their compositional syntax (or, if forms of judgment are not literally syntactic, some structural feature analogous to syntax). Semantic atomists and holists differ over whether semantic value is constructed from the ground up, beginning with atomic concepts, or determined by the entire network of concepts, beliefs, and inferential commitments; but they agree on the basic *units* of analysis, with the sole exception that holists count the entire network as itself the governing unit. Sentence-sized units (beliefs or judgments) are standardly regarded as the basic candidates for warrant as well as truth and hence are treated as the principal units of epistemology, though foundationalists and coherentists differ over whether warrant attaches to some beliefs individually, and the rest through their being the conclusions of valid arguments from other warranted beliefs (foundationalism), or because of a global coherence between beliefs (coherentism).

One type of model—the scientific theory—is often treated as a paradigm case of knowledge. But at least in pre-Kuhnian philosophy of science, theories tended to be treated either as *propositions* (e.g., laws interpreted as universally quantified claims) or as sets of propositions and inferential rules. Since the 1960s, philosophers of science have increasingly regarded entire theories as basic semantic and epistemological units, with theoretical terms interdefined within a theory and the whole theory receiving its warrant as a unit; and more recently, model-based accounts of science have begun to eclipse theory-based accounts in philosophy of science. I argue that several bodies of work in cognitive and developmental psychology (the core knowledge system hypothesis and the theory-theory) and artificial intelligence (explanation of common sense and semantic understanding in terms of frames, scripts, and models) give us reason to conclude that this is not a feature exclusive to scientific theories, but that understanding is *generally* divided into models of different content domains. This suggests that the common division between "central" and "modular" cognition is deeply misleading: a great deal of our learned understanding is based in models and shares a number of features generally associated with modules, such as domain specificity, proprietary representational systems, unconscious and automatic processing, and the speed and cognitive impenetrability characteristic of expert performance.

Part 2, "Models and Understanding," begins with an intuitive development of the notion of a model in general (using examples of external models such as scale models, maps, and flowcharts), which is then used as a basis for understanding the notion of a *mental* model as an internal surrogate that mirrors features of its target domain in a fashion that allows for nondiscursive forms of reasoning about that domain. I then locate my notion of a mental model with respect to several other notions in psychology, philosophy of science, AI, and theoretical cognitive science; argue that extending understanding through domain-specific learned models via a "modeling engine" is a biologically and ecologically plausible strategy for evolving more-intelligent animals; and situate model-based understanding alongside other features of cognitive architecture: developmentally canalized special-purpose modules, learning through conditioning, socially distributed cognition, incorporation of the environment into the extended phenotype, and the special forms of human cognition that build on the capacity for public language and language-like thinking and reasoning.

Part 3, "Epistemology, Semantics, Disunity," explores a number of implications of the cognitive pluralist view. First, because each model employs a particular representational system, is idealized, and is suboptimized for particular practical and epistemic ends, two models can be incommensurable and can license conflicting inferences. I argue that this helps to explain puzzling disunities of knowledge, both within the sciences and in other domains such as ethics, but also gives us reason to doubt that attempts to "unify" the sciences in particular or knowledge in general may be feasible. I then present a model-based account of intuitive judgments, which claims that these are often judgments that can be "read off" the implicit rules of a model and seem plausible and sometimes even necessary because we are in fact committed to them to the extent that we employ a particular model for framing and reasoning about a situation. On the other hand, the intuitive implications of a model are dependable only when the model is aptly applied, and so a model-based account also helps to diagnose several types of cognitive illusion as well. Finally, cognitive pluralism tends to favor a reliabilist epistemology (aptly applied models can be reliable cognitive mechanisms) and a "molecularist" semantics, and thus presents problems for foundationalist and coherentist epistemologies and for atomist and holist semantic theories. However, cognitive pluralism also suggests that we need not view rival epistemological and semantic theories as direct competitors attempting to explain the same phenomena; rather, we might better see them as complementary accounts, each an idealized model of different dimensions of evaluation.

2 A Standard Philosophical View of Cognitive Architecture

If one were to ask philosophers what features of the mind would need to be accounted for in a viable theory of cognitive architecture, *beliefs* and other *intentional states* would surely be at the top of the list. Many researchers in cognitive science (or most, or all, depending on how narrowly you define "cognitive science") would agree.[1] Moreover, this assumption accords well with common sense. In everyday life, we routinely explain people's actions as consequences of what they believe and desire. This style of commonsense belief-desire explanation has been developed extensively in philosophical action theory and extended by theoretical cognitive science into explanations of the *mechanisms* underlying action and other cognitive processes.

But the idea that intentional states, particularly beliefs, play a central role in human cognitive architecture is not a product *only* of reflections on common sense and discussions in philosophy of mind, action theory, and belief-desire explanations in cognitive science. It is also strongly suggested by the ways that philosophers have approached a number of "core" areas of philosophy: truth theory, epistemology, semantics, logic, and theory of reasoning. The most influential approaches to each of these topics have been undertaken in terms of beliefs (or judgments, a distinction I clarify later in the chapter), the concepts that are taken to be the constituents of beliefs, and forms of reasoning that operate over beliefs. Together, these strongly suggest that the basic units of thought are concepts, beliefs (and other intentional states), and inferences over such states. These three units, moreover, form a compositional hierarchy: concepts are constituents of beliefs and other intentional states, and beliefs are constituents of inferences. If these areas of philosophy deal with things that are *psychologically real*, their success would seem to commit us to such entities being important elements of our cognitive architecture. And to the extent that epistemology, semantics, truth theory, and logic can successfully be carried out solely in terms of such units, this suggests either that this segment of cognitive architecture is the whole story about cognitive architecture, or at least that it is a centrally important part of the story and relatively autonomous from everything else. This is a claim that ultimately I wish to dispute. But to do so, the best way to begin is to develop it as compellingly as we can.

2.1 The Central Role of the Concept of Belief

Intentional states, and particularly beliefs, figure prominently in several major areas of philosophy: truth theory, epistemology, semantics, and logic. Intentional states are states that are *about something* in the sense of having a *content*, which is often something that can be formulated as a proposition. I can believe that the cat is at the door, and belief is one kind of intentional state. I can also want the cat to be at the door, hope that the cat is at the door, fear that the cat is at the door, doubt that the cat is at the door, and so on. These would all be intentional states with the same content (the cat is at the door), but different "intentional *attitudes*" of believing, desiring, hoping, and so on. Both belief and other intentional states, particularly desire, are also centrally important to standard theories of action.

Truth is generally conceived as a relationship of *correspondence* between beliefs (or their propositional contents) and states of affairs in the world. There are various accounts of just what this correspondence relationship amounts to, but the relata are the same: the belief (or the proposition believed) and an extramental state of affairs. There is also a minority view to the effect that truth should be understood in terms of relationships *between* beliefs, generally described as "coherence." (This coherence theory of *truth* is not to be confused with the coherence theory of *knowledge* that I will describe shortly.) But both the correspondence theory and the coherence theory of truth treat beliefs as the only *mental* units relevant to truth, the difference between them being that the correspondence theory also regards *extra*mental facts as essential to an account of truth.

Most contemporary epistemologists (theorists of knowledge) share the assumption that *knowledge* is *warranted* (or justified) *true belief*. (This characterization of knowledge was first phrased as "*justified* true belief," but recent accounts increasingly use the word 'warrant' in place of *justification*.) The most influential epistemological theories differ from one another in the accounts they give of what makes a belief warranted. Foundationalists hold that beliefs can be warranted in either of two ways: (1) by being "properly basic"—that is, not requiring a further grounding in anything else—or (2) by being derived in appropriate ways (particularly by valid deductive inference) from other warranted beliefs, ultimately tracing back to a foundation of beliefs that are properly basic. There are a variety of proposals for what might make a belief count as properly basic, such as being self-evident or being produced by a reliable perceptual mechanism; but the basic *units* relevant to the foundationalist account are beliefs and inference patterns ranging over beliefs. Coherentists hold that a belief is warranted by being a part of a coherent set of beliefs. Again, there are various proposals for just what kinds of relations count as "coherence," though consistency is generally deemed to be a minimal necessary condition. But the relevant relata are individual beliefs.

A third and more recent theory of knowledge is *reliabilism*. Reliabilists hold that a belief is warranted if it is produced by a reliable mechanism. (A paradigm case would be the production of beliefs through reliable perceptual mechanisms under standard perceiving conditions.) It is perhaps less clear whether this introduces other types of cognitive phenomena, in addition to beliefs, that might need to be included in a cognitive architecture for knowledge. On the one hand, we might take the basic relata of a reliabilist theory to be beliefs and the states of affairs that produce them. On the other hand, we might take the *mechanisms* of belief production to be among the relata, in which case we are then faced with the question of whether they should be counted as part of *cognitive* architecture, or as something noncognitive or even extramental that is nonetheless relevant to warrant.

We should note, however, that with some beliefs, both their truth and their warrant may not depend on relations to extramental states of affairs. These are beliefs that are true because of the meanings of the concepts and the relationships between them. Propositions like "dogs are animals" and "triangles are three angled" seem to be true *by definition*, and we can *know* them to be true simply by analysis, based in our mastery of conceptual semantics.

Semantics is concerned with meaning, and both concepts and beliefs have semantic properties. Moreover, concepts are constituents of beliefs, and the semantic properties of a belief (or the proposition believed) are generally understood to be a function of those of its constituent concepts plus the logical (or grammatical, or syntactic) form of the proposition (with a few important caveats, such as that reference may also depend on context). There are important disputes about the basis of semantic properties (e.g., that they are based in something internal to the mind, such as sense data, or they latch onto things in the world by direct reference), and also about the level of organization at which they appear. *Semantic atomists* hold that some concepts (semantic atoms) are the most basic bearers of semantic value, and other concepts get their meaning by inclusion of, or definition in terms of, the semantic atoms. *Inferentialists* hold that the meanings of concepts are intertwined, and the meaning of a concept is to be cashed out in terms of the constitutive inference patterns associated with it. The conceptual inclusionist might characterize the relationship between the concepts DOG and ANIMAL by saying that the concept DOG *contains* the concept ANIMAL. The definitionalist might say that the *definition* of DOG involves the information that it is a kind of animal. The inferentialist would say that the semantic value of DOG is partially constituted by the fact that we would be disposed to infer "*x* is an animal" from "*x* is a dog."

A particularly influential version of inferentialism, called *holism*, holds that the semantic value of *every* concept is constitutively interrelated with that of every other concept, and also with every belief and inferential disposition, so that a change in any concept, belief, or inferential disposition ramifies into a change to all the others as well (Quine 1951; Davidson 1967). Holists deny that there can be a principled demarcation between concepts and beliefs, or between semantics and epistemology. Aristotle's discovery that whales are mammals rather than fish might be characterized as a change of the semantic value of the concept WHALE; but it might equally well be characterized as a change in *beliefs* about whales, or as a change in the inferences we are disposed to make about things we classify as whales. But holists, like atomists, still speak of concepts, beliefs, and inferences as basic units of cognition, even though they hold that their content and their warrant are conferred globally rather than locally.

Logic and theory of reasoning are concerned primarily with patterns of reasoning, particularly deductive arguments. Inferences and arguments are composed out of beliefs and propositions arranged in a particular logical form, and logic is particularly concerned with argument forms that are truth preserving—that is, ones in which, if the premises are true, the conclusions must be true as well.

2.2 The Three-Tiered Standard View of Cognitive Architecture

If we were to ask what these standard philosophical treatments of truth, knowledge, meaning, and inference suggest about cognitive architecture, an answer seems readily apparent: thought—or at least the types of thought that those disciplines are concerned with—involves units of three sizes, which stand in a compositional hierarchy: inferences, which are composed of beliefs or judgments, which in turn are composed of concepts. Let us call this the *three-tiered picture* of the mind. The assumption that issues about knowledge, meaning, truth, inference, and understanding can adequately be treated using just the resources of the three-tiered picture I shall call the *standard view*.

One obvious observation about the units of the three-tiered picture is that they clearly correspond to the structural units of language and logic as well. Words in a natural language and predicate letters in logic correspond to concepts, sentences and well-formed formulas correspond to intentional states such as beliefs, and both language and logic have larger units called "arguments" corresponding to inferences, though of course natural languages have many other types of structured discursive units as well (stories, jokes, etc.) that are absent from logic. It seems clearest to characterize the structural *size* of the units with terminology drawn from the description of language: that is, as *word-sized*, *sentence-sized*, and *argument-sized* units.

Moreover, some of the ways theorists have expanded on the three-tiered picture are modeled on things found in linguistics and logic as well: for example, the idea that the meanings of judgments are a function of the meanings of concepts plus compositional syntax, and philosophers' preoccupation with the particular (deductive) forms of inference that predominate in standard logics. Of course, this raises a number of questions about the relationship between thinking, language, and logic that I cannot fully address here. For example, is our capacity for language simply an ability to *express publicly* what was already available *mentally* in the forms of thinking characteristic of human minds, or is the ability to think in these ways a consequence of learning a public language, or at least being a member of a species whose brains have capacities that were originally selected to produce competence in spoken language? Does thinking literally take place in a "language of thought" (Fodor 1975), or is the relationship between thought and language more analogical—say, that thinking and language have structural units with similar sizes and functions? I shall not attempt to settle these questions here. For now, suffice it to say that the theories of a number of core areas of philosophy suggest that thought, or at least a significant portion of thought, is "language-like" in the sense of having structural units analogous to those found in languages.

2.3 Some Philosophical Issues

But *do* truth theory, epistemology, semantics, and logic really commit us to such a thesis about cognitive architecture, or indeed to any theses about cognitive architecture at all? The assumption that they do so presumes that such theories require, or at least suggest, a kind of *psychological realism* about concepts, beliefs, and inferences. There are some important challenges to this assumption, which require some attention and clarification.

Belief and Judgment

First, the word 'belief' is used in several different ways in philosophy (Horst 1995). Sometimes, when we speak of someone "having a belief," we mean that she is having a *psychological episode* in which she is mentally affirming something. The detective sorts through the evidence and suddenly concludes, "The butler did it!" and we report this by saying that she *formed a belief* that the butler did it. But we also speak of people "believing" things that they have never thought of at all. Suppose I ask you if you believe that 119 + 6 = 125, or that dogs have kidneys. You easily answer yes; and of course in the process of thinking about it, you probably actually considered the proposition in question and mentally endorsed it. But suppose I ask you whether you *believed yesterday* that 119 + 6 = 125, or that dogs have kidneys. When I present this question in a classroom, about half of my students tend to say yes and the other half no. I deliberately choose questions that no one is likely to have explicitly thought about before to eliminate the possibility that those who say yes are reporting a *previous* explicit endorsement. Those who say no probably do so because they are restricting the word 'belief' to things we have actually thought about and endorsed. But those who say yes are also using the word 'belief' in a perfectly standard way, to report things we are implicitly committed to or things we are readily disposed to assent to when prompted. And by my reckoning more often than not, philosophers tend to use 'belief' in this broader way, to include what are sometimes distinguished as "dispositional beliefs." Dispositional beliefs are often contrasted with "occurrent beliefs"—those that are supposed to be actual episodes of mental affirmation. Occurrent beliefs are also referred to as *judgments* in some areas of philosophy; and as this seems a more elegant way of referring to mental episodes in which we endorse something, I adopt it here as my preferred terminology, using the word 'belief' in the broader sense that includes dispositions as well.

I have characterized the three-tiered picture in such a fashion as to include both those who are concerned with judgments and those who are concerned with beliefs in the broader sense. To the extent that philosophical and psychological theories of intentionality and reasoning are about *judgments*, they seem to require a realist interpretation. A claim that someone formed a judgment makes a claim about a mental event and can only be deemed true if some event at least roughly corresponding to the characterization actually occurred. But to the extent that such theories are about beliefs but not judgments, they do not. At most, where beliefs but not judgments are invoked, there might be a commitment in the form of a *disposition* to form a judgment or an inference. Dispositions are not themselves elements of a cognitive architecture, though they may point to the need to specify underlying mechanisms that might be parts of such an architecture.

Interpretivism

This leads us to a deeper concern. Some philosophers—call them *intentional realists*—assume that when we talk about beliefs and inferences, we are talking about occurrent episodes of thought involving sentence-like *representations* with propositional content and inferential *processes* that operate on them (Fodor 1975, 1978, 1990). Their reason for doing so is not merely an introspective sense that sometimes we really do think in propositions (or even in natural-language sentences) and follow syllogistic patterns of reasoning. Rather, they tend to appeal to "intentional *explanation*"—explanation of what a person says or does in terms of what she believes and desires—which they view as a form of *causal* explanation in which judgments, occurrent desires, and processes of theoretical and practical reasoning play the role of the explainers. If the claim "Jane went to the refrigerator because she wanted some milk, and believed that the milk was in the refrigerator" is supposed to be a causal explanation, that explanation can be true only if Jane really had the imputed mental states and they played a causal role in producing her action.

But there is an alternative account of what we are doing when we ascribe intentional states, and even when we use them to explain behavior. To use Daniel Dennett's terminology, what we are doing is adopting an *interpretive stance*, which he calls *the intentional stance* (Dennett 1971, 1981/1997). Here is how Dennett explains the intentional stance in one of several papers in which he defends the notion:

> While belief is a perfectly objective phenomenon (this apparently makes me a realist), it can be discerned only from the point of view of one who adopts a certain *predictive strategy*, and its existence can be confirmed only by an assessment of the success of that strategy (that apparently makes me an interpretationist).
>
> First I will describe the strategy, which I call the *intentional strategy* or adopting the *intentional stance*. To a first approximation, the intentional strategy consists of treating the object whose behavior you want to predict as a rational agent with beliefs and desires and other mental states exhibiting what Brentano and others call *intentionality*. … Any *system* whose behavior is well predicted by this strategy is in the fullest sense of the word a believer. *What it is* to be a true believer is to be an

> *intentional system*, a system whose behavior is reliably and voluminously predictable via the intentional strategy. (Dennett 1981/1997, 59)
>
> Here is how [the intentional stance] works: first you decide to treat the object whose behavior is to be predicted as a rational agent; then you figure out what beliefs that agent ought to have, given its place in the world and its purpose. Then you figure out what desires it ought to have, on the same considerations, and finally you predict that this rational agent will act to further its goals in light of its beliefs. A little practical reasoning from the chosen set of beliefs and desires will in many—but not all—instances yield a decision about what the agent *ought* to do; that is what you predict the agent *will* do. (61)

Dennett agrees with intentional realists like Fodor that we can predict and explain people's actions on the basis of their beliefs and desires with striking success—and, moreover, that we have no alternative way of doing so with any hope of similar success. But he does not agree that what we are doing in such predictions and explanations is positing inner causal mechanisms involving *states* or *events* called "beliefs" and "desires" or reasoning processes that operate on them. The intentional stance is a predictive strategy, and it does require that we ascribe beliefs and desires. But what one is doing when one ascribes a belief using this strategy is specifying *what a being must be committed to, given its circumstances, if it is to behave rationally*. Any system that can systematically be predicted in this fashion thereby counts as an "intentional system," regardless of what underlying mechanisms produce those patterns of behavior. And if the intentional strategy works with respect to a particular system, it is getting at something *real* about that system—not (necessarily) real occurrent *states* called "beliefs" (i.e., judgments), "desires," and "reasoning," but real (i.e., empirically robust) *patterns*, patterns that are revealed only when we adopt the intentional stance (Dennett 1991b). Moreover, that such patterns are revealed only from a particular interpretive stance does not render the interpretations fictitious or mere matters of taste: physical and functional properties are likewise revealed only from distinct interpretive/predictive stances, which Dennett calls the *physical stance* and the *design stance*, respectively.

It is important to note that the way Dennett has characterized intentional systems, in a fashion that is *independent* of the *mechanisms* through which a system is able to behave in rational ways, has the consequence that his view is actually *compatible* with what he calls Fodor's "industrial strength realism," which treats intentional states and processes as occurrent states that are central elements of cognitive architecture. But it is also compatible with a denial of that sort of realism. It is perfectly possible that *one* intentional system is correctly ascribed beliefs and desires *because it actually contains language-like mental representations corresponding to each of those beliefs and desires*, while another has no such representations, and a third has representations corresponding to *some* apt ascriptions of beliefs and desires but not to others. For example, some of Dennett's papers include examples involving chess-playing computers, and *some* such computers do what they do by explicitly representing things such as candidate moves and their consequences in something like propositional form. Nothing in Dennett's account should preclude the notion that *sometimes* we have mental episodes involving language-like propositional representations, or that they *sometimes* play a causal role in reasoning. And this is a good thing, as it seems clear that we at least sometimes do so, when we engage in explicit and conscious reasoning.

This actually dovetails nicely with the distinction between belief and judgment in the previous section. Dennett's characterization of belief ascription helps us make sense of ascriptions of beliefs that cannot plausibly be construed as ascriptions of judgments: they are, in effect, ascriptions of *normative commitments*. Of course, judgments involve normative commitments as well, even though we often *do* make judgments that are not ideally rational. Philosophers have raised good and important questions about whether epistemology does or should apply to beliefs that are not judgments and, conversely, whether the question of which beliefs are also judgments is really salient to the concerns of epistemology. (For example, do questions of warrant really make sense if beliefs are imputed from an assumption of ideal rationality?) But I will not address those questions here, as my principal concern is with cognitive architecture.

Another important issue should be addressed. There seem to be clear cases of language-like judgments: if nothing else, when someone mentally endorses a proposition by thinking it to herself *in a natural language*. Likewise, clear cases exist in which we can sensibly ascribe to a person a belief that P in full knowledge that she has never had any *thought* with the content P. For example, I recall one college friend describing another by saying, "He believes he is entitled to half of the speaking time in any conversation, regardless of how many people are in the conversation." It was an apt characterization, but it is unlikely in the extreme that said friend ever formed, much less endorsed, a thought with that content. Indeed, had he been presented with that description of himself, he would doubtless either have disputed it or tried to modify his behavior thereafter. But a wide range of cases also fall in between. The phone rings, and I reach to pick it up. Surely *some* sequence of mental events is going on that we can aptly characterize as "believing the phone is ringing" and "believing that someone is calling me," but (let us assume) I do not consciously think such thoughts in English. Intentional realists are inclined to suppose that these events must be something like judgments, only ones that are not expressed in a natural language (even silently to oneself), and perhaps not consciously experienced at all. But must this be the case? It is certainly not evident from the *phenomenology* of the experiences. It may be that some such cases involve the apt ascription of beliefs using the intentional stance, but the events themselves lack the characteristic propositional structure of judgments. We use the "intentional idiom"—for example, "Bob believed that someone was calling him"—to report such states, but it may be that the assumption that *the states reported* have a propositional structure is an artifact of *how we report them* using a particular grammatical construction.

Sellars

Wilfrid Sellars (1956) proposed that we think and talk about mental states by analogy with public language. He developed this view through a "myth" about how our ancestors came to talk about mental states in the way we do, through three stages of the development of language. In the first stage, people were able to talk about the world around them but did not yet have a way of reporting what others said. The second stage involved the expansion of language to include ways of reporting speech: "Og said, 'The fire is going out,'" or "Og said *that* the fire is going out." Then came a recognition (or at least a hypothesis) that other people's behavior is a result of things going on within them that are not publicly observable, and the development of a way of speaking about these inner states using a variation on the linguistic devices used to report speech: "Og *believes* the fire is going out," or "Og *thinks that* the fire is going out." Sellars's point, as I understand him—the presentation in the form of a myth introduces some interpretive uncertainty—is not that inner episodes are an interpretive fiction but that the way we *understand* and *characterize* such inner episodes (some of which might be quite real) is colored by the fact that we *treat them like inner speech.*

My original reactions to Sellars's myth were of a skeptical sort. This was partly because at first I (wrongly) interpreted it as part of an argument against the reality of intentional states as inner episodes, and partly because I thought there were clearly *other* ways we could first have become aware of, and come to talk about, mental states, namely, through introspection. I think I now appreciate his point better. It clearly has something right about it and raises interesting and important questions. We *do* tend to think of intentional states in a fashion analogous to how we think about speech. And of course in some cases, thinking *takes the form of* subvocalized speech in a natural language. But not all the cases are like that; indeed, the vast majority are probably not. And this raises serious questions about whether the assimilation of thought to language involves a deep *insight*—that thoughts (or at least certain very prevalent types of thoughts) really are structurally language-like even though they are not *in* a *natural* language—or whether the assimilation is an artifact, and potentially a misleading artifact, of using the more familiar ways of reporting discourse to talk about thinking as well.

Introspective evidence clearly indicates that *some* thought takes the form of language-like propositional thinking and reasoning, because sometimes we think *in* a natural language, and sometimes we *apply* argumentative forms such as deductive syllogisms in explicit reasoning. But this evidence extends no farther than conscious judgments couched in language and explicit application of inferential methods. Moreover, we are probably prone to conflate cases in which thinking is *done in* a language with cases in which the actual thinking and reasoning are accomplished in some prelinguistic form and then *rendered in* linguistic form, even if only to oneself. Perhaps this is merely "translation" from one "language" (Fodor's "language of thought") to another (such as English), or perhaps it is a very fast and automatic *conversion* of something that is not language-like in form into language. With reasoning, the situation is perhaps even worse, as numerous psychological studies indicate that when people produce "reasons" for their statements or actions, what they produce are really rationalizations confabulated after the fact that do not reflect the actual *processes* that led to what they said or did.

We thus seem to be faced with the possibility of four distinct kinds of situations, all of which can be reported in the same intentional idiom:

1. Cases in which thinking and reasoning take place in an actual natural language, albeit in an internalized form.
2. Cases in which thinking and reasoning are real events involving language-like inner states and processes that have forms analogous to explicit argumentation but are not in a natural language like English.
3. Cases in which ascriptions of intentional states refer to real events that play a causal role in the production of behavior but are not themselves structurally language-like.
4. Cases in which intentional descriptions pick out real patterns—for example, patterns of normative commitment that are also predictive of behavior—but do not refer to occurrent states or episodes.

I think that it is initially plausible that cases of all four types actually exist. Indeed, I think that evidence for the reality of the first and fourth cases is clear and compelling. Whether the two intermediate categories are exemplified, and how much of our mental life falls into either of them, are the sort of questions that the cognitive sciences attempt to address.

Two Levels of Intentional Explanation

Advocates of language-like mental representations often appeal to "the success of intentional explanation" to justify their position. But it is important to distinguish two very different types of explanation here. The first kind is the sort of explanation of a person's words or deeds in terms of her beliefs, desires, and other intentional states that we find ubiquitously in ordinary life, like the explanation, earlier in the chapter, of why Jane went to the refrigerator. Fodor and other representationalists do indeed appeal to the success of such explanations to justify their position, claiming that such explanations could work only if there really are such things as beliefs and desires with propositional content that play causal roles in the production of behavior, and that the only viable account of how this could work involves treating them as occurrent mental representations. Dennett agrees that this kind of explanation works well, but he denies that it should be taken as the type of explanation that posits causal mechanisms or the existence of occurrent *states* called "beliefs" and "desires." Both of these positions strike me as at least *coherent* accounts of the nature of intentional explanation, and they involve very different commitments to features of cognitive architecture.

Cognitive science also offers a second and deeper type of explanation, involving the postulation of mechanisms that explain not only the resulting behavior but also things like the time it takes to perform a cognitive task, what additional forms of information are sought in the accomplishment of a task, characteristic patterns of error and breakdown, and so on. It is this sort of explanation, not the commonsense explanation of behavior, that is the real work of theoretical cognitive science. Cognitive scientists propose accounts of the types of representations and processes that might underlie different types of competence and performance. There might be any number of possible cognitive architectures involving different types of representations and processes that could, in principle, produce the same end result; but this does not render them indistinguishable, because *how* they produce it can have testable consequences. We can prove, for example, that chess masters do not decide on a move through brute force serial evaluation of the possible moves, because we can calculate how many steps such a process would require, we know the approximate time in milliseconds that a single cognitive operation takes in the human brain, and the result of combining these factors is far longer than the chess master's actual response time.

It would be simply naive to assume that contemporary cognitive scientists who favor a cognitive architecture involving language-like intentional representations do so merely because explaining people's actions in terms of their beliefs and desires succeeds so frequently in ordinary life. Cognitive science is a highly competitive field, in which researchers are exploring alternative cognitive architectures, procedures, and implementations. In fact, it is common to distinguish "cognitive science" from "artificial intelligence" on the basis of whether the research is aimed at an empirically plausible account of human cognition (cognitive science) or the production of intelligent machine behavior through whatever means (artificial intelligence). Theoretical cognitive science is engaged with experimental evidence from cognitive *psychology* and holds itself accountable to explaining such evidence. And researchers in competing labs can be avid in their attempts to show that their approach can accommodate the full range of data in ways that their competitors' cannot. Of course, it is part of any scientific research project that researchers will doggedly try to find ways to make their own approach work in the face of apparent obstacles. But when researchers advocate a cognitive architecture that is committed to language-like propositional representations, they do so generally because that approach has demonstrated some success at producing explanations of a number of psychological phenomena that have been demonstrated in a laboratory, and in a fashion that warrants some hope of a realistic interpretation.[2]

This, however, cuts both ways. Theoretical cognitive science is also home to a number of theories of cognitive architecture in which language-like propositional representations play less of a role or none at all; and a number of aspects of cognition are arguably better explained by alternative cognitive architectures, such as connectionist architectures, Bayesian nets, or models, a form of which I am recommending in this book.

2.4 Alternative Proposals regarding Architecture

Several kinds of potential problems for the three-tiered cognitive architecture of the standard view stem from reasons to favor different views of cognitive architecture. One threat comes from approaches to the mind that seem to oppose the idea that there is really such a thing as *cognitive* architecture at all. A second type of threat comes from concerns that concepts, intentional states, and inference cannot cleanly be separated from a broader notion of *understanding* that includes a great deal that cannot be explained within the framework of the three-tiered picture.

Eliminativism

Some philosophers and psychologists (a minority within both professions), called "eliminativists," completely deny some portion of the three-tiered picture. For example, some claim that there are actually no concepts (Machery 2009, e.g., is sometimes characterized in this fashion), and others that there are no beliefs (Churchland 1981; Stich 1983). Sometimes such claims turn out on closer inspection to be more modest than they might initially sound. Machery's *Doing without Concepts*, for example, is really claiming that the various rival theories of concepts describe separate phenomena that cannot be accommodated within a single theory. It is not really the eliminativist proposal the title might suggest, but a *pluralist* proposal. Churchland's claim that intentional states are "theoretical posits" of what he deems to be a failed theory does not argue that we never make judgments with propositional content, but argues that intentional psychology, viewed as a theory that provides causal explanations of behavior, is a stagnant research program that even at its best would explain only a limited range of behavior. These criticisms are indeed responding to the kind of characterization given to intentional psychology by Fodor, and they raise real (if controversial) questions about its explanatory scope and success. But they leave unaddressed the larger question of whether intentional states are an important structural feature of cognitive architecture (Horst 1995).

Connectionism

Connectionism, which treats cognition as emerging through dynamic processes that take place in interconnected layers of neuronlike nodes, involves a very different proposal about cognitive architecture (Rumelhart and McClelland 1986; Churchland and Sejnowski 1989; Ramsey, Stich, and Rumelhart 1991). It is controversial (among friends and foes alike) whether it should be called a "*cognitive* architecture," as some writers reserve the word 'cognitive' for states, processes, and architectures that operate on units with representational content (e.g., Pylyshyn 1991), while others apply it more broadly to subcognitive implementations of cognitive states and processes, or even to anything that involves information processing, even early processing of perceptual information. (Marr's [1982] book on vision, after all, is generally regarded as a seminal work in *cognitive* science.) Connectionist accounts often *do* have their own characterizations of concepts and intentional states—for example, as activation patterns across nodes in intermediate hidden layers. This approach actually strikes me as being right headed as an account of *judgments* (which are transient mental events) as well as concepts. (It is only when concepts are *activated* or *applied* that we should expect an occurrent state. The notion of *having* a concept or dispositional belief need not require a persistent representation corresponding to the concept or belief.) One can be a connectionist and disavow concepts and judgments altogether; but often the real issues are about how intentional phenomena are *implemented* and whether the dynamic processes that drive thinking operate *over* intentional states or merely sometimes have them as their inputs and outputs.

Embodied Cognition

Advocates of ecological, embodied, and enactive approaches to cognition have likewise positioned their views in opposition to those based around representational intentional states.[3] The core thesis of these views is that cognition is not simply something that takes place within the mind or brain but a kind of skilled practical engagement with the world. Among the sources of this relatively new approach to the mind are the twentieth-century philosopher Martin Heidegger's (1927/1996) claim that intentional states are grounded in and dependent on deeper capacities for skilled bodily activity, and the ecological psychologist J. J. Gibson's (1977) claim that organisms perceive and respond directly to affordances (structures in the world that provide something to the organism or allow it to do something), rather than inferring and representing them on the basis of brute sensory data. Cognition is viewed as a dynamic process that encompasses not only the mind or brain, or even the whole embodied organism, but a dynamically coupled system of interactions between organism and environment (Silberstein and Chemero 2011; Thompson 2007).

Beyond the provocative thesis that cognition, and even "the mind," is not just "in the head," writers who embrace ecological, embodied, and enactive theories of cognition make several additional claims of different degrees of strength. The mildest of these is that there are a variety of processes that do not look much like judgments and inferences that nonetheless play essential roles in intelligent interaction with the world and indeed in sense making. This does not directly threaten commitments to a cognitive architecture that *includes* concepts, judgments, and inferences, but raises a concern, which I explore further in the next section, that this is a seriously *incomplete* description of cognitive architecture, and raises questions about whether the standard view, by itself, can provide an accurate account even of the kinds of states it deals with.

A somewhat stronger claim is that such skilled bodily interaction provides the grounding on which concepts, explicit judgment, and reasoning are based. In something like its Heideggerian form, this is not a repudiation of intentional states, concepts, or inferences but a claim that we need a more fundamental theory to explain how they are psychologically possible at all. (Later I make a similar claim that mental models play such a grounding role, and that the notion of a mental model can be combined with ecological enactive views of cognition.)

Finally, many of the works in this tradition include claims to the effect that some or even all of our intelligent interaction with the world is *nothing but* skilled bodily interaction, with no need to postulate things such as concepts, beliefs, inferences, or mental representations. To my mind, the plausibility of such claims depends on their scope. It strikes me as plausible that much of the online cognition of both humans and nonhuman animals is a form of direct engagement with the world, requiring neither judgments nor inferences, and that we too easily read an interpretive framework involving judgments and inferences onto processes that are in fact accomplished in other ways. On the other hand, as I said earlier, it seems clear that we do sometimes engage in judgments and explicit reasoning. Moreover, it seems necessary to invoke *some* kind of inner processes for off-line cognition in which the mind uses *some* type of "representational" surrogates in lieu of objects in the environment, though that need not imply that they are the types of entities postulated by the standard view.

Intrusions from Broader Cognitive Architectures

This brings us to the broader and older concern that the standard view leaves out a great deal about the mind, and even if the units of cognition it posits are real and play some role in the explanation of behavior, we end up with a seriously misleading picture of how the mind works if we ignore other things. Of course, no one who is even modestly knowledgeable about neuroscience or cognitive psychology believes that the standard view is suitable for a story about *all* the topics one would find in textbooks on those subjects. Autonomic processes, orienting systems, attention, early perceptual processing, proprioception, haptics, motor control, arousal, mood, and emotion are all important elements of our psychology. In many cases, we have a good idea of what sorts of neural processes are involved in their operation, and the neuroscientists' models of those processes are cast in terms very different from intentional psychology; and as often as not, different processes seem to operate in very different ways from one another as well.

None of this is incompatible with the standard view's supplying a useful framework for describing the aspects of our mental life that philosophers have cared most about, which arguably make up a substantial portion of what is distinctive of *human* minds. But it does raise the question of how "open" an architecture it is: whether it is relatively self-contained or cannot really be separated from other aspects of the mind. Perhaps the oldest version of this worry in philosophy concerns the relationship between intentionality and perception. Any epistemological theory faces the question of how sensation and perception provide warrant for beliefs. If we view perception wholly in nonintentional terms, a problem arises for how perceptual information can be translated into a form that provides warrant. Conversely, if it has some element of intentionality within it, the realm of meanings and reasons extends beyond sentence-like intentional states such as judgments. There are similarly old problems about semantics, such as whether intentional states include "nonconceptual content," and to what extent concepts are dependent on (perhaps even constructions out of) sensations or perceptions.

A slightly different sort of issue is occasioned by psychological theories claiming that there are undisputedly intentional phenomena that nonetheless seem to be segregated from the rest of our thinking: for example, special-purpose systems for detecting danger, contagion, or sexual cues, and special forms of reasoning that are applied only in social cognition. Evolutionary psychologists suggest that these are isolated cognitive phenotypic traits that may well have their own neural substrates, adaptive histories, and modes of representation and inference. If this implies that they are separate parts of the mind that work on independent principles, we would need to posit a much more complicated kind of cognitive architecture. In chapter 3, I explore an influential strategy for insulating a domain of "central cognition," characterized by the standard view, from threats stemming from other mental "modules."

Understanding

A related concern can be raised that does not depend on the assumption that different parts of our mental lives are a result of separate *neural* systems or independent evolutionary adaptations. This is the concern that understanding itself involves more than just beliefs or judgments and the kinds of inferences found in explicit argumentation. We speak of "understanding" things such as Newtonian mechanics, consequentialist ethics, the game of chess, the dynamics of free markets, and the Japanese tea ceremony. Understanding of any such domain is clearly not something that can be carved off from the realm of meaning and reasons. It involves an interrelated set of concepts, beliefs, and semantic, inferential, and practical capacities concerning a particular domain of experience. But this seems to go beyond the standard view in two ways. First, understanding also involves abilities to *recognize* events and situations in terms of those concepts, and to *act* in skilled ways within those domains. Second, it at least initially looks as though understanding needs to be approached in a different way from concepts, beliefs, and inferences: as something whose basic units are *domain sized*, where a "domain" is something like the game of chess or the world as described by Newtonian mechanics. As Catherine Elgin notes:

> We understand rules and reasons, actions and passions, objectives and obstacles, techniques and tools, forms, functions, and fictions, as well as facts. We also understand pictures, words, equations, and patterns. Ordinarily these are not isolated accomplishments; they coalesce into an understanding of a subject, discipline, or field of study. (Elgin 1996, 123)

The idea that a content domain might itself be a fundamental unit for understanding was developed most influentially in philosophy of science, beginning in the 1960s, as the idea that scientific theories are not simply combinations of statements or beliefs that can be specified independently of one another; rather, the concepts and law claims that make up a theory are constitutively interdefined, and it is the entire theory that is confirmed or disconfirmed as a unit by empirical evidence. Theories are integral semantic and epistemic units and indeed are in some sense more fundamental semantic and epistemic units than the concepts or claims they employ, as the concepts gain their meaning and the claims their warrant from the larger unit of the theory, not the other way around. As a consequence, we do not fully understand concepts, claims, beliefs, judgments, or the special forms of reasoning distinctive of theories if we do not take this other type of unit into account. And if that is the case, theory-sized units turn out to be eminently relevant to the concerns of epistemology, semantics, and truth theory as well.

Just how broad an implication we should draw from this depends on whether such features are unique to scientific theories, or whether they are a more general characteristic of understanding. In chapter 4, I present a case for the more general thesis that understanding is organized in mental models of content domains, and this thesis is further developed, under the name of cognitive pluralism, in chapter 5.

3 Central and Modular Cognition

In the previous chapter, I claimed that mainstream theories about thinking tend to treat the mind as a single language-like medium with units of three distinctive sizes, related to one another compositionally: word-sized concepts, sentence-sized intentional states (particularly beliefs or judgments), and larger discourse-sized units (particularly deductive arguments). Theories in a number of core areas of philosophy—epistemology, semantics, truth theory, and logic—are at least tacitly committed to this three-tiered picture, though each area has its rival theories about matters such as what makes beliefs warranted, how concepts obtain their semantic values, what relationship between thoughts and the world is needed for beliefs to be true, and what are the most appropriate logical formalization(s) of reasoning. In ordinary life, we often explain people's behavior in terms of their beliefs, desires, and other intentional states and how they reason on the basis of these. And influential mainstream projects in cognitive science have attempted to provide more explicit and exacting theories of such processes through formal and computational models.

While philosophers have often managed to ignore the fact that there is also much more going on in mind and brain than intentional states and processes, this option is hardly available within the cognitive sciences. Even if we were to stipulate that only mental states involving concepts, intentionality, and reasoning are to be called "cognitive," these still make up only a part of what goes on within some larger architecture that is needed to account for things such as sensation, perception, motor control, proprioception, and other topics studied in the cognitive sciences. And this raises the question of whether, even in this narrow sense of the cognitive, a *cognitive* architecture can really be specified in a fashion that is independent of the rest. Alternatively, if we use the word 'cognitive' more broadly, can the architecture for what we might call "the space of meaning and reasons" be specified independently of a broader cognitive architecture?

This chapter describes an influential proposal for how such a space of meaning and reasons, conceived as "central cognition," might effectively be distinguished from other aspects of the mind, conceived as "modular." Ultimately, I argue that this bifurcation of central versus modular cognition is not a helpful one. Even so, it will prove useful to develop and examine it, as the terms in which it is framed, once held up to scrutiny, will help us see more clearly an alternative to the standard view.

3.1 The Mind in Philosophy, Psychology, and Neuroscience

While philosophy may have been the first discipline to study the mind, it has been joined by a number of scientific disciplines, sometimes collectively referred to as the cognitive sciences. Not only has each of these disciplines taken on a life of its own, with professional societies, degree programs, conferences, journals, and monographs, but the overarching disciplines like psychology and neuroscience have each branched into a number of subdisciplines, to such an extent that keeping up with the literature in any one of them could be a full-time occupation. Some of the topics studied in these disciplines would be familiar to philosophers and touch in different ways on topics of central philosophical interest, such as concepts and reasoning. Others lie far afield from philosophy, such as the more biological aspects of neurophysiology and neuroanatomy. And others lie somewhere in between, dealing with topics like perception and emotion that philosophers have long been aware of, and many philosophers have written about, but which have tended to remain separate from discussions of epistemology, semantics, truth, and logic.

It is useful to do at least a brief survey of the sorts of topics that one finds in textbooks in psychology and neuroscience and compare these with those discussed by philosophers. In compiling these lists, I have followed the simple procedure of looking at the tables of contents of some contemporary textbooks in neuroscience and in cognitive and developmental psychology, doing some minor rearranging, and omitting some things like neurophysiology and neurochemistry that lie far afield from mainstream philosophical interests.

Topics in Textbooks in Cognitive and Developmental Psychology

- Perception
  Perceptual organization in vision, object recognition, face recognition, eye movements, event perception

- Attention
  Perception and attention, spatial attention, disorders of attention, automaticity, unconscious processes
- Memory
  Long-term, short-term, working, episodic, semantic
- Executive processes
- Emotion
  Theories of emotion, emotion perception, emotion and memory, emotion regulation, aggression and antisocial behavior
- Decision making
- Problem solving
- Reasoning
  Induction, judgment under uncertainty, analogical learning, affective forecasting, practical reasoning, moral reasoning
- Motor cognition and mental simulation
- Language
  Language acquisition, syntax, semantics, speech perception, spoken-word perception, reading, discourse comprehension
- Expertise
- Creativity
- Consciousness
- Social cognition
  Self-knowledge, person perception, theory of mind, attitude change, cultural differences
- Concepts and categories
  Nature of concepts, culture and categories, mental images
- Theories of intelligence
- Cognitive style
- Personality

Topics in Textbooks in Neuroscience

- Neural signaling
- Sensation and sensory processing
  Somatic sensory system, pain, vision: the eye, central visual pathways, auditory system, vestibular system, chemical senses
- Perception
  Somatosensory system, touch, pain, constructive nature of visual processing, low-level visual processing: the retina, intermediate-level visual processing and visual primitives, high-level visual processing:

cognitive influences, visual processing and action, the inner ear, the auditory central nervous system, the chemical senses

- Movement
  Organization and planning of movement, motor unit and muscle action, spinal reflexes, locomotion, voluntary movement: the primary motor cortex, voluntary movement: the parietal and premotor cortex, control of gaze, vestibular system, posture, cerebellum, basal ganglia, genetic mechanisms and diseases of the motor system
- Unconscious and conscious processing of neural information
- Development and the emergence of behavior
- The changing brain
  Development, plasticity

These lists contain some topics that have little directly to do with the mind as philosophers conceive it. In some cases, this is for perfectly ordinary reasons that can be found whenever we compare the topics discussed by two disciplines with overlapping interests. Neuroscience is concerned with some activities that are paradigmatically mental, but frames its questions in terms of the role of the brain, or more broadly the nervous system; and there are important things to say about the brain and nervous system that have nothing to do with the mind. (The processing of visual information in the retina, for example, has little to do with how we think, and spinal reflexes are a result of activity of the nervous system but do not involve mind *or* brain.)

In other cases, psychology and neuroscience investigate topics that have figured in philosophical discussions of the mind, sometimes providing perspectives that challenge traditional philosophical assumptions and offering new resources for philosophers to work with. Examples of these include the distinctions between several types of memory, theories of concepts that treat them as prototypes or exemplars, theories of attention and its relationship to consciousness, theories of multiple intelligences, and discoveries of species-typical patterns of reasoning (such as Gigerenzer and Todd's fast and frugal heuristics [1999]) that are invalid and do not correspond to good Bayesian reasoning but have been shown to be adaptive, at least when applied in particular contexts. Of particular interest, however, are discussions of topics that have always been regarded as relevant to thought, knowledge, and reasoning, but whose philosophical treatment has always left important open questions, such as perception and action/motor control.

Perception can produce beliefs and other intentional states. Indeed, when philosophers speak of "perception," they are often speaking of things like perceptual gestalts—seeing a dog or seeing something *as* a dog—that are reasonably regarded as *being* intentional states. At the least, they involve the application of concepts and the formation of assumptions about perceived states of affairs; and some classifications of intentional states (e.g., those of Brentano [1874] and Husserl [1900/1973, 1913/1989]) treat such states as forming a class of intentional states alongside of beliefs, hopes, and desires. The bulk of what one finds about perception in a neuroscience textbook, however, is concerned with "early processing" or "preprocessing" that takes place in the brain before application of concepts, for example, the separate visual processing pathways in the brain for color, form, and motion, and the organization of the visual cortex. Most of this processing not only is unconscious but cannot be made into an object of conscious awareness through introspection. Moreover, perceptual mechanisms can operate adaptively without the formation of conscious intentional states. (And, depending on one's views of what species of nonhuman animals have intentional states, it may be possible for perception to drive adaptive behavior in animals that lack even the capacity for intentional states, or at least the kind that can enter into reasoning and produce knowledge.) Even if the philosophers' paradigm examples of perceptual states are intentional, supply the warrant for perceptual beliefs, and can enter into reasoning, the mechanisms that produce these states operate on different, nonintentional, nonrational principles, and they can produce adaptive behavior even when they do not produce intentional states. Moreover, there are separate systems for each perceptual modality, which operate in different ways, and at least in the case of vision, there are several distinct subsystems that process different kinds of visual information.

We find much the same thing with motor control. When philosophers think about the control of bodily activity, they are primarily concerned with *action*, in which motor control is driven by intent, often intent that involves various beliefs and desires and practical reasoning. But what one finds in a neuroscience textbook is largely about the unconscious, nonintentional, nonrational, and mechanistic operation of circuits in the motor and premotor cortices and associated noncortical areas. There is a useful and suggestive distinction between "motor" and "premotor" areas of the brain—the latter being involved in motor planning and the organization of behavior into particular sequences of bodily movements—but even this seems something of a bridge between the "inner" space of conceptual thought and reasoning and the more mechanism-like operations of the motor areas. Motor behavior, moreover, can take place without the experience of intentional states or reasoning, not only in the case of true reflexes but also, importantly, in the case of expert performance, such as that of an athlete "in the zone" or a chess master.

Another topic with a similar profile is the comprehension of language. Phenomenologically, we simply hear sounds as meaningful speech. But there are complicated mechanisms underlying the processing of sounds into phonemic, syntactic, and lexical units, and the integration of these units into an interpretation as meaningful speech via sentence parsing. Both the operation of these capacities and their acquisition seem to require special mechanisms, which operate on nonintentional, nonrational principles (Chomsky 1965, 1966). Unlike the mechanisms for perception and motor control, those for comprehension of language seem likely to be unique to the human species.

In each of these cases, we seem to have phenomena that have one foot in the world of concepts, intentionality, and the space of reasons, and the other in the world of neural mechanisms that operate on different principles—principles that are in a broad sense "mechanical." And the relation between the intentional states and the mechanistic neural processes in this case is not between a state at a high level of organization and its "realizing substrate" but more like different links in a causal chain. When we perceive, early perceptual processing *causes* conceptually laden perceptual gestalts and subsequent intentional states and reasoning. When we act, intentional states and practical reasoning *cause* the processes that direct bodily movement to come into play.

This, in turn, suggests a basic way of extending the standard view to accommodate phenomena such as perception, action, and language parsing. There are, as it were, two importantly different types of cognition. On the one hand, there is "central" cognition, which is the subject matter of standard philosophical theories and described by the three-tiered picture, involving concepts, intentional states, and reasoned inference. On the other hand, there are a variety of "peripheral" processes, operating on different, nonintentional, nonrational principles, which can either produce outputs into central cognition (in the cases of perception and language parsing, which indeed might be seen as a specialized form of perception) or take inputs from central cognition (in the case of the motor systems) and produce behavior as a consequence. In the next section, I describe a familiar approach to this division, which contrasts central and peripheral processes by treating peripheral processes as "modular."

3.2 Fodor's *Modularity of Mind*

Jerry Fodor's *Modularity of Mind* (1983) has become a kind of locus classicus for contemporary discussions of modularity and central cognition. I do not mean to suggest that everyone agrees with the conclusions Fodor draws there. Indeed, for every subsequent writer who treats *Modularity of Mind* as a definitive text, there is another who pins it to the wall as a target to shoot at. However, one aspect of Fodor's treatment of modularity has gained a great deal of currency among friends and foes alike: the criteria he lists for what features are typical of modular systems.

Actually, Fodor supplies two such lists, which resemble each other only roughly. The first list emerges from a series of questions he poses about cognitive systems in mind and brain at the end of part 1 (36–37):

1. Is [a cognitive system] **domain specific**, or do its operations cross content domains? This is, of course, the question of vertical versus horizontal cognitive organization; Gall versus Plato.
2. Is the computational system **innately specified**, or is its structure formed by some sort of learning process?
3. Is the computational system "assembled" (in the sense of having been put together from some stock of more elementary subprocesses) or **does its virtual architecture map relatively directly onto its neural implementation**?
4. Is it **hardwired** (in the sense of being associated with specific, localized, and elaborately structured neural systems) or is it implemented by relatively equipotential neural mechanisms?
5. Is it **computationally autonomous** (in Gall's sense), or does it share horizontal resources (of memory, attention, or whatever) with other cognitive systems?

Prospectus: I now propose to use this taxonomic apparatus to introduce the notion of a *cognitive module*. (Fodor 1983, 36–37; boldface emphasis added)

The features that I have rendered in bold type are supposed to be characteristic of modules: "Roughly, modular cognitive systems are domain specific, innately specified, hardwired, autonomous, and not assembled" (37). Fodor immediately goes on to caution (37) that "each of questions 1–5 is susceptible to a 'more or less' sort of answer": that is, these are features that can be had by matter of degree rather than being an all-or-nothing affair. This, of course, raises the question of what to say about systems that possess such features in some intermediate degree, and to this Fodor never gives an answer. Similarly, he leaves it unclear how these features are jointly to provide a litmus for modularity. Does the conjunction of them provide a necessary and sufficient condition for modularity? It seems clear that Fodor does not mean anything so strong (or so precise; recall the "roughly" earlier). But no clear answer ever emerges. Indeed, Fodor seems to think no exact criterion is needed, because the cases fall into two clearly separated groups. But I will explore problems with this assumption anon.

Parts 3 and 4 of *Modularity of Mind* are devoted to making the case for two major claims. First, part 3 argues that "input systems"—including both early sensory processing and the mechanisms underlying language parsing—are modular. (A similar claim could probably be made for "output" systems involved in motor control.) In the course of arguing this claim, Fodor explores a number of features that he finds in such systems, and these features can be viewed as a second list of features associated with modules. Since they are spread across the section headings of part 3, I will simply list them here:

1. Domain specificity
2. Mandatory operation
3. Limited central access (cognitive impenetrability)
4. Fast processing speed
5. Informational encapsulation
6. "Shallow" outputs
7. Fixed neural architecture
8. Characteristic breakdown patterns
9. Characteristic ontogenetic pace and sequencing

Domain specificity appears at the top of each list, but beyond that, the reader is left to figure out how to map the two lists onto each other. I will sidestep this issue by simply treating the second list as canonical.

The Modularity of Input Systems

Fodor regards what he calls "encapsulation" as the most important typical feature of modules. Actually, Fodor sometimes seems to conflate encapsulation and cognitive impenetrability. Both of these ideas are concerned with what kinds of information from other parts of the mind are available to a system. At a first approximation, we might think of cognitive impenetrability in the following way: higher cognitive processes (including, but not limited to, conscious introspection) have access to *some* but not *all* of what goes on in the mind. In vision, I am aware of objects, and sometimes shapes and colors, and perhaps even that my vision is a bit fuzzy this morning. But I am not aware of the early visual processes that lead to visual experience, like edge detection or the operations performed in the ganglion cells or visual cortex. Indeed, it is not only that I characteristically am not aware of such processes; even if I try as hard as I can, I cannot get access to these stages of visual processing. At some point, the early visual processors produce output (say, a perceptual gestalt) that is introspectable and can be used in thinking; but *how* it was produced, and whatever operations and representations were used in producing it, are not passed on for further use (though they can be reconstructed through neuroscientific research). The systems that perform such operations are thus "cognitively impenetrable" in the sense that higher thought cannot "see into" them but has to make do with their outputs.

The opposite question about the information sharing of such systems is what kind of information they can use. Some types of mental processes are quite open-ended in what information they can use. My understanding of cooking and my understanding of thermodynamics might seem like separate things. But if, for example, I am cooking at high altitude, I may have to adjust my cooking in light of what I know about how water boils at different altitudes. That is, my culinary prowess is, in principle, capable of using knowledge that is not itself culinary in character. Likewise, it can use the input of the senses, proprioceptive feedback, and so on. By contrast, early perceptual processing seems largely insulated from almost everything else I know. Conceptual and contextual knowledge *do* play a role at *some* point in perception—say, in forming beliefs about what I am seeing. But they do not affect how the visual system detects edges or constitutes figures, as evidenced by the persistence of the standard visual illusions even after we know them to be illusions. In this way, the early perceptual processing systems are "encapsulated" in the sense of being insensitive to information that might exist elsewhere in mind and brain.

Encapsulation, in turn, is plausibly linked to other features Fodor lists as characteristics of modules. Encapsulated processes have homologues across a great deal of the animal kingdom and are likely to be products of evolution, at least weakly nativistic (in the sense of being species typical, early appearing, and developmentally canalized), and subserved by standard neural structures. They are also likely to be *fast*, both because they are (plausibly) "hardwired" and because they do not waste time gossiping with the rest of the brain. Because they operate largely on their own, without input from higher cognition, they are mandatory. And because they are subserved by standardized species-typical neural mechanisms, they have characteristic breakdown patterns.

In short, there is a prima facie case that we have a set of processing systems in the brain that share a certain set of features that give us reason to group them as a class. Moreover, we can see how at least some of these features might quite naturally be related to one another. Indeed, it is reasonable to read Fodor's argument here as being example driven. Early perceptual mechanisms form most of the class of examples, and the list of features associated with modularity is patterned on what they seem to share in common.

An important caveat here is that Fodor also includes an additional type of process that might *not* initially seem to have much in common with sensory input: linguistic processing. More exactly, he suggests that the mechanisms that lead up to the constitution of sounds as a sentence in a language—that is, a token of a grammatical structure with certain lexical items filling its slots—should be viewed as a module. While a great deal of what we think of as "linguistic thought" (in the sense of thinking *in* a language) is eminently cognitively penetrable, voluntary, and sensitive to a broad swath of knowledge, the mechanisms by which the brain transforms sounds into sentences do seem to be fast and automatic. They are certainly not "nativistic" in the sense of being present from birth: it takes a several years to learn a native language. But since Chomsky's work in linguistics in the 1960s (1965, 1966), if not Broca's and Wernicke's discoveries that injuries to particular areas of the cortex result in the loss of abilities to produce or comprehend language, the idea that there is a nativistic specialized language "unit" has been commonplace in the cognitive sciences. Indeed, Fodor is not the first to think of linguistic comprehension as a kind of perception.

Central Cognition

In part 4, Fodor turns to a second claim that provides a counterpoint to his modularism. Whereas input systems are modular, what we normally regard as "thinking"—more or less, anything involving *beliefs*—is not. The main argument for the claim that most higher cognition is nonmodular goes something like this:

1. Scientific discovery and confirmation are the only higher cognitive processes that we have any real understanding of.
2. Scientific theories are "isotropic" and "Quinean."
3. All thinking is plausibly of a kind with scientific discovery and confirmation: thinking is all about deciding what to *believe*.
4. Therefore thinking is (plausibly) isotropic and Quinean.

Fodor explains the terms 'isotropic' and 'Quinean' as follows:

> By saying that confirmation is isotropic, I mean that the facts relevant to the confirmation of a scientific hypothesis may be drawn from anywhere in the field of previously established empirical (and, of course, demonstrative) truths. Crudely:

everything that the scientist knows is, in principle, relevant to determining what else he ought to believe. In principle, our botany constrains our astronomy, if only we could think of ways to make them connect. (105)

By saying that scientific confirmation is Quinean, I mean that the degree of confirmation assigned to any given hypothesis is sensitive to properties of the entire belief system; as it were, the shape of our whole science bears on the epistemic status of each scientific hypothesis. (108)

These points may perhaps be made more clearly without the neologisms. To take a previous example, one might be tempted to think of culinary understanding as a "knowledge domain." But my culinary beliefs are, at least potentially, subject to challenge and revision from an open-ended set of sources. They are, of course, subject to challenge from the senses. (I *thought* this would be enough salt, but upon tasting the results, I decide otherwise.) But they are also subject to challenge from other domains of knowledge: for instance, I may have to revise the cooking time for a dish when cooking at high altitude based on what I know about how altitude affects cooking, or I may revise a recipe in light of what I have learned about how different cooking techniques affect nutrient content. Or in science, what we discover about electromagnetism (say, when it is incorporated into quantum theory) may eventually pose a problem for a corpuscularian view of matter.

In interpreting the claim that central cognition is "Quinean," it is helpful to distinguish a mild Quineanism from a more radical variety. The more radical variety says that every change in concepts or beliefs necessarily affects *all* my concepts and beliefs. To put it crudely, my understanding of gravitation undergoes subtle changes when I acquire the concept ARCHDEACON or learn that Attila the Hun turned back from his siege of Rome after a conversation with Leo the Great. I have always found such claims wildly implausible. But a more prudent Quineanism goes something like this: we hold our concepts and beliefs only tentatively. They are subject to revision in light of further evidence and integration with other concepts and beliefs. And there is no a priori boundary to where the reasons for revision might potentially come from. This milder Quinean thesis makes an interesting and important claim and is directly relevant to claims about encapsulation, for if all beliefs are Quinean, they are all, in principle, subject to influence from all the rest of our beliefs.

We might do well to distinguish this from another kind of argument for the nonencapsulation of concepts and beliefs. Isotropy and Quineanism are concerned with how concepts and beliefs (recall that the two are inseparable for Quine) are *constituted* and how they may be *revised*. But knowledge domains are also porous in another way. We can take concepts and beliefs from different content domains and combine them through logical machinery. We can string beliefs about different subjects together with logical connectives. And we can combine them in syllogisms. In this sense, both thinking and language are "promiscuous": they are abilities that allow us to combine beliefs about different content domains, both in the simple sense of stringing them together with logical connectives and in the fuller sense of creating new beliefs as a product of deductive reasoning.

In short, neural localists like Gall had it right about input systems, but neural globalists like Flourens had it right about thinking. The mind—at least the human mind—has modular elements, primarily at the sensory periphery. But all the things we are accustomed to regard as "thinking" are nonmodular. And given Fodor's roots in the computationalist tradition, the natural term for such processes is "central." A computer may use peripheral units like a keyboard, scanner, mouse, and printer, and even a separate graphics card, each of which contains its own special hardware circuitry; but most of the *computation* is done by a single "central processing unit" using programs and data stored in a common memory space.

The book ends on a note of epistemological pessimism. Fodor points out that whereas the cognitive sciences have been producing impressive explanations of processes at the sensory and motor peripheries of the brain, we understand almost nothing about how thinking takes place. We do not know of neural localizations for various forms of thought like concepts or judgments the way we do for, say, detection of color and form. And computational models of thinking founder on the frame problem—the problem of setting algorithmic (or even heuristic) boundaries around what a program should treat as the *relevant* information in performing and monitoring a task. This, Fodor points out, has an eerie resonance with isotropy and Quineanism, and he suggests (plausibly) that isotropic and Quinean processes may not be susceptible to modeling by traditional computational approaches in artificial intelligence, and (a bit more speculatively) there may be no other way to gain epistemic traction on thinking, either.

3.3 Motivations, Criticisms, and Alternatives

Fodor's presentation of the distinction between central and modular cognition at least appears to have been intended primarily as a thesis about cognitive architecture, beholden to empirical data about the mind. Moreover, it seems largely to have been case driven: perceptual systems and language processing (and one might reasonably add motor control to the list) seem to have a set of distinctive features that differentiate them from conceptual thinking and reasoning. Whatever the initial motivations, however, the bifurcation between central and modular cognition also seems to carve out a space for the kinds of mental phenomena (concepts, intentional states, reasoning) studied in several core areas of philosophy (epistemology, semantics, truth theory, logic). Moreover, it presents central cognition as a single representational medium that has a common set of types of units (word-sized concepts, sentence-sized intentional states, discourse-sized forms of inference), where thoughts about any subject matter can be brought into contact with one another, and where they are potentially mutually relevant. But if central cognition is architecturally unified in this way, modular cognition, by contrast, is quite disunified: each module has its own characteristic domain, representational system, and inference patterns, and plausibly its own distinctive neural realization and natural history in a process of selection.

While Fodor's notion of a module is widely used in philosophy and the cognitive sciences, it has also received a great deal of criticism, much of which can in some measure be traced to the looseness of his characterization. Recall that Fodor does not tell us how his various criteria for modularity are supposed to fit together—for example, whether they are supposed to jointly supply necessary and sufficient conditions. Some things he says—such as that some of the features are not all-or-nothing affairs but admit of degrees—might suggest a weaker interpretation; but his conclusion that the number of modular systems is quite small, limited more or less to processing of perceptual inputs, motor outputs, and language comprehension, suggests that he intends the criteria to be fairly stringent, and he clearly resists the tendency of some other writers to speak of "modules" wherever a functional decomposition of cognitive abilities occurs (cf. Fodor 2001, 56–57).

Criticisms have therefore come on a number of fronts. Some critics, such as Jesse Prinz, claim that Fodor's criteria are too stringent jointly, and in some cases even individually, to be applied even to the few things that Fodor considers to be modular. On the other hand, there are any number of cognitive abilities that share some but not all of Fodor's criteria.

> Fodor's criteria for modularity do not carve out interesting divisions in the mind. Systems that have been alleged to be modular cannot be characterized by the properties on Fodor's list. At best, these systems have components that satisfy some of Fodor's criteria. There is little reason to think that these criteria hang together, and,

when considered individually, they apply to a scattered and sundry assortment of subsystems. (Prinz 2006, 33)

In subsequent chapters, I make a case that a subset of these features can be applied equally well to a great deal of what Fodor would count as central cognition, and they are quite relevant to the question of what units are needed for a theory of understanding.

The other main objection to Fodorian modularity has come from advocates of the massive modularity thesis (e.g., Carruthers 2006a, 2006b; Tooby and Cosmides 2005). Fodor claims that while the mind has modules, their number is comparatively small, largely limited to input and output processors and a language parser. Advocates of massive modularity argue that the number of mental modules is actually much larger—perhaps in the hundreds or thousands—and, perhaps more importantly, they are not limited to "peripheral" processes of perception and motor control.

The massive modularity thesis is generally developed in tandem with views from evolutionary psychology. We seem to have little problem thinking of *animal* minds as including a grab bag of useful instincts that are products of millions of years of natural selection—that is, of seeing the cognitive abilities of nonhuman animals as distinctly biological traits. Indeed, philosophers have often doubted that any nonhuman animals possess reasoning and other features of centralized cognition. But when we think about *human* minds, we tend to have the opposite bias. In spite of believing that we too are products of natural selection, and indeed that the genetic distance between us and other primates is slight, we seem to have a hard time believing that we too possess many psychological traits that are biological adaptations, and try to attribute as much as possible about our minds to reasoning.

From a biological standpoint, these biases seem misguided. It is true that the tiny genetic differences between humans and chimpanzees support enormous differences in intellectual capacity: the capacity for language, a greatly expanded ability to create and use tools, more complex social cognition, and most of what philosophers have lumped under the headings of "thought" and "reason." And we are right to be interested in these things, as they are important components of who we are, both as individuals and as a species. But biological mutation does not operate by wiping clean the slate of millions of years' accumulation of adaptive traits and designing a new type of mind from scratch. New traits may be built on the old and transform them in many ways; but many of the old traits are sure to still exist, and indeed to still be adaptive in many cases. Moreover, new species-typical psychological traits are themselves likely to be products of selection, and perhaps even adaptations.

Evolutionary psychology thus biases the investigation of divisions of cognitive labor in a direction different from Fodor's. Fundamental to the constitution of units of cognition is their selection history. A capacity is a *trait* because it is a product of selection. Moreover, evolutionary psychologists have often assumed that products of selection will be *adaptations* rather than exaptations or evolutionary by-products, and tend to classify the *function* of a trait in terms of the adaptive advantage it conferred in an ancient ancestral environment.

Viewing cognitive capacities as grounded in evolutionary processes also connects nicely with some of the features attributed to modules. Adaptations are encoded in the genome, and this explains their species typicality. They can thus be expected to have a characteristic developmental sequence that is strongly canalized. And because they are thought of in biological terms, it seems plausible that in many cases they will have standard neural localizations. Both localization and the ability to perform fitness-enhancing functions suggest that they will be fast and automatic. If we assume that selection has operated on various traits individually, it seems likely that the traits will be domain specific, and at least plausible that many of them will be functionally autonomous from one another, including being to some extent encapsulated. And if evolution selects "quick-and-dirty" ways of solving particular problems, and these ways take the form of special-purpose mechanisms in the brain, it seems likely that each will have a representational system that is tooled toward a specific type of problem and response.

Evolutionary psychology thus makes sense of why a number of Fodor's features of modules might go together. It also provides its own criterion for what is likely to be a module. For a trait to be a module, not only must it be species typical, but it must be something that could have been produced through natural selection sufficiently far back in our lineage that the population it appeared in included common ancestors of all contemporary human beings. Anything that has appeared during recorded history could not be a biological adaptation common to the species and hence could not be a module, *even if* it had spread to all human cultures through learning and cultural transmission. (And, indeed, such an ability would be "species typical" only in a statistical, not a biological, sense.) There is thus an important role for something like Fodor's "central cognition" as well, though the principal criterion for it lies not in cognitive architecture but in whether an ability is a product of evolution or of culture and learning. This way of looking at modules greatly expands the number of cognitive capacities that could plausibly count as modular. To name a few: folk physics, folk psychology, folk biology, facial recognition, social contracts, cheater detection, "mind reading" (assessing the psychological states of others), storytelling, contagion avoidance, music, dance, empathy, kindness, practical reasoning, assessment of kinship, incest avoidance, and navigation by dead reckoning.

Evolutionary psychology is itself a highly controversial approach to the mind and has received a number of withering critiques from philosophers of science (Kitcher 1985; O'Hear 1997; Dupré 2001a). Two criticisms stand out as having special importance. First, evolutionary psychologists have been criticized for assuming that products of evolution must be adaptations. Critics point out that many heritable traits are evolutionary by-products and exaptations as well. Second, the treatment of traits as adaptations is linked to highly speculative "just-so stories" about what adaptive advantage a mutation might have conferred in an imagined ancestral environment, and this adaptational story is then used as an account of the *function* of the presumed module.

These criticisms are no doubt justly applied to some of the publications of evolutionary psychologists. There is a tendency toward "strict adaptationism" within evolutionary psychology, and any account one might give of what an ancestral environment was like, or the adaptive advantage a trait conferred in such an environment, will of necessity be highly speculative. But to my mind, the force of these criticisms is often overestimated. Many proponents of evolutionary psychology acknowledge that some traits are exaptations or by-products (cf. Tooby and Cosmides 2005; Carruthers 2006a, 2006b). And the real damage done by this criticism is to the attempt to ground an account of the function of a module in its ancient adaptive role, and not to the broader thesis that many special-purpose traits exist that are products of selection. Likewise, if we abandon the commitment to strict adaptationism, the speculative "just-so stories" play a much weaker role in the justification of the massive modularity hypothesis. We have additional forms of evidence for species-typical traits, such as cross-species comparisons, and for separation of traits, such as functional and developmental dissociability. To the extent that we are content to speak of heritable psychological traits in nonhuman animals, there seems little reason to be skeptical about them when the species in question happens to be our own. Indeed, even psychologists *not* inclined to evolutionary reasoning tend to recognize the existence of things like mechanisms for contagion avoidance and detection of the mental states of conspecifics.

For our purposes, however, it is perhaps more useful to note that the massive modularity thesis is, in important ways, a fairly conservative revision of Fodor's thesis. The criteria for modularity may differ considerably from Fodor's (Carruthers [2006b], for example, stresses dissociability, domain specificity, mandatoriness, neural localizability, and central inaccessibility), but they are not so broad as to include understanding that is accomplished through specialized learning: there are not modules for Newtonian mechanics or chess, though these are ways of understanding particular domains that involve highly specific proprietary rules and are largely dissociable from other kinds of understanding. Moreover, in many cases, the postulation of additional modules does not threaten the modular/central cognition framework. If a module exists for face recognition, contagion, or cheater detection, it may be encapsulated and cognitively impenetrable and function largely to provide something like a propositional input for further centralized cognition and behavioral control. (Indeed, many such purported modules might easily be viewed as specialized perceptual processing mechanisms for things like faces, cheating, and sources of contagion.) Likewise, it is perfectly compatible with the modular/central cognition framework that modules, operating individually or in tandem, may sometimes guide behavior in ways that bypass central cognition entirely. For many animal species, they would probably *have to* do so, as we must draw the line somewhere as to what species possess central cognition at all.

More generally, massive modularity threatens the *unity* of central cognition only to the extent that the modules postulated actually operate *within* the realm of concepts, intentional states, and reasoning. If they merely supply *inputs* to central cognition, or execute motor routines in a fashion *caused by* central cognition, or bypass it entirely, it really does not matter how many modules there are. In the next chapter, I look at a number of types of mental phenomena that clearly involve intentionality and inference but seem to display some of the features Fodor attributes to modules, particularly domain specificity, proprietary representational systems, and a certain degree of automaticity, impenetrability, and encapsulation.

4 Beyond Modularity and Central Cognition

In chapter 2, I noted that several core areas of philosophy concerned with the mind and thinking—semantics, epistemology, truth theory, and logic—treat thought as having units of three distinctive sizes: word-sized concepts; sentence-sized beliefs, judgments, and other intentional states; and argument-sized inferences. I dubbed this assumption about the structural units of thinking the "three-tiered picture" and claimed that the standard view in philosophy seems to be that the resources provided by this picture are adequate for an account of understanding: understanding something is a matter of having the right set of concepts, beliefs, and inferential dispositions, related and deployed in the right ways.

Of course, many subfields within neuroscience and even cognitive psychology deal with cognitive systems that certainly do not involve concepts, beliefs, or argument-like inferences, and this might at first glance seem like an obvious objection to the standard view. However, in chapter 3, I explored an influential strategy for preserving the autonomy of the sphere of meaning and reasons from the rest of the phenomena studied by psychology and neuroscience: the distinction between central and modular cognition. According to this view, many neural systems are modular and as such are domain specific, have proprietary ways of representing their domains and rules for processing information about them, and operate quickly, automatically, and largely independently of our beliefs and intentions. In many cases, they are species-typical traits that might be products of natural selection and have standard neural realizations and characteristic developmental sequences. Central cognition, by contrast, is a single language-like system involving concepts, intentional states, and forms of inference. These concepts, beliefs, and inference methods are primarily learned and vary with culture and individual learning history, operate slowly and laboriously by contrast with modules, and are open to conscious inspection; and central cognition is "domain general" in the sense that it can contain concepts and beliefs about anything we are capable of having concepts of and entertaining beliefs about, and mixing them together in thought and inference. Central and modular systems can interact, but only in the limited sense that modular perceptual systems can provide inputs to central cognition, and central cognition can drive modular motor outputs. In short, the things that philosophers have traditionally classified as "thinking," and perhaps even as "mental," fall in the sphere of central cognition.

In this chapter, I present several lines of research from the cognitive sciences that suggest that this bifurcation between central and modular cognition is not realistic. A great deal of our thinking and understanding, ranging from an infant's earliest ways of understanding the world to scientific theory, with all adult commonsense understanding in between, *also* seems to be organized around particular content domains, and this seems to require that we posit units of thinking of corresponding size—considerably larger than individual concepts, beliefs, and inferences, but considerably smaller than the entire web of concepts, beliefs, and inferential dispositions. Moreover, to make sense of this, we would seem to need to conclude that the cognitive systems that deal in these units—mental models, if you will—must each have their own internal ways of representing their own domains (often quite different from model to model) and rules for producing inferences about them. And because expert performance is expertise *in* such domains, and the judgments of experts are often, in contrast with those of novices, fast, automatic, and not reached through explicit reasoning, we must conclude that model-based understanding *can* have these other features associated with modules as well, even if they are not species typical and lack standard neural localizations.

The upshot of this is not that thinking is never explicit reasoning in a language-like system, or that there are not any systems in the brain that meet the conditions for a Fodorian module. The implication, rather, is that the bifurcation of the mind into central and modular systems is not a good or useful one and obscures a much more general feature of cognition: that both within the sphere of meaning and reasons and outside it, it seems to be a general principle of our cognitive architecture that we deal with the world through many special-purpose systems that represent different features of the world in distinct ways, and we use them *in conjunction with one another* rather than integrating them into a single representational system.

4.1 Core Systems

In recent years, a number of developmental and cognitive psychologists, such as Elizabeth Spelke (2000) and Susan Carey (2011), have argued that the human mind possesses several cognitive systems called core knowledge systems, which are species typical, strongly canalized in development, and already evidenced at the earliest ages at which we have ways of testing for them. The case for core systems is primarily empirical. It consists in evidence that there are particular ways of understanding the world, retained in adulthood but different from full adult understanding, that appear extremely early in development. Indeed, they seem to appear too early to be products of social learning mediated by language, and in some cases, they are evidenced so early that it is difficult to regard them as products of learning at all. These include ways of understanding both the physical and the social world. In their excellent review article, Spelke and Kinzler (2007) summarize it thus:

> Studies of human infants and non-human animals, focused on the ontogenetic and phylogenetic origins of knowledge, provide evidence for four core knowledge systems. ... These systems serve to represent inanimate objects and their mechanical interactions, agents and their goal-directed actions, sets and their numerical relationships of ordering, addition and subtraction, and places in the spatial layout and their geometric relationships. Each system centers on a set of principles that serves to individuate the entities in its domain and to support inferences about the entities' behavior. Each system, moreover, is characterized by a set of signature limits that allow investigators to identify the system across tasks, ages, species, and human cultures. (Spelke and Kinzler 2007, 89)

This is an elegant and compact summary, but precisely for that reason, there is much to unpack within it. I continue by examining their description of each system, and the evidence used to argue for its existence.

The Core Object System

The first candidate for a core system involves an understanding of contiguous solid Objects with distinct and enduring boundaries. (I will capitalize the words 'Object' and 'Agent' when they are used in the special senses that are internal to the core systems hypothesis.) Very early on, children show evidence of expecting such Objects to have a certain set of characteristic properties: they will preserve their boundaries, move as a unit, interact with one another only through contact, and be set into motion only when acted on through direct contact. In addition, children show abilities to track such Objects visually, even when occluded, well before the age that Piaget had supposed that children "discover" object constancy. The system seems to have built-in resources for a very limited form of counting as well, or at least a sensitivity to low numerosities.

It is important to note that the notion of "Object" here is inextricably linked with the ways things are constituted *as* Objects. Philosophers are used to applying the word 'object' much more broadly—say, to abstract objects, to persons, to Cartesian souls, to sets. The notion of the "Object System" is used in explicit contrast with systems oriented toward intentional agents or sets. The relevant notion of "Object" is further contrasted with things to which it is not applied, like liquids and heaps. In other words, the claim is more precisely that there is a psychological system that is applied to spatiotemporal stimuli that are perceived as being cohesive and continuous. Indeed, even this is not completely perspicuous, as it suggests that *first* there is some *independent* test for properties like cohesion and continuity, and *then*, if these are met, the Object System is applied. But in fact the applications of principles of cohesion and continuity are supposed to be themselves part of the Object System—indeed, one might venture to say that if there is such a core system, *this is where our ideas of spatiotemporal cohesion and continuity, as well as Objecthood, originate from.*

The Object System is not supposed to be *universal* in its application. Rather, it applies only to a subset of things we experience. Indeed, that subset is more or less defined by the system itself. The "rules" that govern which experiences activate this system as an interpretive tool—in Kantian terms, their "schematism"—must appear as early in development as the system itself. In these regards, the Object System is importantly different from some other ways that we come to understand the world later in childhood, such as foods, artifacts, piles, and liquids.

There also seem to be *signature constraints* on the ways in which the system can be used.

> Infants are able to represent only a small number of objects at a time (about three ...). These findings provide evidence that a single system, with signature limits, underlies infants' reasoning about the inanimate world. By focusing on these signature limits, investigators of animal cognition have discovered the same core system of object representation in adult non-human primates. ... Like human infants, monkeys' object representations obey the continuity and contact constraints ... and show a set size limit (of four ...). Investigators of cognitive processes in human adults have discovered that the same system governs adults' processes of object directed attention. ... Human adults are able to attend to three or four separately moving objects, for example, when the objects' boundaries and motions accord with the cohesion and continuity constraints. Adults fail to track entities beyond this set size limit, and

they fail to track entities that do not obey the spatiotemporal constraints on objects. (Spelke and Kinzler 2007, 90)

These latter studies suggest that the core object system is *not* simply a developmental stage, like the succession of schemas hypothesized by Piaget, which were understood to be temporary constructs to be discarded when a more adequate schema became available. Rather, they seem to involve enduring capacities that continue to exist in adulthood alongside more sophisticated adult modes of cognition. And since some of those more sophisticated forms of adult cognition—such as particular ways of typifying Objects into more exact categories, and generalized arithmetic abilities—are subject to cultural variation and transmission, adult cognition should include the core abilities even if a person has no understanding of arithmetic operations like addition and multiplication, or even a generalized notion of the counting numbers.

The Core Number System

Proponents of core systems claim that there is a second system that is also concerned with numerosity. Whereas the core object system is supposed to involve abilities involving exact small numerosities, the core number system involves abilities to detect and compare significantly larger numerosities, such as might be presented by two stimuli, one displaying ten objects and another thirty. This ability is, additionally, nonmodal, in the sense that it does not operate only on stimuli presented through one sensory modality. And, significantly, its application is not limited to contiguous Objects but includes elements such as sounds and actions in its domain. Again, such abilities appear very early, endure into adulthood, and are found in nearby species.

The Core Agency System

From early on, children treat people, animals, and other stimuli that display signs of agency differently from inanimate objects, liquids, and things like piles of sand. The behaviors of the former are treated as goal directed, and infants tend to mirror them in their own behavior. Some such mirroring, such as mimicry of four stereotypical facial gestures, has been observed very early indeed, perhaps as early as hours after birth.

Spatio-temporal principles do not govern infants' representations of agents, who need not be cohesive ..., continuous in their paths of motion ..., or subject to contact in their interactions with other agents. ... Instead, the intentional actions of agents are directed to goals ..., and agents achieve their goals through means that are efficient. ... Agents also interact contingently ... and reciprocally. ... Agents do not need to have perceptible faces. ... When they do, however, infants use their direction of gaze to interpret their social and non-social actions ..., even as newborns. ... In contrast, infants do not interpret the motions of inanimate objects as goal-directed ..., and they do not attempt to mirror such actions. (Spelke and Kinzler 2007, 90)

Interpreting agents differently from inanimate objects is clearly a feature of adult cognition as well, and a similar profile of agent-directed cognition has been evidenced by studies of nonhuman animals.

The core agency system is more controversial than some of the other core systems. One reason for this is that a module for interpreting Agency (capitalized to reflect its special usage within the theory) requires encoding of specialized ways of detecting some highly sophisticated phenomena. An innate ability to detect contiguous solids requires much less sophistication of design than one to detect purposeful action. But by the same token, Agency should be a difficult notion for humans (or other animals) to construct out of a prior inventory of notions confined to spatiotemporal properties of stimuli. (Indeed, one might argue that such notions are insufficient to the task. Agency-concepts are not a conservative extension of Object-concepts.) In other words, there is reason to think that this is exactly the sort of case where one might most strongly expect to find special-purpose ways of understanding the world, on grounds similar to Chomsky's "poverty of the stimulus" argument for an innate grammar module. People and tree leaves both move, and perhaps even an empiricist mind could detect that they display different patterns of movement. However, it is less clear that, just on the basis of the objects and their movements, such a mind could come up with the hypothesis of Agency. And even if it could do so, there are an open-ended number of ways of grouping different types of motion that would be hypotheses equally consistent with the data. But in fact we all employ more or less the same framework for interpreting Agents, and we show evidence of doing so very early. Since the stimuli underdetermine this interpretation, we have reason to think that there is at least some innate bias toward it. In addition, it is not clear that we could function as members of a *social* species, or engage in particular types of early social learning, without the very early introduction of something that allows us to treat conspecifics as having typical features of Agency, and so it seems likely that any social species must possess developmentally canalized species-typical cognitive abilities for the detection of Agency.

The core agency system would seem to be constitutively tied to our notions of Agency. That is, if the hypothesis is correct, we may be unable to get outside our innate understanding of Agency to analyze it or its domain in nonintentional terms. We can, to be sure, come to understand that some things that initially *seemed* to be Agents—say, characters in cartoons—are not really bona fide Agents at all. But it seems unlikely that we can cash out the phenomena that are interpreted by this system, or the things that make up its domain, in anything but an intentional vocabulary. Such notions as we have in this region of our conceptual space would seem to be tied to our developmentally canalized endowment of core capacities, even if they can be expanded on, enriched, and adumbrated. In this respect, Dennett's (1971, 1987) idea of an intentional stance—a special framework for interpreting things as acting on the basis of their beliefs and goals—seems well conceived. Likewise, the domain specificity of the core agency system seems constitutive in nature. That is, one cannot specify the class of things to which the system is applied *except* through the system itself, or perhaps some more sophisticated descendant of it.

The Core Geometric System

The fourth core system is devoted to an understanding of the geometric properties of the environment, "the distance, angle, and sense relations among extended surfaces in the surrounding layout" (Spelke and Kinzler 2007, 91). It does not represent nongeometric properties such as color or odor (91). The evidence for such a system is drawn from studies of how children, adults, and nonhuman animals orient themselves spatially. Adults use at least two orienting systems, one guided by surface geometry and the other by landmarks. The landmark-based orientation system develops later, whereas the geometric orientation system is evidenced very early. The Geometric System, moreover, seems to undergo an important type of generalization as it becomes dissociated from the need for a fixed origin and angle of orientation.

The existence of such a system may seem either obvious or unnecessary. The geometrical character of the physical world seems so patently obvious that it may seem unclear why we should need a special system to understand it. Yet this very obviousness may simply be a product of how deeply ingrained our geometric assumptions in fact are. (As Kant suggested, we may actually be unable to perceive space except as structured by a highly specific geometry.) We can see this most readily by considering artifacts that have some features of human perception while lacking others. A camera receives visual information about the environment but does not interpret that information as being a product of a three-dimensional world. To get a computer to infer such a world of objects from optical input devices such as cameras, it is necessary to endow it with a representational system for the geometry of such objects, and ways of constructing representations of three-dimensional objects from two-dimensional views. More than one such system is available. For example, one could do so using spherical coordinates that orient objects only to a particular position and angle of gaze. Such a system does not guarantee that there will also be a way of extracting perspective-invariant representations of geometric layouts or predicting how objects will appear after transformations of the origin or the direction of gaze. To achieve this, the system must be endowed with a particular type of geometric representation, in which objects are treated as located with respect to one another in an observer-independent frame of reference. There are, of course, still multiple geometries that could be applied for this purpose—Euclidean, spherical, hyperbolic, and so on. But any particular representational system employed for this purpose must employ some particular type of geometry.

Characteristics of the Core System Hypotheses

As empirical claims, each of the hypotheses about particular core systems should be taken individually. However, that they share some common features is of both scientific and philosophical interest. (Indeed, with such issues, it is hard to find a clear dividing line between philosophically oriented science and scientifically oriented philosophy.) A number of such features are attributed to all four core systems:

1. *Species typicality*. The core systems are found in both children and adults.
2. *Nativism*. The systems are developmentally canalized and early appearing.
3. *Analogs*. Analogs are found in closely related species.
4. *Domain specificity*. The systems are applied to particular types of stimuli and not to other types.
5. *Proprietary representational system*. Each system represents its subject matter in a particular fashion, affording particular types of information and expectations about it.
6. *Characteristic breakdown patterns*.
7. *Signature limits*.

The similarities between this list and Fodor's criteria for modularity are substantial. Indeed, advocates of the core systems hypothesis tend to regard core systems as modules. Significantly, however, they are not simply systems for preprocessing perceptual inputs but ways of conceptualizing and thinking about the world. They are used in interactions with the world *as* a world of Objects and Agents existing in a geometrically characterized space, and they allow for anticipations, predictions, inferences, and motor planning. In this respect, they seem much more like Fodor's central cognition and in particular have similarities to the other, more sophisticated, ways that we acquire of thinking about objects, agents, space, and numerosity through later development and specialized learning, like the folk theories described in the next section. Both the core systems and these later-acquired ways of conceptualizing the world are "domain specific" in that they are ways of thinking about particular parts of aspects of the world. Yet at the same time, they do not allow for cross-domain reasoning *within* themselves, a defining feature of central cognition.

An important question about the core systems as a group is how they are related to one another. In particular, the two systems that provide ways of categorizing types of things—the Object and Agency systems—are functionally dissociable from each other and have contradictory rules, yet one and the same thing can trigger either or both systems. The psychologist Paul Bloom (2004), for example, emphasizes that children (and adults) can think of something as an Agent without thinking of it as an Object, and argues that this dissociability might be what lends intuitive plausibility to the idea of a soul that is independent of the body. When the Agency system operates without the Object system, one conceives of a thing as an Agent but not as an Object, and such a thought is not counterintuitive to the child because only the Agency system is in play, and it has no rules that require Agents to also be Objects. Indeed, some of the rules of the two systems are logically incompatible: Agents move on the basis of their own goals, whereas Objects are incapable of self-initiated motion. Objects must move from place to place in spatially continuous paths, and infants show surprise if things that have shown the profile of Objects seem to disappear from one place and reappear in another, but do not show the same surprise when things showing the profile of Agents do so. Nothing could really correspond to the rules of both the Agency and the Object systems, because some of the rules of the two systems contradict one another.

At the same time, a single stimulus might be able to trigger either system, perhaps even both at once. A toy that is made with features such as a face, like Thomas the Tank Engine, might at one time be regarded as an Object, which the child expects to move only when someone pushes it down the tracks. But at another time it might be regarded as an Agent, especially when the child has reached an age when she can watch videos depicting it as doing agential things or hear stories about its adventures. Indeed, even without the deliberately ambiguous features, a model train that is first observed inert might be assumed to be an Object and then, to the child's surprise, be reconstituted as an Agent when the power switch is thrown and the train begins to move without any obvious external cause.

Of course, as I have described the examples, the identification of the Agent and the Object as "the same thing" is supplied from the perspective of the adult observer, and in the examples described, the stimulus is perceived first through one core system and then through the other. It is less clear whether (a) a single stimulus can *simultaneously* trigger both systems, and (b) if it can do so, whether the child can bind the two representations together as representations *of* the selfsame object and regard one and the same thing as being both Agent and Object. Bloom seems to think not and claims that we naturally think of human beings as consisting of two things—a body (an Object) and a soul (an Agent). I am skeptical of Bloom's conclusion. Adults, at least, seem to be quite capable of triangulating a single object through multiple cognitive lenses, even if the two ways of regarding it are logically incompatible. Of course, it is possible that we do this through some mechanism other than the core systems. And even if adults can triangulate through the core systems, it is also possible that the ability is acquired only long after the core systems are in place. While the specific question about how the core systems can be combined is an empirical question requiring more research, there is nonetheless an important point here about *some* forms of cognition: that we understand the selfsame object through more than one way of thinking about it, and incompatibilities between the two ways of thinking about it do not seem to present an obstacle to the psychological possibility of doing so.

4.2 Folk Theories

A second and slightly older hypothesis in cognitive and developmental psychology is the claim that ordinary human understanding involves implicit theories about the natural, biological, mental, and social world, acquired over the course of early to middle childhood and persisting through adulthood. These implicit theories are commonly referred to by names such as "folk physics," "folk biology," and "folk psychology." They consist of whatever species-typical assumptions there may be about matters such as how objects move when thrown or dropped, what features we can expect of any animal species (e.g., that individual animals of a species are offspring of parents of the same species, have a species-typical physiology and diet, etc.), and how we interpret people's actions as consequences of their beliefs and desires.

Developmentalists who study folk theories note that they are not completely mastered until sometime in mid- to late childhood (the time frame varies with the folk theory) and have some typical developmental sequencing (cf. Gopnik and Meltzoff 1997; Gopnik 1996). For example, children attribute beliefs to other people before they understand that other people's beliefs are different from their own. This latter capacity is generally gained at an age of around four or five years in human children, and it is a matter of some controversy which, if any, other animal species share it. Advocates of the view that commonsense understanding is theoretical—sometimes called "the theory-theory"—tend to characterize this understanding in terms of general *beliefs*, like "animals have parents of the same species" and "people tend to do what they believe will bring about the things they desire." The different forms of folk understanding (e.g., folk biology and folk psychology) also seem to have their own central concepts, like SPECIES and BELIEF, and characteristic inference patterns. Folk-theoretic understanding is thus described in the vocabulary used for what Fodor calls central cognition, though folk theories themselves have some characteristics of modules, such as being domain specific, appearing in a characteristic developmental sequence, being applied automatically, and having their own special categories, forms of representation, and inference rules.

Some cognitivists also suggest that these somewhat abstract central concepts also form a kind of *template* for more specific concepts. Pascal Boyer (2001), for example, calls the concepts ANIMAL and PERSON "ontological categories." When a person learns a new animal concept—say, ZEBRA—she does not need to learn that members of the kind will have a characteristic diet, physiology, and means of reproduction. The ANIMAL category is used as a template, with "slots" for certain types of information that are assumed to be available for all species and can be filled in with more specific information about a particular species as one learns more about it. Thus even children know many of the right questions to ask to learn more about a new animal kind they have just encountered or been told about: What do they eat, and how do they get their food? How big do they get? What kind of environment do they live in? How do they move around? Thus something about the *structure* of the folk theory encodes a certain kind of understanding—often left tacit—that guides the formation of more particular beliefs.

While the acquisition of folk theories seems to be strongly developmentally canalized—they are species-typical features of how adults and older children think about physical objects, persons, and animals, and these appear at characteristic ages and in characteristic developmental sequences—they are clearly a feature of everyday thinking of the sorts that, under Fodor's classification, would count as central cognition. But they also have several features characteristic of Fodorian modules: they pertain to particular domains (physical objects, persons, animals), employ distinctive categories for thinking about their domains, encode particular assumptions about these domains, and provide a framework for more specific sorts of representations of things falling within them.

4.3 Scientific Theories

The expression "folk theory" was explicitly intended to convey the hypothesis that what developing children are doing in constructing an understanding of the world shares important commonalities with what scientists are doing in constructing more specialized and exacting theories in disciplines such as physics, biology, and psychology. Indeed, one of the leading advocates of this "theory-theory," Alison Gopnik (1996), titled one of her articles "The Child as Scientist." Important differences, of course, distinguish the child's cognitive development from the processes of theory formation and testing in the theoretical sciences. Personally I should prefer to reserve the term 'theory' for the paradigmatic scientific instances; but the basic point, that the child's construction of ways of interpreting particular domains in the world and the scientist's enterprise are in significant ways members of a common class of cognitive undertakings, seems plausible. Scientific theories, of course, stand at the opposite end of the spectrum of intellectual sophistication from our early acquired "folk" understandings of the world. They do, however, have at least one feature that recommends them for consideration in any discussion of cognition: namely, that they have been extensively studied and analyzed.

Before about 1960, the prevailing school of philosophy of science, logical empiricism, was committed to the view that the vocabulary of science must be in a theory-neutral observational language, and that theories are propositions, or sets of propositions, using that vocabulary and terms constructed from it. (Theories and law claims were additionally widely assumed to be universally quantified claims. For a critique of this assumption, see Horst 2011.) Disagreements over theories and theory change were thus to be seen as differences concerning what propositions should be held true; but the propositions were understood to be framed in a common vocabulary, or at least one that could be specified independently of the competing theories. All of this began to change in the 1960s, as a number of philosophers of science began to see individual theories as tightly interconnected units, and to see competing (or successor) theories not so much in terms of making incompatible claims in the same vocabulary as offering alternative ways of conceptualizing their subject matter. Competing (or successor) theories came to be regarded not so much as *contradictory* as *incommensurable*, and theory change not as piecemeal change in particular scientific beliefs but as revolutionary change in paradigms for understanding particular aspects of the world.

A number of different philosophers played roles in developing a new consensus understanding of scientific theories—Thomas Kuhn, Imre Lakatos, Paul Feyerabend, and David Lewis all certainly deserve special mention—and entire books have been written about their works, individually and collectively. As my aim here is merely to pick out certain consensus views, I will not even attempt to do justice to all the contributors to this important period in the philosophy of science. I will cite Kuhn as a main contributor, though ultimately my interest is as much in his later views (around the time of his 2000 APA presidential address) as in his seminal *The Structure of Scientific Revolutions* (1962).

Kuhn's initial approach was as a historian of science, concerned with the nature of theory change. As an introduction, here is an extended quote offered retrospectively in a later publication:

> A historian reading an out-of-date scientific text characteristically encounters passages that make no sense. That is an experience I have had repeatedly whether my subject is an Aristotle, a Newton, a Volta, a Bohr, or a Planck. It has been standard to ignore such passages or to dismiss them as products of error, ignorance, or superstition, and that response is occasionally appropriate. More often, however, sympathetic contemplation of the troublesome passages suggests a different diagnosis. The apparent textual anomalies are artifacts, products of misreading.
>
> For lack of an alternative, the historian has been understanding words and phrases in the text as he or she would if they had occurred in contemporary discourse. Through much of the text that way of reading proceeds without difficulty; most terms in the historian's vocabulary are still used as they were by the author of the text. But some sets of interrelated terms are not, and it is [the] failure to isolate those terms and to discover how they were used that has permitted the passages in question to seem anomalous. Apparent anomaly is thus ordinarily evidence of the need for local adjustment of the lexicon, and it often provides clues to the nature of that adjustment as well. An important clue to problems in reading Aristotle's physics is provided by the discovery that the term translated 'motion' in his text refers not simply to change of position but to all changes characterized by two end points. Similar difficulties in reading Planck's early papers begin to dissolve with the discovery that, for Planck before 1907, 'the energy element *hv*' referred, not to a physically indivisible atom of energy (later to be called 'the energy quantum') but to a

mental subdivision of the energy continuum, any point on which could be physically occupied.

These examples all turn out to involve more than mere changes in the use of terms, thus illustrating what I had in mind years ago when speaking of the "incommensurability" of successive scientific theories. In its original mathematical use 'incommensurability' meant "no common measure," for example of the hypotenuse and side of an isosceles right triangle. Applied to a pair of theories in the same historical line, the term meant that there was no common language into which both could be fully translated. (Kuhn 1989/2000, 9–10)

While scientific theories employ terms used more generally in ordinary language, and the same term may appear in multiple theories, key theoretical terminology is proprietary to the theory and cannot be understood apart from it. To learn a new theory, one must master the terminology as a whole: "Many of the referring terms of at least scientific languages cannot be acquired or defined one at a time but must instead be learned in clusters" (Kuhn 1983/2000, 211). And as the meanings of the terms and the connections between them differ from theory to theory, a statement from one theory may literally be nonsensical in the framework of another. The Newtonian notions of absolute space and of mass that is independent of velocity, for example, are nonsensical within the context of relativistic mechanics. The different theoretical vocabularies are also tied to different theoretical taxonomies of objects. Ptolemy's theory classified the sun as a planet, defined as something that orbits the Earth, whereas Copernicus's theory classified the sun as a star and planets as things that orbit stars, hence making the Earth a planet. Moreover, not only does the classificatory vocabulary of a theory come as an ensemble—with different elements in nonoverlapping contrast classes—but it is also interdefined with the laws of the theory. The tight constitutive interconnections within scientific theories between terms and other terms, and between terms and laws, have the important consequence that any change in terms or laws ramifies to constitute changes in meanings of terms and the law or laws involved with the theory (though, in significant contrast with Quinean holism, it need not ramify to constitute changes in meaning, belief, or inferential commitments outside the boundaries of the theory).

While Kuhn's initial interest was in revolutionary changes in theories about what is in a broader sense a single phenomenon (e.g., changes in theories of gravitation, thermodynamics, or astronomy), he later came to realize that similar considerations could be applied to differences in uses of theoretical terms between contemporary subdisciplines in a science (1983/2000, 238). And while he continued to favor a linguistic analogy for talking about conceptual change and incommensurability, he moved from speaking about moving between theories as "translation" to a "bilingualism" that afforded multiple resources for understanding the world—a change that is particularly important when considering differences in terms as used in different subdisciplines.

Scientific theories are thus like modules in having specific domains, which are represented in a proprietary fashion. There are as many scientific theories as there are well-understood phenomena, and each theory is constitutively intertwined with the ways phenomena are understood through it. Theories likewise employ proprietary taxonomies of object kinds, properties, relations, and transformations and use proprietary representational systems with their own inference rules. But theories are unlike Fodorian modules in that they are products of learning, and subject to scrutiny and revision. Indeed, scientific theories are a case par excellence of intellectual products that require explicit representation. Moreover, they are like Fodorian central cognition in that they can be combined with other forms of thought in reasoning: they can be combined with other theoretical claims and with commonsense observation and thinking in reasoning to a conclusion, their appropriate use can be monitored through ways of evaluating context, and their content can be challenged through observation and dialectic.

4.4 Intuitive Reasoning, Semantic Reasoning, and Knowledge Representation

We have thus far looked at several kinds of cognition that do not fit comfortably into the bifurcated framework of central and modular cognition. Core systems, folk theories, and scientific theories are domain specific, and the understanding they underwrite comes in units larger than the size of individual concepts and beliefs, but smaller than the entire web of concepts, beliefs, and inference patterns. Core systems, like modules, are unlearned and cannot be altered; folk theories and scientific theories are products of learning, and the learning and alteration of them involve wholesale alterations in units the size of a domain-centered theory because the concepts and inference patterns distinctive of them are constitutively interdefined. They involve implicit divisions of their domains into what we might call *implicit ontologies*, with proprietary categories for kinds of things, properties, relations, and transformations. And once learned to the level of expertise, they can operate more or less automatically, quickly producing thoughts and inferences that are "intuitive" in the sense that they are not based on explicit conscious reasoning. In these ways, they have much in common with Fodorian modules.

With respect to some of Fodor's other criteria, they differ from one another: core systems are highly developmentally canalized, folk theories are canalized as well but require a great deal of training and bootstrapping, while scientific theories are products of specialized learning, usually requiring invention or instruction. Core and folk systems, but not scientific theories, are plausibly products of evolution and may have species-typical neural realizations. It would seem that the mind has *some* uses for domain-specific understanding that is not modular in Fodor's sense. But perhaps it is too early to draw general conclusions from this, as core systems, folk theories, and scientific theories, while clearly "mental" in ways that perceptual input processors are not, are each rather special forms of cognition, in the first two cases distinguished by their early appearance and developmental canalization, and in the latter by the specialized learning required.

But in fact we have good reason to think that domain-sized units of understanding, with proprietary ways of representing domains and tight connections between concepts and inference patterns within the domain but much looser connections across domain boundaries, are found broadly within what Fodor would classify as central cognition. They would seem to be, among other things, the basis for semantically based reasoning and intuitive inference. One reason for thinking this might be so is that researchers in artificial intelligence, who started out trying to simulate human cognitive abilities on the model of explicit reasoning in something like central cognition, ended up having to turn to theories positing domain-specific understanding to explain everyday intuitive reasoning abilities.

Artificial intelligence emerged as a discipline fast on the heels of Turing's seminal discussion of digital computation. The first projects in AI were still closely tied to computation's mathematical heritage. They were automated theorem provers like Newell and Simon's (1956) Logic Theory Machine. The viability of theorem-proving computers was already assured in principle by Turing's theoretical work. A computer can execute any formalized algorithm, and so any proof that can be carried out by formal means can in principle be proved by a computing machine. Researchers then attempted to model human formal reasoning more generally, in systems like General Problem Solver (Newell, Shaw, and Simon 1959). Such work is often viewed as the first generation of AI.

In attempting to model human reasoning more generally, however, AI researchers quickly hit an important roadblock. Precisely because formal reasoning techniques abstract away from the semantic values of the symbols, they are unsuited to underwriting the many semantically based inferences human beings routinely perform. Formal techniques treat predicates like 'dog' as lexical primitives. They underwrite valid syllogisms involving such lexical units, but not the knowledge that is present in human semantic competence. For example, we all know that if Lassie is a dog, then Lassie is an animal, she has a weight, she has bones and kidneys, and so on. But to get a computer to simulate this type of understanding, we must equip it with more than the ability to perform valid inference techniques on symbolic structures like "Lassie is a dog." In short, we must encode human-level semantic understanding of the concept DOG.

One way to do this is to supply the computer with a set of explicit propositional representations, like "all dogs are animals," "dogs have bones," and so on. If this could be done with everything we know about dogs, it might allow formal inference techniques to generate a set of inferences equivalent to those a human being would make. And if our only concern were to get a computer to reach the same conclusions humans would reach, regardless of *how* they reach them, this might be enough. But if we are also concerned to get the computer to make such inferences in a fashion that resembles the way human beings make them, encoding semantic understanding in the form of explicit propositional representations seems psychologically unrealistic. We understand things we have never put into explicit propositional form. You may never have considered the proposition "dogs have kidneys" before, but in some sense you understood it to be true. You may never have entertained the proposition "119 + 7 = 126" before, but it was in some sense implicit in your mathematical understanding. And indeed we are inclined to say that we *knew* those things yesterday, even though we had never thought of them. That is, we didn't just *learn* them overnight, and had we been asked, we would have given the right answer without any difficulty.

A second way of approaching the problem is to model semantic understanding, not in the form of explicit propositions but in terms of a *data structure*—say, like that of a relational database. The second generation of AI was largely devoted to exploring the way semantic understanding is structured as *data*. The appeal of this strategy lay not only in its greater ability to produce simulations that could not be achieved through propositional representation, but also because it seems the more psychologically realistic, natural way to model semantic relations that are couched at the conceptual, rather than the propositional, level. The work of the models is done, however, not at the level of the semantics of individual conceptual or lexical units, but in explicating the systematic relations between them. Treating semantic relations as data structures also has the advantage that many such relations are not well captured by standard logical machinery. Our understanding, say, that dogs are four-legged is *not* accurately captured by the proposition "All dogs have four legs," as we understand that some dogs have lost one or more legs, and indeed a few dogs have extra legs. Some sort of semantic link clearly exists between the concepts DOG and FOUR-LEGGED, and perhaps one could devise a logical operator that would track this link extensionally. But as it is not clear from the outset how to do this, we are well served to find a way to model such links, if we can, in a fashion that avoids the problem of how to model their precise relation to extensional logic.

Information Chunking in Knowledge Representation

The second generation of AI, in the 1970s, was marked by efforts to explore such an approach. Marvin Minsky, writing in 1974, sums up the general concerns as follows:

> It seems to me that the ingredients of most theories both in Artificial Intelligence and in Psychology have been on the whole too minute, local, and unstructured to account—either practically or phenomenologically—for the effectiveness of common-sense thought. The "chunks" of reasoning, language, memory, and "perception" ought to be larger and more structured; their factual and procedural contents must be more intimately connected in order to explain the apparent power and speed of mental activities.
>
> Similar feelings seem to be emerging in several centers working on theories of intelligence. They take one form in the proposal of Papert and myself ... to substructure knowledge into "micro-worlds"; another form in the "Problem-spaces" of Newell and Simon ...; and yet another in new, large structures that theorists like Schank ..., Abelson ..., and Norman ... assign to linguistic objects. I see all these as moving away from the traditional attempts both by behavioristic psychologists and by logic-oriented students of Artificial Intelligence in trying to represent knowledge as collections of separate, simple fragments. (Minsky 1974)

It is useful to look at a few examples of such proposed structures for modeling understanding and knowledge deriving from these projects.

Semantic Networks

Semantic networks are data structures for encoding the semantic relations between lexical or conceptual units. The structure of semantic networks is best conveyed in diagrammatic form. Conceptual or lexical units are represented by nodes (diagrammatically, by words or by boxes or other closed figures), and semantic relations between them by links between nodes (diagrammatically, by lines and arrows) (fig. 4.1). One important relationship between sortal concepts is that of category subsumption: for example, that dogs are animals. In a semantic network diagram, this could be represented by nodes for DOG and ANIMAL connected by a directional link (an arrow) with a label representing the fact that the particular semantic relation in question is a subsumption relation. (This was often expressed with the label "IS-A," which was unfortunately used for a number of conceptually and ontologically distinct relations.)

Of course, explicating the semantics of these nodes would require us to make explicit their relations to many other nodes as well, thus resulting in the characteristic network structure of this type of representation.

Semantic networks afford us not only an approach to encoding semantic relations in digital computers, but also a way of understanding how they

Figure 4.1
A simple node-and-link structure in a semantic network encoding the information that dogs are animals.

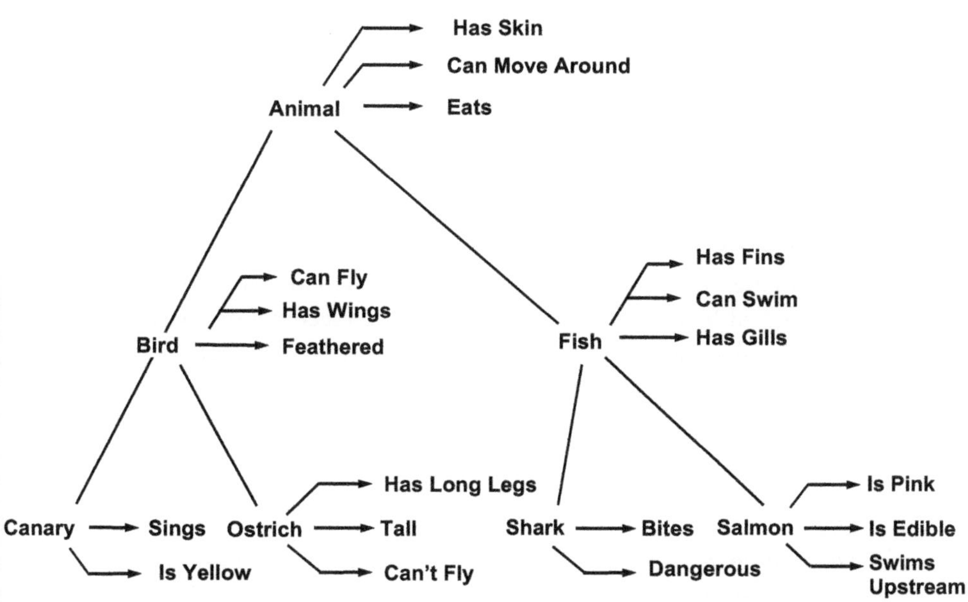

Figure 4.2
An example of a network memory structure from Collins and Quillian 1969. Reprinted from Cohen and Feigenbaum 1982, 40.

might be structured in human beings (the "structure" in question being a kind of abstract formal/functional structure, not an anatomical structure). It is an approach that seems more psychologically realistic than treating our semantic understanding as a matter of having an enormous number of explicit propositional beliefs like "dogs are animals." Our understanding of dogs seems to involve a vast network of things that we seldom if ever explicitly think about, but which are in some sense "there" for us should we ever need to access them. It is tempting, in trying to characterize what we do when we employ semantically based understanding, to turn to metaphors like tracing down the strands of a web until we find what we need. Of course, neither introspection nor the formalism of semantic networks can tell us how—through what mechanisms—this is accomplished in the human brain. But it seems, at some level of abstraction, to get something right about our psychology that is not captured by treating semantic understanding in terms of formal inference: there is a great deal of our understanding that is packed into semantic relations, and our way of accessing this understanding is different from the retrieval of stored propositional representations.

Another advantage of the semantic network approach is that it leaves the question of what types of links there might be as one that can be guided by empirical research, rather than forcing it into the mold of whatever logic a theorist happens to be most familiar with. Indeed, research in knowledge representation has led to an expansion of our understanding of the types of logics that might be relevant to modeling human thought.

Semantic networks in AI bear important resemblances to network theories of meaning in philosophy of language and provide a machinery for testing and refining such theories. But they also raise an important question that philosophers often overlook. Given that our semantic understanding is characterized, at least in part, by links between conceptual or lexical units, are *all* such units connected into a single network, or are there perhaps a variety of disjoint networks whose units can be connected nonsemantically through language and syntactically based inference techniques? In philosophy, advocates of network views tend also to be holists. But semantic network modeling does not require holism, though it is compatible with it. The topology of semantic connections is ultimately an empirical question, and one that bears heavily on the topics of unities and disunities of mind and understanding.

Frames

Semantic networks are structures for representing and encoding semantic understanding at the level of semantic relations between conceptual or lexical units. Semantic understanding and logical deduction, however, do not exhaust the types of cognitive skills that are needed for understanding and interacting with the everyday world. When we interact with a new object of a familiar type, or walk into a situation of a type we have encountered before, we do not have to start from the ground up in assessing the situation or knowing what to expect. In interpreting an object as a ball or a cube, or a room as the dining room of a restaurant, we automatically have a whole set of expectations about what we will encounter, what sorts of actions we might perform that are relevant to that situation (and perhaps even stereotypical of it), and what sequence of events we would expect to ensue if we were to perform one or another of those actions. We automatically assume a great deal about the object or situation that is not directly perceived: that the cube will look a certain way if we were to go around to the other side, that the hostess will seat us, that there is an unseen kitchen where food is being prepared, that we will be expected to pay for our meal, and so on.

In short, we seem to have *mental models* of various familiar types of objects and situations. Such models allow for a great deal of variety of detail to be filled in and actively explored, but they will also often involve default expectations. For example, unless one sees a cafeteria line or a sign saying "Please Seat Yourself," we assume that someone who works for the restaurant will come and take us to a table. Marvin Minsky (1974) writes of this:

> When we enter a room we seem to see the entire scene at a glance. But seeing is really an extended process. It takes time to fill in details, collect evidence, make conjectures, test, deduce, and interpret in ways that depend on our knowledge, expectations and goals. Wrong first impressions have to be revised. Nevertheless, all this proceeds so quickly and smoothly that it seems to demand a special explanation.

The "special explanation" he offers is that such abilities are underwritten by a particular type of knowledge structure that he calls a "frame."

> A *frame* is a data-structure for representing a stereotyped situation, like being in a certain kind of living room, or going to a child's birthday party. Attached to each frame are several kinds of information. Some of this information is about how to use the frame. Some is about what one can expect to happen next. Some is about what to do if these expectations are not confirmed.
>
> We can think of a frame as a network of nodes and relations. The "top levels" of a frame are fixed, and represent things that are always true about the supposed situation. The lower levels have many *terminals*—"slots" that must be filled by specific

> instances or data. Each terminal can specify conditions its assignments must meet. (The assignments themselves are usually smaller "sub-frames.") Simple conditions are specified by markers that might require a terminal assignment to be a person, an object of sufficient value, or a pointer to a sub-frame of a certain type. More complex conditions can specify relations among the things assigned to several terminals.
>
> Collections of related frames are linked together into *frame-systems*. The effects of important actions are mirrored by transformations between the frames of a system. These are used to make certain kinds of calculations economical, to represent changes of emphasis and attention, and to account for the effectiveness of "imagery." (Minsky 1974)

Frames, like semantic networks, are thus data structures involving a node-and-link architecture. But whereas semantic networks are cast at the level of semantic relations between lexical or conceptual units, frames represent the space of possible objects, events, and actions stereotypically associated with a particular type of context or situation.

Minsky (1974) explores a diverse set of examples of how the frame concept can be applied, ranging from visual perception of an object and spatial imagery to meaning structure of a discourse and understandings of social interactions; late in the article, he assimilates frames to Kuhnian paradigms, though without elaborating on the nature of the relationship (e.g., whether he thinks that scientific understanding is frame based). He also cites the work of several other researchers whose work, although employing other labels, like "model" or "script," is proceeding essentially along similar lines. Relating all of these, not to mention other applications of frame-based knowledge representation subsequent to Minsky's article, would unduly extend this chapter. But it is difficult to get the idea across adequately without attending to at least a few examples that give some sense of what a frame is, and how flexibly this way of representing knowledge can be applied.

Vision and Perspective

Consider the common experience of becoming acquainted with an object visually. At any one time, one sees the object from a single viewing angle, from which only a portion of its surface is visible. One knows, however, that there will be views from additional angles as well, and may anticipate some of the visual changes that will occur as one moves, say, around the object to the right (or, in the case of a small object, as one moves it around in one's hands). As one does so, new information becomes available: the shape of the formerly unseen parts of the object may meet our expectations or surprise us, and the colors, textures, and patterns on the additional surfaces come into view. At the same time, we do not *forget* the sides we saw before, which have now passed out of view. Rather, as we move around, we are building an ever-more-comprehensive understanding of the spatial and visual properties of the object. To think about features that were visible before, we may not need to go back and look again but may be able simply to call them to mind. Moreover, when we encounter the object again or encounter a similar object, we can imagine features of the unseen sides (perhaps incorrectly, of course) without directly reacquainting ourselves with them by moving ourselves or the object so that they are in view.

The implication of this seems to be that, in seeing an object, we are not simply experiencing something like a photograph (or a binocular interpolation of photographs) of the object from a given angle, or even seeing a succession of photographs taken from different vantages. We do indeed always see an object from an angle; but we also have some kind of mental model of the object that either consists in, or is generative of, understanding of what it looks like from multiple perspectives. And this is not simply a feature of how we see individual objects. We gain templates for the understanding of new objects of familiar kinds: if you are handed a new pair of dice at the casino, you do not need to examine them to see what is on the sides (unless you have reasons for suspicion), because your mind already fills in default information that each side will have one to six pips. On the other hand, some aspects of what one will find are specified only at a generic level: picking up a book, you expect to see text, but the content of the text requires empirical examination, as does the question of whether there will also be pictures.

The question Minsky essentially poses is, What kind of information structure could have these features? His answer is that it is a particular type of frame structure. In his own example, an observer is looking at a cube from multiple angles. Minsky develops the data structure diagrammatically, using a system of nodes and links (fig. 4.3).

To the left is a representation for the reader of how the cube looks from an angle, with sides labeled with letters. To the right is a diagram intended to represent the data structure used to encode such a view *as* a view of a cube. The circular node at the top of the diagram represents the type of the object, in this case a cube. This node has three arrows going down that are labeled "region-of"—that is, the three nodes to which they connect are represented as being regions of (the surface of) the cube. These are labeled in the diagram with arbitrary letter designations of faces of the cube, A, B and E. The region-nodes have internal links that are labeled for the spatial relations between regions. Each region-node also has an additional outbound

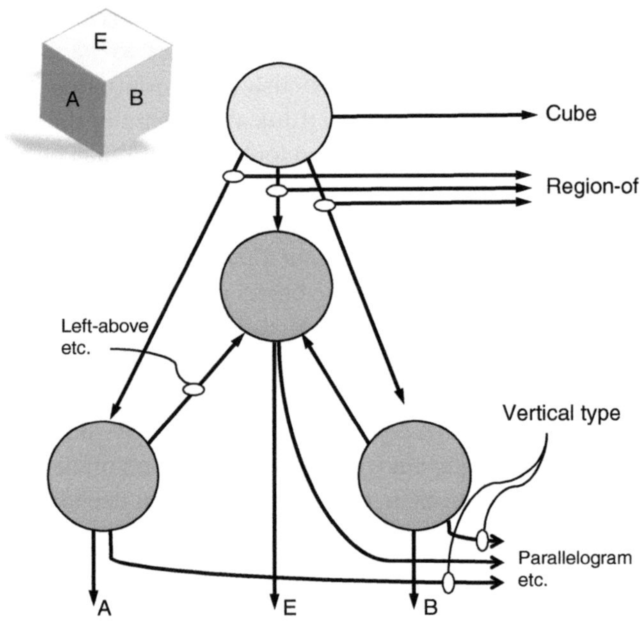

Figure 4.3
Frame structure for views of a cube. Based on figure in Minsky 1974 (electronic, n.p.).

link that indicates its internal spatial configuration (e.g., it appears as a particular type of parallelogram from this angle).

Now suppose I move around to the right (or turn the cube correspondingly to the left). Face A disappears from view, a new face C appears, and I learn about the surface features of C that were previously unavailable to my view. This new view could be represented by a new node-and-link structure that is structurally similar to the previous one, but with nodes for different sides of the cube. But my visual understanding of the cube in fact involves more than simply having one view after another. To retrieve information about the first view, I do not need to go back to my original position, because it is still represented in my mental model of the object. (As Shepard and Meltzer [1971] had already shown, we are able to rotate 3-D images in visual imagination.) Minsky points out that we can posit a data structure for our frame that can encompass more views than we can actually see at a given time:

> Since we know we moved to the right, we can save "B" by assigning it also to the "left face" terminal of a second cube-frame. To save "A" (just in case!) we connect it also to an extra, invisible face-terminal of the new cube-schema as in figure [4.4].

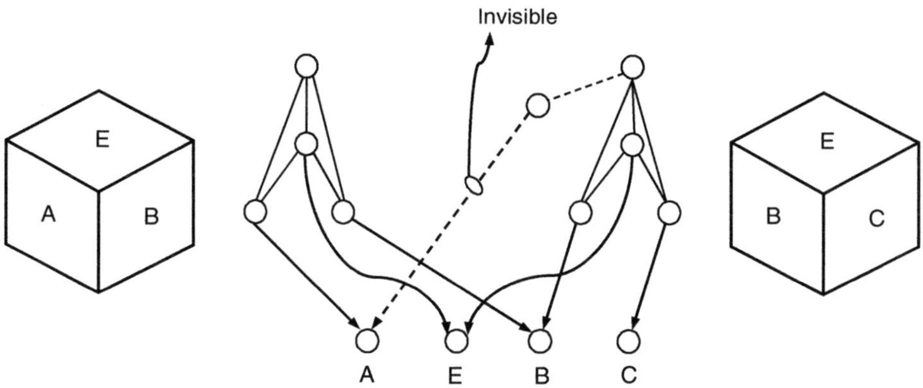

Figure 4.4
Based on fig. 1.2 from Minsky 1974.

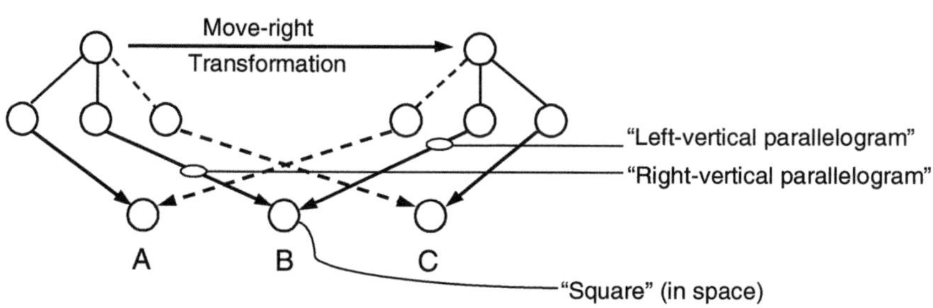

Figure 4.5
Based on fig. 1.3 from Minsky 1974.

If later we move back to the left, *we can reconstruct the first scene without any perceptual computation at all*: just restore the top-level pointers to the first cube-frame. We now need a place to store "C"; we can add yet another invisible face to the right in the first cube-frame! [See figure 4.5.]

We could extend this to represent further excursions around the object. This would lead to a more comprehensive frame system, in which each frame represents a different "perspective" of a cube. In figure [4.6] there are three frames corresponding to 45-degree MOVE-RIGHT and MOVE-LEFT actions. (Minsky 1974; figure numbers amended to correspond to those in this chapter)

In this example, a multiview representation of the object takes the form of a *frame system*, composed of individual frames corresponding to views, and linked by spatial operations. Such a frame system, moreover, is useful at two levels of abstraction. First, it is used as a representation of the (seen

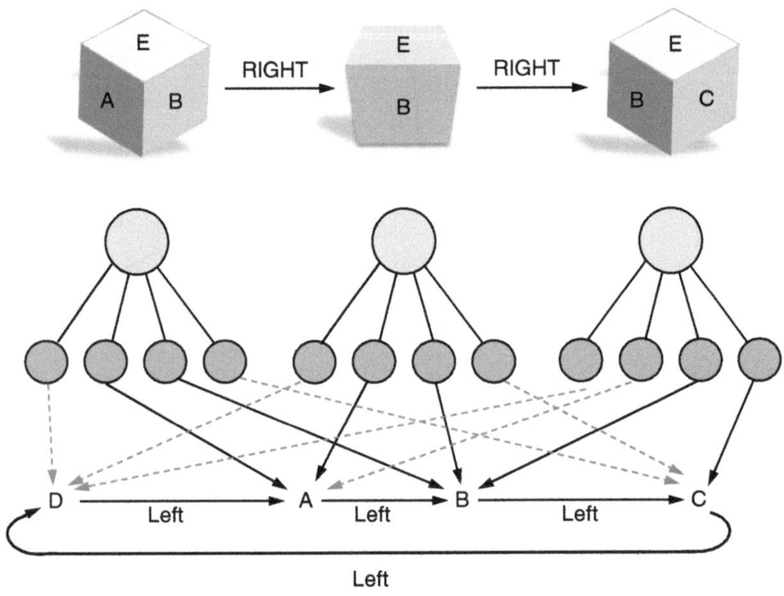

Figure 4.6
Based on fig. 1.4 from Minsky 1974.

and unseen) visual features of a particular object. But once such a frame system has been constructed, it can then be *reused* when one encounters other objects that are similar to the original. That is, such a frame system becomes a way of representing *cubes in general*. And this, in turn, supplies us with knowledge of a way of interacting with new cubes we encounter, including what kinds of information to look for, and how to manipulate the object to find it.

It is thus crucial that frames and frame systems have a combination of information that is specified and variable-valued "slots" that are left to be "filled in" by exploration of each particular object. This affords an organism that possesses frame-based cognition a powerful tool in dealing with new objects of familiar types. It can search for a frame that seems to be a match with a new stimulus, and then efficiently explore the stimulus, looking for just the information that will fill in the slots specified by the frame (say, the surface features of a new cube). In the course of this process, it may turn out that the chosen frame structure is not a good match for the new stimulus, and then the organism must either find another existing frame or else construct a new one based on the new stimulus as a paradigm case.

Visual Imagination

Most humans are capable of imagining objects visually. (Degrees of proficiency in visual imagery seem to cover quite a spectrum, ranging from individuals who apparently have no visual imagery at all to people with extremely vivid visual imagery that can interfere significantly with online visual perception.) I can picture an object, and in doing so, I do not simply apprehend a snapshot image. Rather, I can do things like *rotate* the image in my mind and intuit the shape of its unseen sides (Shepard and Meltzer 1971; Shepard and Cooper 1982). Thus at least some sorts of imagining involve not simply the mental equivalent of two-dimensional pictures but robust and manipulable three-dimensional models of objects. This will be familiar to many readers from the exercises in spatial reasoning found in some standardized tests, in which one is presented with a picture of a convoluted object from one perspective and then asked to determine what it would look like from another angle. In visual imagination, objects are generally not presented with all the visual detail that one would find in visual perception. They may, for example, be more like wireframe models of objects, with borders and surfaces, but no visual detail like color or texture on the surfaces.

Both of these features of visual imagination can be accounted for by the hypothesis that the mind contains something like frame systems that are permanently stored templates for visible objects. The templates themselves are by nature *abstract*, in the sense that they specify some, but not all, of the visual features that we might find in an individual object. This allows them to be reusable and to be decoupled from perception. Such structures afford us a number of important capacities. They provide templates for the classification of visual stimuli. They then allow for efficient information gathering about objects we encounter, by limiting the search space of relevant questions to resolve. But they also allow for offline reasoning about objects. In actual perception, we can extrapolate unseen features. And even in the absence of the object, we can figure out things about it, and about ways of interacting with it. (I recall that when I was learning the cello, I would often think through fingering patterns in imagination.)

Social Interactions and Scripts

This type of abstract informational structure has uses outside perception and imagination as well. For example, every human being has an understanding of a stock of social situations and interactions. Some of our practical and social knowledge involves an understanding of stereotypic sequences of events or actions. A recipe, for example, involves a sequence of actions on a set of objects that must be performed in a certain order. A social situation like dining at a restaurant involves a standard sequence of events like being taken to one's table, being given a menu, choosing items from the menu, having the courses brought in order and consuming them, being brought the bill, paying it, and tipping. Roger Schank explored a particular framelike structure called a *script* for encoding such situations, his most noted example being a script for restaurant dining (fig. 4.7; Schank and Abelson 1977).

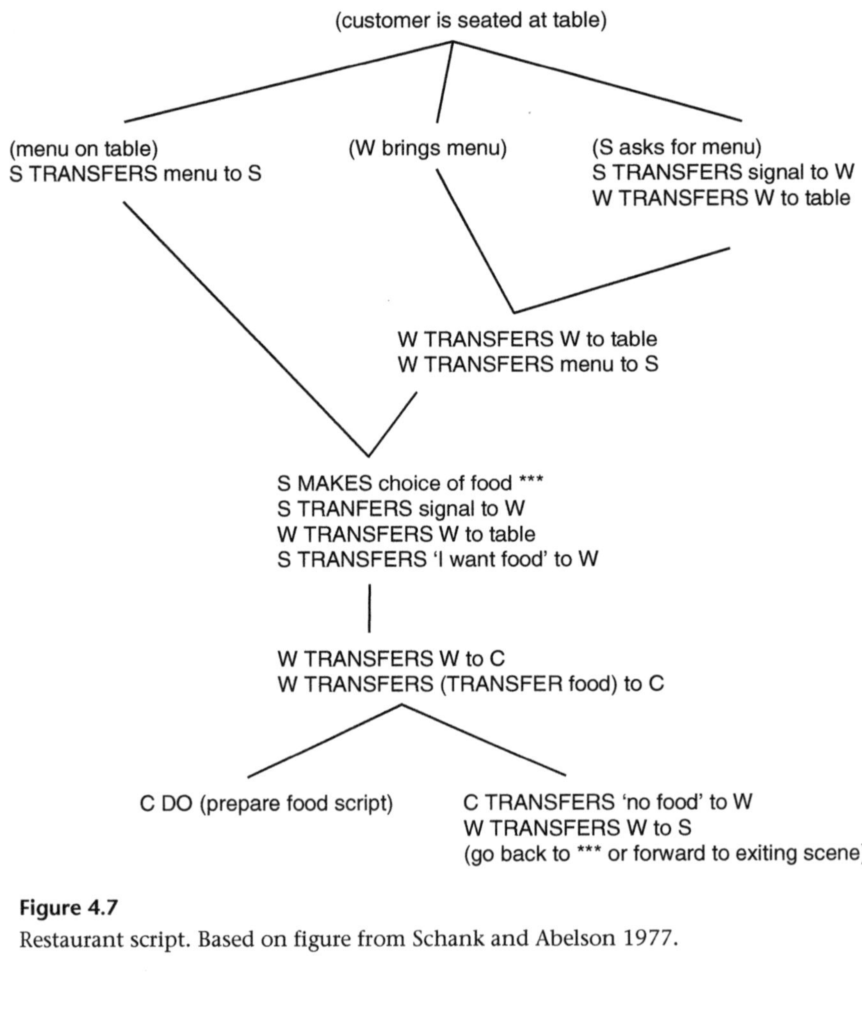

Figure 4.7
Restaurant script. Based on figure from Schank and Abelson 1977.

Reflections on Frames

Knowledge theorists like Minsky and Schank took themselves to be in the business both of finding ways that machines could replicate human competences and of postulating how such competences are possible in human thought. The criteria for success in these two enterprises are, of course, quite different. As projects in artificial intelligence, they are successful to the extent that they endow machines with particular types of competence, regardless of whether this is achieved in the same fashion that it takes place in human beings. As projects in cognitive science, they are successful to the extent that they postulate processes that are psychologically and neurally plausible in humans (and other species), regardless of how closely the models that researchers actually program can replicate actual human performance.

As artificial intelligence, projects of interpreting understanding in situation-specific representational systems ran into an important barrier. While researchers like Minsky and Schank achieved some success in simulating competence in carefully constrained situations through algorithmic techniques, the same techniques do not seem to be suited to the task of replicating our ability to assess what is relevant to a situation and choose a frame to use in novel or complicated situations, or when to shift frames. (For example, we know when to shift out of the restaurant script if, say, the building is on fire or another diner is choking.) Some writers, like Hubert Dreyfus (1979), have argued that this broader task of frame choice and relevance assessment is not something that can be reduced to an algorithm, and hence is not suitable for treatment by standard (algorithmic) techniques in artificial intelligence. This may be an important limitation for at least one form of AI. But it need not be a problem for cognitive science. If knowledge structures like frames exist in the mind, *they need not be implemented algorithmically*. And even if they are implemented algorithmically, other brain processes underwriting the assessment of relevance may be nonalgorithmic.

Minsky seems clearly to be onto something in holding that there must be some type of knowledge structures that are chunked at the level of situation-specific or domain-specific understanding. Human cognition seems to make heavy use of models that encode knowledge very particular to a situation or domain. And these seem to require representation of abstract properties with further "slots" left to be filled in by the particular cases one encounters. It is a further, and more dubious, question whether a given *particular* theoretical model of how knowledge of a particular domain is represented is a good one. Frames, however, strike me as providing a useful and flexible tool for empirical research that aims at making explicit what is involved in situational understanding.

Assuming that there are things like frames that organize our understanding at all, we must have a great many of them. But beyond this, we will often apply several of them to the same situation. The birthday cake on the table in front of me may be both dessert and a cube. I admire the artistic icing by turning it around like any other cube. I know to use the dessert fork because of my understanding of etiquette. As Minsky points out in examples of his own, this applicability of multiple frames is found in quite a variety of situations.

> Sometimes, in "problem-solving" we use two or more descriptions in a more complex way to construct an analogy or to apply two radically *different* kinds of analysis to the same situation. For hard problems, one "problem space" is usually not enough! The context of the von Neumann quotation [which introduces the section in Minsky's article] is a proof that the two early formulations of quantum mechanics, Heisenberg's matrix theory and Schrodinger's wave mechanics, could be seen as mathematically identical, when viewed from the frame of Hilbert Space. Here, two very different structural descriptions were shown to be very similar, but only by representing both of them from a third viewpoint.
>
> But we do not have to look to mathematics for such examples; we find the same thing in this everyday scenario: Suppose your car battery runs down. You believe that there is an electricity shortage and blame the generator.
>
> Seen as a mechanical system, the generator can be represented as a rotor with pulley wheel driven by a belt from the engine. Is the belt still intact and tight enough? The output, seen mechanically, is a cable to the battery. Is the cable still intact? Are the bolts tight, etc.?
>
> Seen electrically, the generator's rotor is seen as a flux-linking coil. The brushes and commutator (in older models) are seen as electrical switches. The output is current that runs through conductors.
>
> We thus represent the situation in two quite different frame-systems. In one, the armature is a mechanical rotor with pulley, in the other it is a conductor in a changing magnetic field. The same—or analogous—elements share terminals of different frames, and the frame-transformations apply only to some of them. (Minsky 1974)

This will turn out to be an important aspect of cognition. Just as we (and other animals) are able to unite perceptions of an object *as* perceptions of a single object through various sensory modalities, so we are also able to think about what is in some sense a single object, situation, or subject matter through multiple frames. This in turn raises the questions of how cross-frame reasoning and understanding are possible, whether the use of multiple frames is a *necessary* feature of cognition, and whether it may introduce any unwanted artifacts into our reasoning.

4.5 Mental Models

The Scottish psychologist Kenneth Craik suggested that we understand and interact with the world by having a "small-scale model" of the world.

> If the organism carries a "small-scale model" of external reality and of its own possible actions within its head, it is able to try out various alternatives, conclude which is the best of them, react to future situations before they arise, utilize the knowledge of past events in dealing with the present and the future, and in every way react in a much fuller, safer, and more competent manner to the emergencies which face it. (Craik 1943, 61)

Craik was prevented by his untimely death from working out this idea in detail, but one key component seems to have been that models reflect systems of relations in the phenomena of which they are models. While the foregoing quote may suggest that Craik thought that each of us has a single, comprehensive model of "external reality" and our own possible actions, other passages suggest that there are separate models of different domains. He writes of models more generally:

> By a model we thus mean any physical or chemical system which has a similar relation-structure to that of the process it imitates. By "relation-structure" I do not mean some obscure non-physical entity which attends to the model, but the fact that it is a physical working model which works in the same way as the process it parallels, in the aspects under consideration at any moment. (Craik 1943, 51)

And of mental models:

> My hypothesis then is that thought models, or parallels, reality—that its essential feature is not "the mind," "the self," "sense-data," nor propositions but symbolism, and this symbolism is largely of the same kind as that which is familiar to us in mechanical devices which aid thought and calculation. (Craik 1943, 57)

Just what this "nonpropositional symbolism" might amount to is unclear, and we should probably resist the temptation to read later associations the word 'symbol' has from within the computational approach to thought. Johnson-Laird has suggested possible connections to Wittgenstein's pictorial or isomorphism view of thought and Peirce's iconic symbols, which represent by dint of structural similarity. Craik's interest, however, seems to be not with semantics but with adaptive cognition: if structural similarities between a model and the system modeled are important, it is because this allows the model to function as a mental surrogate for the system it models in off-line cognition through simulation or nonpropositional reasoning. Craik's suggestion that mental models include representations not only of features of the world but also of our own actions is also highly suggestive of the idea that models of this sort might bridge the gap between theoretical and practical reasoning and provide a basis for an account of understanding.

The language of "mental models" lay somewhat dormant between Craik's death and the reintroduction of the terminology by Gentner and Stevens (1983) and Johnson-Laird (1983), apart from occasional informal usage by writers like Minsky who had their own technical terminology. It is not clear to me that there is a substantive difference between the *general* notions of frames and models, though more recent advocates of things called "mental models" have developed the metaphor in several different ways. The term 'mental model' has been used more widely than 'frame' in work in cognitive psychology, AI, cognitivist philosophy of science, and exploration of human–computer interfaces. As it is my own preferred terminology, I will defer a discussion of the range of model-based accounts of cognition until chapter 5.

4.6 Moving beyond Central and Modular Cognition

The examples surveyed in this chapter cut a wide swath across our mental lives. They range from our earliest ways of thinking about the world as infants, through commonsense understanding of the physical and social world, to our most sophisticated forms of understanding, scientific theories. All involve concepts, beliefs, and inferential dispositions. But in each case, the concepts, beliefs, and inferential patterns involved seem to be tightly bound together and centered on particular content domains. In this respect, each of these types of understanding resonates more closely with Fodor's characterization of modules than of central cognition. With the exception of core systems, they are unlike Fodorian modules in being products of learning and in most cases can be revised through further learning. Many of them, moreover, are not universal or even typical of human cognition: you need special types of learning and a particular type of social context to learn restaurant etiquette or Newtonian mechanics. On the other hand, once acquired to a level of expertise, even the more specialized ones can produce intuitive judgments far more quickly than can be done through stepwise explicit reasoning, and the processes through which they produce such judgments are cognitively impenetrable.

In short, these examples suggest that understanding does not look much like the way central cognition is generally characterized. Does this mean that the space of beliefs and reasons is also modular? The answer one is inclined to give to that question will depend heavily on just how one decides to use the word 'module'. Folk theories, scientific theories, frames, and models clearly do not meet all the criteria for Fodor's modularity, but one could decide that this calls for a far looser notion of modularity. At one time I was tempted by this strategy—to fight over the definition of 'modularity' and claim that understanding is modularized as well. Now, however, I think this is the wrong approach, for two reasons. The first is that Fodor's definition is sufficiently entrenched in the professional literature that one might as well concede the definition of that particular word. The second and more important reason is that the distinction between modular and central cognition just does not seem to be all that appropriate or useful if human understanding, from infancy through scientific theory, shares so many properties that are supposed to be distinctive of modules.

The better conclusion to draw is that *some* of the features Fodor assigned to modules actually seem to be much more general features of human cognition, appearing in cognitive systems that are deeply canalized and ones that require special learning, in systems that fall into place at different developmental stages, in systems that are consciously accessible and ones that are not, in systems that employ language and ones that do not, in systems that are shared with many nonhuman species and ones that are uniquely human. We need to move beyond the bifurcated framework of modular versus central cognition and develop an account of these shared features of very different cognitive systems, particularly of their implications for an account of understanding. This does not require that we deny that there are systems that answer to Fodor's strict characterization of a module, and it certainly does not require that we deny that we can think in a language (and perhaps in language-like ways even in the absence of a natural language). But it does require that we give a more general account of how cognition in general and understanding in particular make heavy use of domain-sized units, which I call "mental models." To this task I turn in the next chapter.

5 Cognitive Pluralism

In chapter 4, we saw evidence that understanding is organized into units that are domain specific, relatively self-contained, and possess distinctive representational systems and inference patterns. This characterization has much in common with the way cognitive "modules" are typically described, but is intended to describe a much more general feature of cognition. The organization of understanding into domain-centered units that can operate semiautonomously on their own principles is found in nativistic and highly encapsulated perceptual Fodorian modules, in nonperceptual cognitive systems (such as core knowledge systems) that are canalized early in development and might be products of natural selection, in the organization of acquired understanding in semantic networks and frames, and in scientific models, our most exacting and regimented form of understanding. This would seem to be a feature of the architecture of understanding that is independent of whether a particular mode of understanding is "nativistic" or learned, neutrally localized or distributed, species typical or dependent on individual learning history or acculturation. Indeed, I will claim that it is a basic design principle for the architecture of understanding.

We need a nomenclature both for this view of cognitive architecture and for the domain-specific units of understanding that it posits. I will refer to the view I am espousing here as *cognitive pluralism* and will use the expression 'mental model' as a generic term for the domain-sized units of understanding we have observed in modules, core knowledge systems, folk theories, scientific theories, and Minskian frames. Settling on this terminology for domain-sized units of understanding was in some measure a matter of choice, as there were several obvious candidates, none of them perfect, and each having preexisting usages: 'model', 'schema', 'frame', 'framework', 'theory'. Short of inventing entirely new terminology, there was really no way of finding a term that would not risk some degree of confusion with other writers' uses of familiar words. In the end, I settled on 'model' largely because it seemed to cause fewer problems and misunderstandings than the others. Psychologists balked at 'schema' because of confusions with a well-known "schema theory" in their discipline (Bartlett 1932). Minsky's notion of a frame is very close to what I mean by a mental model, but is likewise associated with a particular theory, one whose associations with strong AI (the assumption that the mind is literally a computer and that mental processes are computational) I do not wish to endorse. (I do not actually think that Minsky's frame theory requires an endorsement of strong AI; however, given that it originated in this context, in the end I decided not to use his terminology to avoid potential misunderstandings.) The term 'theory' I prefer to reserve for a specialized form of cognition found paradigmatically in the sciences. And so I was left with 'model' by a process of elimination. Of course, the term 'model' also has a number of existing uses, including an importantly different use in cognitive science explored by Philip Johnson-Laird (1983) and a philosophical use in logic, and I will attempt to situate my notion of a model in relation to these and other existing uses in chapter 9, and to develop at greater length what I mean by a "mental model" in chapters 6 and 7.

The claim that mental models are fundamental units of understanding does not require us to deny that we also possess a more ecumenical form of language-like thinking, in which concepts and beliefs related to different domains can be brought together through syntactic construction and formal inference. Indeed, this would be a strange thing to deny. Even if you are skeptical about how much of our thinking takes the form of occurrent sentence-like thoughts, there is no way around the fact that we think in this fashion, at the least, when we think in a natural language such as English. Language-like thinking and model-based thinking are good for different things and are complementary to each other. Models provide a basis for semantic understanding and intuitive inference, but each model provides insight only into its own domain. In my view, semantic understanding and semantic inference are not really a part of language (or language-like thinking) at all. Our concepts and the things they allow us to represent and reason about are primarily a product of the models we possess. Language allows us to take these and combine them regardless of their models of origin and to perform types of inference that are content neutral. Language also allows us to make our thoughts explicit, and thereby to examine, test, and refine them, both individually and in interpersonal dialectic. Much of what is distinctive about human cognition would be impossible without language or language-like thought. But language would be largely devoid of content without models. (I develop these themes further in chap. 11.)

5.1 What Is Cognitive Pluralism?

The basic thesis of cognitive pluralism is that the mind employs multiple special-purpose models of parts, aspects, and features of the world, rather than (1) a single, consistent, and integrated model of everything or (2) a long inventory of more specific and independent individual beliefs. If a "worldview" is construed as a comprehensive and consistent model of the world, then we possess nothing that answers to the description of a worldview. But our understanding is at the same time far more systematic than simply a list of beliefs, even beliefs that happen to cohere into a consistent set.

This is, first and foremost, a claim about cognitive architecture: that domain-centered mental models are fundamental units of understanding, distinct from, and irreducible to, beliefs or judgments. In fact, I argue that models provide much of the basis for the semantic properties of concepts and judgments, and for a psychologically realistic account of dispositional beliefs. This, and other implications for semantics and epistemology, is the focus of part 3. One of the main questions I address there concerns whether the representational disunity that comes with a multitude of different mental models presents principled problems for various concepts of the unification of knowledge and understanding.

The type of "plurality" at the heart of cognitive pluralism is primarily a *representational* plurality, and the type of "unity" that is denied is a representational unity. Cognitive pluralism is the thesis that our ways of understanding the world are all partial, idealized, and cast in individual representational systems and perhaps cannot be reconstructed into a single representational system that is at once comprehensive and consistent. These different representational systems are attuned to particular phenomena in the world and weakly optimized for pragmatic goals in interacting with them. Some of these representational systems are weakly nativistic and take species-typical forms. Others are acquired through trial and error, social learning, and the special processes involved in learning technical theories like those found in mathematics and the sciences. Which models are acquired may vary widely between individuals and over the course of a lifetime in a single individual. Through them, we "triangulate" a common reality without the construction of a comprehensive and consistent worldview.

Cognitive pluralism need not be committed to denying other, nonrepresentational types of cognitive unity, such as personal identity, the transcendental unity of apperception, the unity of individual intentional states or perceptual gestalts, or the ability to combine insights originating in separate models through logic and language. Nor need the cognitive pluralist be hostile to the project of unification as a regulative ideal. Moreover, while cognitive pluralism holds that mental models are a fundamental unit of understanding distinct from, and arguably prior to, language-like thinking in concepts, judgments, and inferences, it does not deny that there *is* such language-like thought. Indeed, the addition of capacities for language-like thinking to a model-based architecture confers distinct and substantial cognitive advances that make up a great deal of what is distinctive about human minds.

5.2 Modules and Models

I propose that a pluralistic cognitive architecture can be found at a number of biological, evolutionary, and cognitive levels. It is found as a design principle in many animal species in the form of developmentally canalized modules. This modular architecture arguably becomes more weakly nativistic (i.e., is increasingly open to developmental variables and learning) in animals with more complex neural structures but is conserved even in human beings. In humans and numerous other species, pluralistic architecture takes a new turn in the ability to acquire knowledge of the world and of how to interact with it in the form of domain-specific *learned models*. *Scientific* models are a special case of learned models, whose special features lie in their regimentation and minimization (though not elimination) of features peculiar to the cognizer. In humans, this pluralistic architecture is supplemented by special capacities for logical and linguistic thought, which permit concepts and insights originating in modular and model-based understanding to be combined in a domain-general representational medium. However, this domain-general medium does not thereby become a universal and domain-general super model. In humans, moreover, a great number of learned models are socially shared or even socially distributed and are transmitted through language and social learning.

I will treat the notions of "module" and "model" as overlapping categories, typified on different grounds. If a cognitive system is a module, then it must have a strongly or weakly nativistic (i.e., canalized) etiology and a species-typical relation between functionality and neural localization (broadly understood to include distributed localization [M. Anderson 2014]). By contrast, I will speak of a "model" as a unit of understanding that encodes systematic relational information about a domain and thereby affords the possibility of representations of particular states of affairs within a state-space of situations in the world, in the organism, or at the interface between organism and world. (I explore the nature of models at greater length in chaps. 6 and 7.) "Models" in this sense can be nativistic or products of learning, and thus some modules may be models as well. However, further examination of some modules may lead to the conclusion that they drive adaptive response without representing features of the world, the organism, or the interface between the two. It is thus an empirical and theoretical question whether all modules are also models. (I apologize for the difficulties of parsing that may be caused by the phonetic similarities of the words 'model' and 'module'. Unfortunately, each seems the most apt word to use for its own subject matter.)

We may, however, contrast modules, including those that are also models, with *learned* models, whether the latter are acquired through trial-and-error learning or through social transmission. It is possible that this will prove to be a continuum, rather than a clean partition, as weakly nativistic structures often require training through experience. It is largely an empirical question whether there will prove to be a continuum of ways that learning is implicated in the formation of mental models, or whether there will be a natural line of demarcation.

5.3 Models and Representation

To model something is necessarily to model it *in a particular way*, employing a particular representational system. For example, classical mechanics modeled space under gravitation using a Euclidean metric, whereas relativistic mechanics models space under gravitation using a Lorentzian metric. One map models the surface of the earth using a Mercator projection with lines representing roads, while another uses a polar projection with lines representing altitudes of landforms. Fechner modeled psychophysical data using a logarithmic scale; Stevens used a power function.

A model is characterized by

- the types of objects, properties, relations, states of affairs, and affordances it models;
- the representational system it uses to model them;
- the set of operations or transformations among these it models;
- the space of possible representations of states of affairs that is generated by its representational system.

For example, a model of the game of chess must have ways of representing different types of pieces, the space of positions on the board, the moves possible for each piece, captures, and the overall game state. A model of space under gravitation must contain resources for representing objects (bodies), their relevant properties (mass, position, momentum), and laws governing their gravitational dynamics.

A model is *apt* to the extent that its representational system tracks the salient properties of the phenomena that it is its function to model. Aptness need not be an all-or-nothing matter. Classical gravitational models are quite apt for modeling the influence of gravity in most of the cases human beings encounter, but not for relativistic cases. Indeed, aptness may have multiple components. It may involve both *fit with a range of data* and *suitability for things like computation.* Classical models approach relativistic models asymptotically with respect to aptness of fit in low-mass, low-velocity situations and exceed them with respect to computational simplicity. Aptness is a pragmatic matter, and there are multiple dimensions of pragmatic virtue.

5.4 Representation

The notion of representation has a checkered history in philosophy. Indeed, in my first book (Horst 1996), I took computationalists' theories of the mind to task for their use of a notion of representation. Readers of that book may thus find it surprising that I should use a notion of representation here. My concern there, however, was that a particular familiar notion of representation was being put to an illegitimate use. Computationalists rely on a notion of representation grounded in the paradigm of symbols. Paradigm examples of symbols, like those in written or spoken language, have both syntactic and semantic properties. Computationalists posit that "meaning" may be attributed univocally to symbols in a natural language, mental states, and the hypothetical symbols of a language of thought. They then suggest that the meanings of mental states can be explained by the supposed meanings of symbols in the language of thought, on the grounds that we already know (on the basis of symbols in a public language) that symbols are the sorts of things that can have meanings.

But the sense in which mental states are meaningful cannot be explained by appeal to the type of meaning attributable to public-language symbols, as the latter needs to be cashed out in terms of (1) the conventions of a public language and/or (2) the intentions of their authors and/or (3) the actual interpretations of hearers. To say that an utterance or inscription "means-X" is just to say something about its conventional interpretation, the intentions of its author or speaker, or the interpretation given by its reader or hearer. But this notion of "symbolic meaning" is not suitable to underwrite the meaningfulness of mental states. To posit *this* sort of meaning for symbols in a language of thought, and then to use this to explain the meaningfulness of mental states, is to fall into circularity and regress, as each mental state would require explanation in terms of symbolic meaning, and the meaning of each symbol would require explanation in terms of a prior mental state and convention. (For a longer version of this argument, see Horst 1996, esp. chaps. 4–5.)

While this criticism undercuts a quick and easy way of explaining the meanings of mental states, and a too-close assimilation of whatever underlies mental states to public-language symbols, it was never meant to imply that there is no important use of the notion of representation in the cognitive sciences. However, it is necessary to try to make explicit the notions of "representation" and "representational system" that are relevant to the current enterprise.

First, I view the principal form of the family of words 'represent', 'representation', and their cognates to be the verb form: that is, *to represent things as being thus-and-so*. Various senses of *representation* are derivative from this.

1. A "representation" in the primary sense is thus an *act* of represent*ing* (that things are thus-and-so) (e.g., "Smith objected to Hill's representation of the facts of the case").
2. In a secondary sense, we may speak of the *vehicle* through which this representing-as-thus-and-so takes place as a "representation." It is in this sense, for example, that we speak of token symbols and diagrams as representations.
3. To the extent that there are features of the vehicle that allow vehicles of that kind to be used systematically to represent things as being thus-and-so, we may apply the term 'representation' to the type as well as the token. (This distinction is necessary because some kinds of representation tokens, such as scribbled diagrams, may not fall into representational types, while others, such as symbols, typically do so.)

We may use the notion of "representing things as being thus-and-so" so broadly that it can apply both to cases that are communicative (A represents things as being thus-and-so to B by uttering sentence S) and to cases that are purely mental (A represents things as being thus-and-so [to herself] in the sense of, say, judging that they are thus-and so). The notions of "representing" and "representation," one should note, are mental and intentional through and through. These notions of "representation" can be applied only to beings that think and to the representational tokens and types that they employ as aids in thinking and communication.

But how broad a category "representation" turns out to be will then depend on how broadly we construe "thinking" and "intentionality." Should we count the cognitive states of nonlinguistic animals as thinking, and hence potentially as representational? And if so, *which* nonlinguistic animals? Chimps? Cats? Lizards? Flatworms? As I have argued elsewhere, I think there are significant issues here (Horst 1996). We conceive of "the mental" through the lens of our own mental life, drawing on paradigm examples that are conscious, available for scrutiny and reasoning, and articulable in language. A kind of terminological choice is involved when we decide either (a) to apply terms like 'mental' and 'representational' to creatures without consciousness, reasoning, or language or (b) to restrict those terms to cases closely resembling our own. I think we have good reason to treat the primary uses of words like 'mental' and 'intentional' as closely and perhaps constitutively linked with consciousness, reasoning, and language. In the present enterprise, however, our concern is not with consciousness, nor is it immediately with reasoning or language as special features of human minds. Rather, it is with cognition. And we have good reason both to use the word 'cognition' so that it covers processes that take place in nonhuman animals, and to stress the continuities between human and animal cognition.

I thus propose to use 'representation' in a stipulative and technical way, in which representing-things-as-being-thus is something that is situated in the internal economy of a cognizing system, paradigmatically a living organism. It may be that the cognitive capacities of some species are so simple that we cannot plausibly apply this notion to them. Likewise, it may be that this usage is naturally applied to certain nonbiological homeostatic or autopoietic systems as well. But it is applied only *analogously* to things like computer simulations of cognition that are not embedded in homeostatic devices where the "representing" plays an adaptive function in those devices maintaining their characteristic form of being. I stress that this is a technical and stipulative usage of the word 'representation'.

To represent things as being thus-and-so, an organism requires two types of things. The first is a *representational system*; the second, a particular representation, in the sense of a *representational vehicle* through which the organism counts as representing things as being A rather than as being B. The space of possible representational-vehicle types is generated by the representational system. A representational system consists of elements that have the function of tracking invariants in the world, the organism, or their interface (i.e., affordances). These may include resources that track:

- Kinds of objects or states of affairs
- Properties of objects
- Individuals
- Relations
- State transitions

A representational system may include some of these elements without others. For example, it is likely that some organisms can track particular objects in perception but cannot reidentify them as the selfsame object over repeated encounters, thus lacking the kind of resource needed to track individuals, as opposed to kind-instances. A representational system may also include some analogue of nonsemantic machinery such as logical connectives and modal operators. (By "nonsemantic" I mean that they do not have the functions of tracking sense or reference, though of course they contribute to the semantic properties of complex representations.)

Let us take some examples with different levels of cognitive sophistication.

Scientific models: A scientific model, such as a theory of gravitation, may be viewed as a representational system, in the sense of being a system that provides resources for representing and reasoning about particular types of states of affairs. A particular gravitational model will have some particular domain, such as bodies with mass, and will model its objects in some particular way—for example, as point masses or as extended bodies with distributed mass. It will track particular properties like mass and perhaps shape and size, but not others, like charge or color. It will employ a particular geometry (e.g., Euclidean or Lorentzian) defining a space of possible locations. And it will possess dynamic laws whose function is to track state changes due to gravitation over time. Other scientific models will have different internal ontologies (e.g., living organisms), different properties and relations (e.g., kinship relations), and different dynamic principles (e.g., principles of selection).

Commonsense models: A frame-type model of a social situation, such as behavior at a restaurant, will have an ontology consisting of things like persons in roles (customer, waiter, host), menus, dishes, utensils, furniture, bills, and the like. Its stock of properties, relations, and transformations will include things like being seated at a table, ordering, and paying the bill. The "dynamic principles" will take the form of something like Schank's scripts, which track characteristic temporal patterns of states and transitions.

Models in perceptual modules: Modules used in perception, such as layers of V4 involved in color perception, may also be viewed as representational systems. A system for color, for example, produces arrays of color representations in a retinotopic map, tracking arrangements of chromatic properties within the visual field. Such a model involves a spatial arrangement of units, with an internal geometry and topology, and each unit is capable of a range of states modeled by the color solid. Areas later in the perceptual stream used to represent spatial geometry provide a model describable in terms of geometric rules that is used for representing objects in space.

A representational system defines a possibility space for possible token representations. A particular gravitational model rigorously defines a way of thinking or speaking about the locations, motions, and dynamic interactions of bodies in space over time. This, in turn, provides the resources for describing a particular set of bodies (e.g., the sun and the earth) in terms of position, mass, and inertia, and for calculating the resultant gravitational component of their dynamics. A restaurant script defines a set of possible events and actions (and thoughts and utterances about them) that are on script. This provides the resources for describing, understanding, and acting appropriately in a variety of particular occasions of dining in which different specific things happen within the constraints of the script. The architecture of a layer of V4 defines a possible set of visual representations of colored environments, and perhaps of chromatic experiences. Particular chromatic experiences can then be located with respect to this space. A particular utterance or judgment about a gravitational situation picks out one possibility from such a possibility space, and likewise a judgment, utterance, or perception about events in a restaurant or a chromatic experience. In some cases, a representational token will pick out a possibility at the most fine-grained level: for example, at any one time, a person's subjective color array has fully determinate values. In other cases, a representational token will narrow the possibility space while leaving some things indeterminate: for example, a description of the sun and the earth within a particular gravitational model need not affirm or deny the existence of other bodies with mass, and the description of position and mass may be given to a finite approximation, although the model theoretically allows for indefinitely more exact descriptions.

Some neurally realized representational systems, such as those involved in regulating autonomic processes and the proprioceptive systems, have the function of tracking things going on within the organism. Others, like physical models, have the function of tracking things in the world, though the organism may itself be represented as an object in the world. Still others may have the function of tracking affordances—that is, organism-salient features of the environment to which it is related. (The cognitive systems of simpler organisms like flies seem mostly to be of this type, and likewise human systems for things like tracking the position of head, eyes, and torso used in orienting to things in the world and modulating perception.)

While each representational system has its internal ontology, these vary in their degree of "objectivity." By this I mean that there are varying degrees to which the internal ontology is an artifact of things about the organism itself. The human color space, for example, does not neatly map onto electromagnetic frequencies. It is heavily conditioned by the properties of human cone and ganglion cells in the retina and is presumably a product of selection that has been shaped over evolutionary history to track a particular set of visual invariants that have been relevant to the adaptive performance of the species in an ancestral environment or a succession of such environments. At the opposite extreme, the ontology of objects and properties in contemporary physics has been carefully developed to exclude, so far as it is possible, artifacts of human perception and cognition, and to track properties that objects have independently of how we perceive or think about them. Such "objectification" is never entirely complete, as we must always think about objects in some particular way, and the available ways are constrained by our cognitive architecture. (For example, our tendency to think of the world in terms of discrete objects with properties is probably due to a deep bias of human cognition, one that proves problematic when we turn to quantum phenomena. But even our representations of quantum phenomena reflect human concepts imported from other domains by metaphorical transposition (e.g., "wave"), and a stock of mathematical concepts that the human mind is suited to dealing with. And, indeed, many people seem to be unable to comprehend a system without classical objects at all.)

Each representational system has its own formal shape. In some cases, such as theories in physics, the formal shape is made explicit, and the system is partly defined by mathematical equations. In other cases, it is implicit and is made explicit only by formal modeling, such as is undertaken in computational psychology and computational neuroscience.

Some representational systems can be modified or replaced, while others cannot. We can progress from folk to classical to relativistic and quantum physics or make adjustments to the gravitational constant within a relativistic model. We can learn new rules of restaurant etiquette. But we probably cannot make adjustments to our subjective color space short of surgical (or perhaps biochemical) interventions. Indeed, even when we learn scientific models, they replace our folk physics only in the sense of providing a more rigorous alternative, and not in the sense that we forget how to think in folk-physical terms.

5.5 Models and Idealization

Models are abstract and idealized. A model is (generally, and perhaps universally) a model of some particular subset of the features, objects, processes, and events that make up the world. It abstracts away from most of the features that make up the world to attend to a few. This confers understanding of particular invariants, generally in accordance with some pragmatic goal, whether explicit or implicit. Such abstraction is necessary given facts about both perception and cognition. On the one hand, an organism must gain information about the world through processes of transduction, and transducers are sensitive only to particular types of information. On the other hand, cognitive processing capacities are limited and thus unsuited to grasping everything at once. Moreover, it is by taking the particular and organizing it under more general patterns that we become sensitive to the various types of order in a complex and messy world.

Idealization is a special type of abstraction. I will use the term 'idealization' for types of abstraction that result in models that can mischaracterize how the world actually is in ways that can matter. Of course, what matters is relative to particular pragmatic goals. Treating objects as point masses, as we do in some physical models, may matter for some problems but not for others. Elsewhere (Horst 2007), I have distinguished several types of idealization:

- *Bracketing idealizations*: These idealize by ignoring features that are present in the real world. For example, a gravitational model brackets other forces like charge and wind resistance, which most bodies with mass are also subject to. Hence bodies generally do not move exactly as described by gravitational models. A restaurant script brackets many other things that may in fact happen in restaurants—a customer choking, or the fire alarm going off.
- *Distorting idealizations*: These idealize by modeling their subject domains in ways that are not fully accurate, such as treating bodies as point masses, or employing classical gravitational models that do not reflect the curvature of space or temporal frame relativity.

- *Finite approximations*: When mathematically exact models are applied in calculation, constants are often replaced by finite approximations of their values.

5.6 Two Types of Alethetic Virtue

Because models (a) define a possibility space of representational vehicles and (b) are idealized, we must speak of two types of alethetic virtues,[1] one applying to representational systems and one to token representational vehicles. On the one hand, we may evaluate how good a model is for its task. Classical gravitation is good enough for some types of problems but is not a good model for objects moving at relativistic velocities. Descartes's understanding of gravity as an effect of particles in the plenum pressing down is not a good model at all. The restaurant script a person has learned from going to Denny's may be inadequate to model what he finds when he goes to a state dinner at Buckingham Palace. I will speak of the degree of fit between a model and the phenomena it models as its *aptness*. Aptness is always judged relative to particular cases or classes of cases.

A model (or its representational system) defines a space of possible representations of how things might be. (As a result, many and perhaps all representations of how things are will be framed within the context of particular models.) It is to such token representations that we apply the honorifics "true" and "false." Models and representational systems are apt or inapt (to particular problems). Individual representations are (or at least are capable of being) true or false. When we speak of something like a scientific theory being "true," what we really mean is that it is *apt*, or perhaps that the claim *that* it is apt is a true claim. Evaluations of truth or falsehood are generally claims about whether a given representation of how things are is the best representation available from a given model and for a particular set of purposes, and furthermore are normally offered without qualification only when *some* available representation afforded by the model is adequate. When *none* of the canonical ways of representing things supplied by a particular model is up to the task, we generally qualify truth claims or else point out the need for a more adequate way of framing the situation. For example, when two women came to King Solomon, each claiming to be the mother of a particular baby, there was a single true answer to the question of who was really the baby's mother given their shared understanding of motherhood. But in contemporary cases where one woman supplied the egg, the second the womb, and both did so at the behest of a third woman who wished to become the legal mother, our erstwhile understanding of maternity relations is inadequate to the problem. Likewise, if one asks where an electron is within an orbital, the question can no longer be answered without qualification, given contemporary physics.

Often (and perhaps always) the truth of a claim can be assessed only once one has fixed what model is in play. Suppose that Smith witnesses a traffic accident and testifies that the sedan was standing still and the bus ran into it. In court, it will not count as a refutation of Smith's testimony if the opposing counsel points out that, according to our best physical models, there is no such thing as absolute rest, and hence the sedan could not have been standing still. What "standing still" *means* depends on whether we are adopting a commonsense model in which motion is judged relative to the face of the earth or a model suited to physics and cosmology. It is the former that the judge is likely to count as relevant to matters of traffic law, and likely this was the type of model that Smith had in mind when giving his testimony.

5.7 Types of Error

Corresponding to the two levels of alethetic assessment (aptness and truth), there are two ways that representation may go awry. One of these involves the tokening of a representation that is not the best match available within the limits of the operative model for tracking whatever it is its function to track. Asked to give the value of π, I say that it is 17. Or I know about both cows and horses and misidentify a cow as a horse, perhaps because of bad lighting. Or I say that the two lines in an illusory figure are of different lengths when their lengths are equal. In such cases, the representation is false.

But there is another sort of case, in which a given model does not possess adequate resources to produce an apt representation at all, and perhaps I lack (and may not be capable of grasping) a model that can do so. For example, I can produce only a finite approximation of the value of π, but a fully accurate representation requires an infinite decimal sequence. Or my subjective zoological ontology differentiates cows from horses but does not extend to differentiating different breeds of cattle, so I cannot identify this cow as a jersey or a Black Angus. Or I find myself unable to represent the world except in terms of classical objects and thus find the problems and models of quantum mechanics unfathomable.

Both of these types of errors can be *forced errors*—that is, errors that one cannot avoid. If I do not possess a representational system that affords an apt representation of a situation, I cannot represent it aptly. The error is *strongly* forced if I am incapable of possessing such a representational system (say, I cannot represent an infinite decimal sequence), and *weakly* forced if I simply do not possess one but could learn one. The first type of error, the kind that results in *false* judgments, can also be forced. For example, in standard visual illusions, like the Müller-Lyer illusion, in which two equal lines appear unequal because they are terminated in arrows pointing inward and outward, the problem is not that I cannot represent lines as of equal length but that my visual system is wired in such a way that *this* type of stimulus systematically produces an incorrect representation of comparative length, and production of the erroneous representation is encapsulated and cognitively impenetrable. The error is forced *for the system*, and for *me when I use only that system*; but it is not forced for me full stop, because I can also assess length in additional ways. (This is one of several ways that the possession of multiple models can aid us in getting a good epistemic grip on the world.)

5.8 Knowledge and Understanding

The distinction between models and their representational systems, on the one hand, and token representational vehicles, on the other, has thus far provided further distinctions between different types of alethetic virtues (aptness and truth, respectively) and corresponding types of error. It is also useful to distinguish the kind of epistemic grip on the world that is afforded, on the one hand, by occurrent cognitive states that are the epistemic "size" of single representations—that is, sentence sized—and afforded, on the other hand, by a systematic model of some set of features.

I will speak of the representation-sized or proposition-sized units as *judgments* or *beliefs* and (when the occasion befits) *knowledge*, and the model-sized units (when they are apt) as *understanding*. This usage is, once again, stipulative. These are all terms that have been used in a variety of technical ways by philosophers for centuries (and if we count equivalent words in Greek and Latin, for millennia). My preference for applying the terms 'belief' and 'knowledge' to proposition-sized units is largely influenced by recent epistemology, which has treated individual propositions or propositional contents as the units of belief and knowledge. (If one were to follow ancient, medieval, or early modern usages, 'knowledge' might have been aligned with differently sized units.) Indeed, even many coherentists are content to use the justified-true-belief formula for knowledge, though what they count as "justification" is holistic. But both coherentists and foundationalists alike appeal to something more systematic when they turn to what confers justification or warrant. We are not yet in a position to assess the implications of cognitive pluralism for questions of warrant, but models clearly provide a kind of epistemic unit that is relevant to epistemological questions, one in which the basic units are neither so small as individual propositions, nor so holistic as the entire epistemic web, but model sized. As it is our grasp of models that allows us to do much of the reasoning we are able to do about a domain, it seems appropriate to use the term *understanding* for it.

5.9 Looking Forward

In the chapters of part 2, I expand on this brief introduction to cognitive pluralism. I begin with a more concerted intuitive development of the notion of models in general and mental models in particular, followed by observations about relationships between mental models. Part 2 concludes with two chapters that consider why a cognitive architecture using a large number of mental models is a good strategy for developing intelligent organisms, and how mental models might be related to other capacities of the mind, particularly public language and language-like thinking. Part 3 then turns to implications of cognitive pluralism for the unity of understanding, intuitive reasoning, cognitive illusions, epistemology, and semantics.

II Models and Understanding

6 Models

One of the crucial theses of cognitive pluralism is that a fundamental unit for the analysis of understanding is a *mental model*. Chapter 5 presented a summary overview of what I mean by a mental model and its role in cognitive pluralism: that models are idealized domain-sized units of understanding, each of which has its own internal ontology for representing the objects, properties, relations, events, and processes of its content domain and its own rules for inferences about the properties and transformations of these. Perhaps some readers will have found these to be clear and intuitively plausible claims in the form in which they were initially presented. But the notion of a mental model may also need more deliberate explication, not only because it is of such central importance to cognitive pluralism, but also because the notion I wish to develop could easily be confused with other notions that have gone by the same name.

Over the past thirty years or so, model-based theories have grown in popularity in both the cognitive sciences and philosophy of science, even if they have made curiously little impact on other areas of philosophy, including philosophy of mind. But if their advocates agree on one thing, it is that many different uses of the expressions 'model' and 'mental model' are at work in their respective literatures. Some of these are quite different from the idea I have in mind, some roughly consonant with it, and some would count, by my standards, either as *types* of models or as narrower theoretical claims about how models are *implemented* in human cognitive architecture.

It thus seems prudent, on a number of grounds, to explore the relevant notion of a mental model more methodically. I will try to develop the idea intuitively, first examining several familiar instances of things called "models" in this chapter before proceeding to look at specifically mental models in the next. The examples in this chapter are of things that we might call *external* models: scale models, maps, blueprints, computer flowcharts, and computer models. These will be used to illustrate what I take to be important features of models and modeling, and will serve as intuition pumps for what it would mean to call something a *mental* model. I explore the relationship between my notion of a mental model and other uses of the word 'model' in chapter 8.

6.1 Scale Models (Target Domains, Idealization, and Aptness)

We have all interacted at one time or another with scale models, ranging from children's toys to models of molecules used in a chemistry class. Scale models are typified, first and foremost, by their structural similarity to what they are models of, which we may refer to as the *target* of the model. A model of a 1964 Mustang has approximately the same shape and proportions as a genuine 1964 Mustang, only much smaller. In some such models, the similarity is limited to the surface contours and perhaps the colors. Other models may reproduce other features in miniature—the doors, trunk, and hood may open realistically, the steering wheel may spin and turn the wheels, and so on. A very fine model may even be composed of parts matching those of the real car: little pieces shaped like the engine, the catalytic converter, the struts and beams, and so on. Some may be powered, so that the model can move on its own power.

But, of course, many other features will not be modeled. Even a powered model car will likely run on batteries (of a shape and relative size unlike those of an automobile battery) and not on a tiny internal combustion engine. The parts will generally not be made of the same materials, and hence the balance and the ratio of size to weight will be different, and so on. In models designed for different purposes, some of these features may in fact be modeled as well. Engineers designing a car or an airplane may build a scale model first and use it to test properties like wind drag and stability. If a model is to be used for such purposes, it will need to model more properties of the target than will a child's toy. Design models are also different from models of production-model cars or airplanes in that they are typically built *before* their targets.

Scale models are also used in the sciences and in science classrooms. There are, for example, kits for building models of molecules, using balls of different sizes and colors for the atoms and tubes for the bonds. These are used to roughly approximate the geometric configurations of molecules, and as such, they allow both students and researchers to understand structural properties of the molecules modeled and to explore what kinds of molecular configurations are structurally possible. They serve as easily perceived and readily understood surrogates for the real thing, fulfilling heuristic, explanatory, and predictive functions. Of course, some of the relationships between elements of the model and their targets are purely conventional. The balls that represent carbon atoms may be black, those that represent hydrogen atoms red, and so on, but the atoms themselves do not possess color at all. The size ratios between the balls may not match those between atoms, the atoms themselves are not homogeneous spheres with holes for connectors, and the bonds are not physical solids. The modeling kit affords modeling only of one set of properties of the molecules it is used to model—the physical, structural, and geometric properties, and those only roughly.

Idealization

However realistic the model, it is made in such a way that it models only a subset of the features of the target system. (If it is *exactly* like a '64 Mustang, it is the real thing, or a working prototype, or a reproduction, not a model. If we build something isomorphic to methanol using atoms of carbon, hydrogen, and oxygen, we are building an actual molecule of methanol, not a model.) We may say that in this sense the models are *idealized*. They have elements corresponding to some elements of the system modeled and permit relations and transformations corresponding to some of those characteristic of the target system while differing in other properties. We can use such models as a kind of stand-in for the real thing for various purposes. But just what set of purposes they can effectively be used for depends on what features are modeled. And hence, to use the model effectively, one must understand how it is idealized and what it reproduces faithfully. A good scale model of a train may provide information about the relative size of the engine and caboose but cannot be used to accurately predict the real train's structural integrity in a crash. A classroom model of an organic molecule may provide information about its shape and what it can bond with, but not about how the bonds of real molecules are formed or broken.

Aptness

We need a term for the suitability of a model to serve as a stand-in for the real thing with respect to a given task. As noted in chapter 5, I will speak of this as the *aptness* of a model. Aptness is always relative to a purpose. A model train is apt for modeling features like appearance and dimensions, but not for features like structural integrity. A classroom model of a molecule is apt for modeling features such as the rough geometry of molecules, and for determining what molecular configurations are possible, but not for other features, like the nature or strength of the bonds. Indeed, the model may itself permit configurations or transformations that are impossible in the target system, or prohibit ones that are really possible. For example, it may be possible to connect two balls with tubes representing a particular type of bond when the molecules they are used to model cannot actually be bonded in that way. This, too, is a limit on the aptness of the model, a limitation on the purposes for which it may aptly be employed. Insofar as the user understands these limitations, this amounts to a further way the model is idealized. But insofar as the user is unaware of the potential problems, the model may be capable of misrepresenting what is modeled. Indeed, if the model is badly constructed, it may even be forced to misrepresent what is modeled.

Multiple Models and Incommensurability

There can, of course, be multiple scale models of the same type of object. For example, model trains come in different gauge sizes defined by different scales. The engineer's model of a car or airplane will model different features than will a toy model. Two kits for modeling molecules may be in different scales, use different materials, employ different coloring conventions, and use different types of connectors. The kinds of information available from two models of the same type (say, two HO-gauge train models or two models of organic molecules using the same kit) will afford information about comparisons (one engine is longer than another) or combinations (the two molecules can or cannot be combined into a larger molecule). But the use of models of different types is more complicated. One cannot directly compare the length of two train engines using an HO-scale model of one and an O-scale model of another, though if one knows the scale of each, one can calculate and compare the sizes of the originals by using the models, measuring, and applying algebraic calculations. Model molecules built from different kits may allow some types of comparisons (the geometry, the number of carbon atoms) but not allow the models to be used for other purposes: the two models' failure to fit together into a larger molecular model may be due to different scales or connectors that do not match and tell us nothing about what chemical combinations are possible; or if the connectors do fit together by chance, the ways they do so may not correspond to actual physical possibilities, and the geometry of the resulting model will be incorrect if the components are of different scales. A mismatch between the types of models combined can thus result in artifacts, and two models may even be incommensurable in the sense that they cannot actually be combined.

6.2 Maps

Scale models, as *scale* models, share the physical dimensionality and geometric proportions of their targets. One can also have scale models of physical environments (e.g., people who reenact historical battles with model soldiers often build scale models of the terrain of the battle they are reprising) or even the whole Earth (a globe). More often, however, we use a different sort of model to represent geographical information: the map. Or perhaps better, as we shall see, we use *various types* of maps. For while maps share as a class the property of being two-dimensional representations of broadly geographical features, they come in a bewildering number of varieties. But we can bring some order to the variety in terms of just two variables: *what* features a map represents, and *how* it represents them.

Representational Systems and Uses

Consider the maps in figures 6.1 through 6.4, all of which represent approximately the same geographic area in and around Boston. The first map is a map of roads and light rail lines, drawn to scale. The reproductions here are grayscale, but in its original form, the map uses particular color conventions to represent land, water, roads, and each of the MBTA train lines. It also employs *labels* in plain text for minor roads, and special conventional symbols for state and federal highways. Such a map affords the ability to measure distances (at least roughly) between points in Boston, either by a particular road route or as the crow flies. If the map indicates which roads are one-way, it also allows the user to plot driving routes. (Having lived in Boston, I can vouch for the fact that a street map of that city *without* such indications is largely useless for driving there.) And it allows one to see, very easily, what surface points are accessible by rail, and to plan rail routes involving a change of train.

The second map (fig. 6.2) is much like the first in reproducing the rough spatial features of the city and the rail lines but does not include streets other than major highways. It does, however, note rail stations with labels, whereas the first map indicates only a few major stops and does not label these. Such a map is intended for use in rail navigation by a person who already knows where the stations are, or the name of the station that is her destination, but not how to get there. Precisely because the map omits other information, it is easier to use if one does not need this other information. On the other hand, it is less useful if one needs, say, to determine which station is closest to a known street intersection.

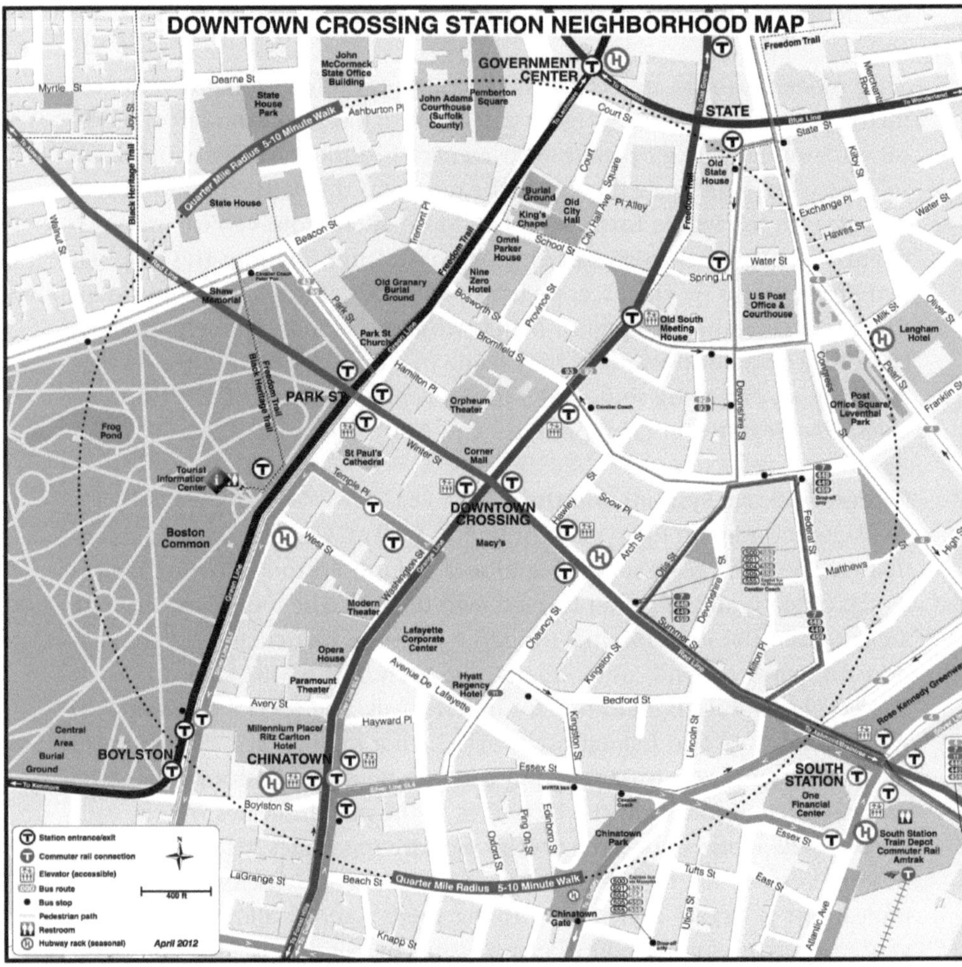

Figure 6.1

Map of streets and MBTA stops in downtown Boston preserving surface geometry. (Source: Boston MBTA, April 2012.)

Figure 6.2

Map of Boston MBTA network with station stops preserving rough surface geometry without streets. (Source: Boston MBTA, October 2009.)

The third map (fig. 6.3), an updated version of one I recall seeing on trains when I lived in Boston, does not preserve the spatial relationships between stations. Train lines are represented (in the original) by distinctive colors as before, and stops with labeled dots. But the lines do not preserve the geometric properties of the routes—the north–south and east–west directions or the comparative distances between stops. The map represents the rail system *topologically* rather than geometrically. This is actually an elegant way of presenting the information needed to quickly determine how to get from one rail station to another. In navigating by rail, one in fact needs to *think topologically*.

The fourth map (fig. 6.4), of a more limited portion of Boston, is topographic. It represents isotopic elevations with specially colored lines and marked with elevations, types of landform with colors, and landmarks with labels. Topographic maps are particularly useful for reconstructing three-dimensional features of a landscape from a two-dimensional map. While their use is something of an acquired skill, they are useful for orienting oneself to one's environment, and for calculating elevation-dependent affordances, like the steepness of a route, optimal locations for radio and microwave towers, and patterns of water runoff.

Each of these four maps represents certain features of the area it models and abstracts away from others. Just *what* is modeled differs from map to map, and the choice of what to include and what to leave out has consequences for the purposes for which each map may aptly be employed. Indeed, it is safe to assume that each map was carefully designed with a particular set of practical purposes in mind, and the mapmakers sought to optimize each for use in a particular set of tasks. The kinds of information needed by the driver, the rail passenger, the cross-country hiker, and the city planner are not the same. And a given type of information is made most readily available by excluding from the map all unnecessary information.

Consider now a second set of maps (figs. 6.5 and 6.6). These are both maps of the surface of the globe. Their most obvious difference is in how they project the surface of a sphere onto a flat surface. The skin of a spherical surface cannot simply be flattened onto a plane without distortion. One must either stretch it, or make cuts in it, or both. The first map, a Mercator projection (fig. 6.5), is a result of something like stretching the surface. It results in a rectangular map (or, more precisely, the surface of a curved portion of a cylinder flattened onto a rectangular plane), in which each latitude line is equally long, stretching from left to right. Since different latitude lines are in fact of different lengths on the surface of a sphere, this results in a geometric distortion. The map does not preserve the comparative areas of

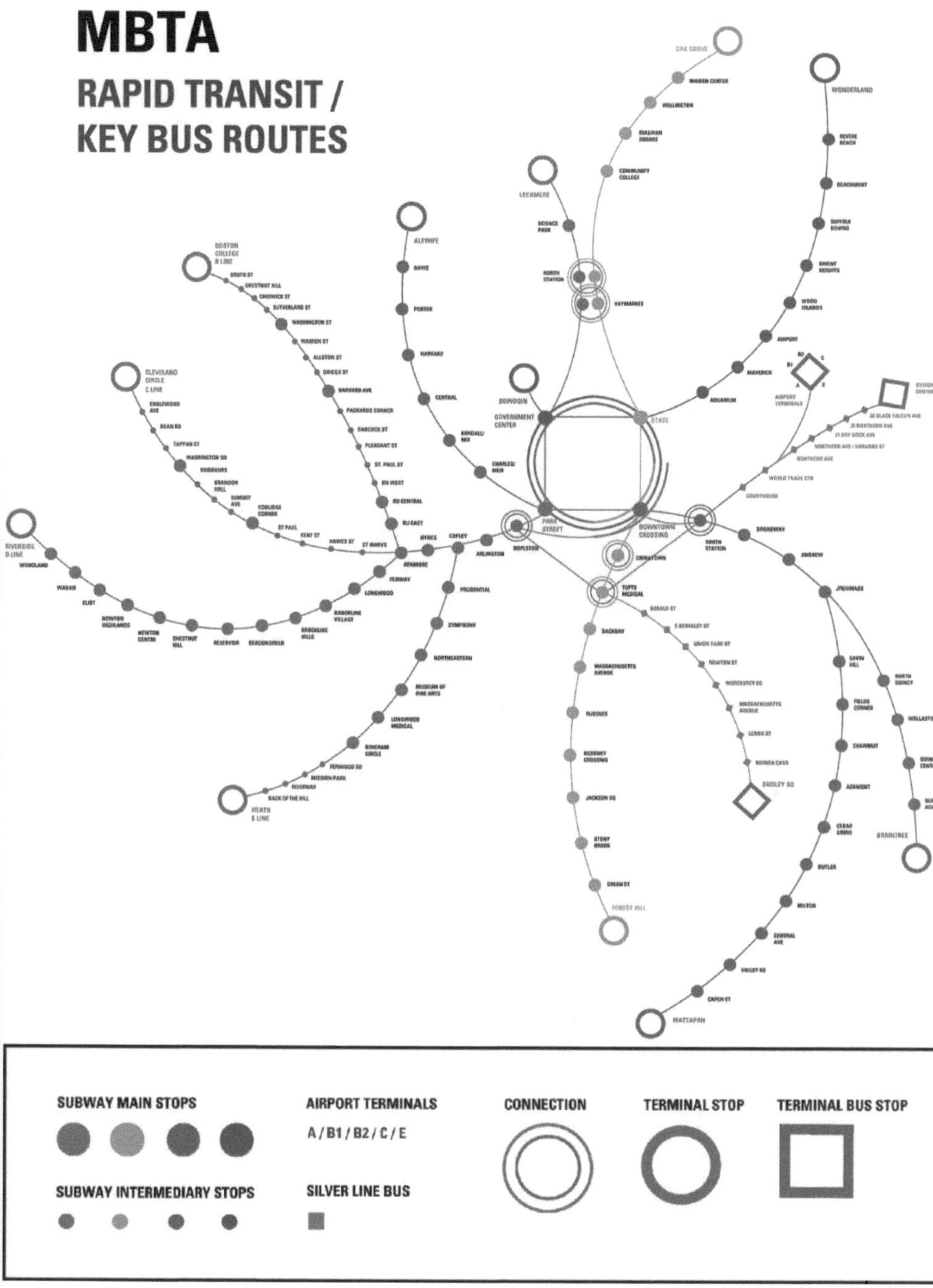

Figure 6.3

Topological map of Boston MBTA system. Copyright 2012 Samantha Havas, used by permission.

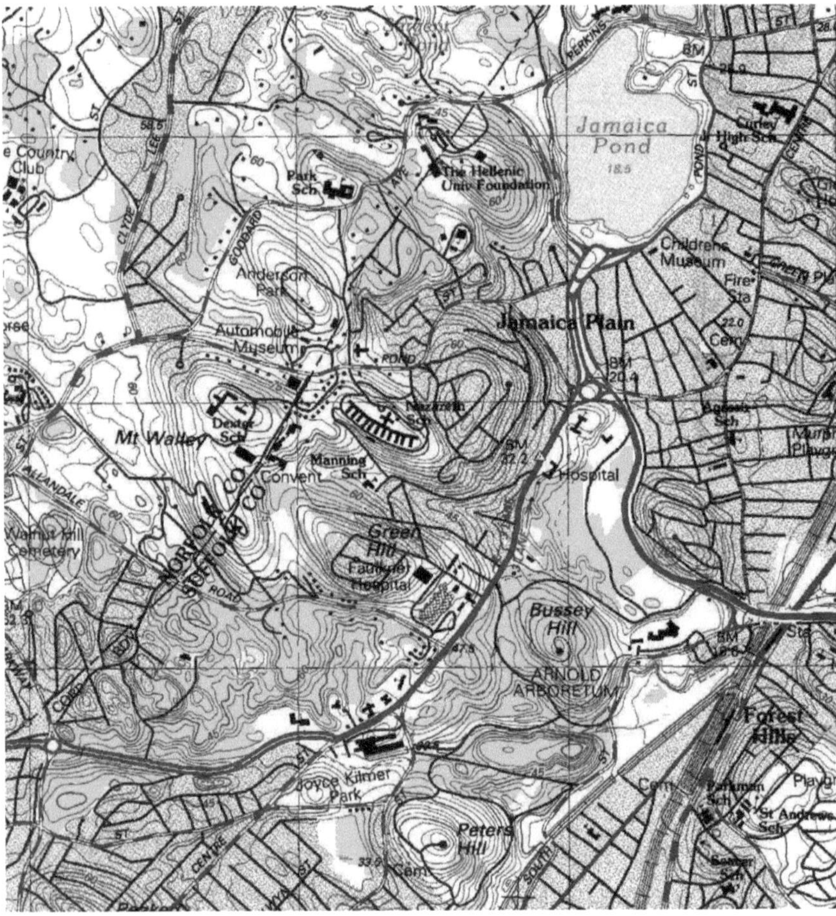

Figure 6.4
Topographic map of Jamaica Plain area of Boston. (Source: USGS info.)

Figure 6.5

Mercator projection of globe. Geordie Bosanko, August 30, 2011. Via Wikimedia Commons, http://commons.wikimedia.org/wiki/File%3AMercator_Projection.svg.

land masses at the poles and at the equator, and hence Greenland appears much larger in relation to other land masses than in fact it is. A Mercator projection cannot be used directly to calculate distances (except along a north–south axis) or areas. But it does preserve all contiguous relations between points on the globe, except at the points lying along the edges of the map. Moreover, it is useful for nautical navigation, as it makes it easy to plot lines of constant course, called rhumb lines.

The second map, a sinusoidal projection (fig. 6.6), preserves proportions of areas on the globe, essentially by slicing up the surface of the globe rather than stretching it. It results in a figure that is initially difficult to orient to, because of the many discontinuities, which seem counterintuitive. The sinusoidal projection distorts distance measurements between two points except those located on the same parallel. However, for purposes of preserving information about areas, it has advantages over the Mercator projection.

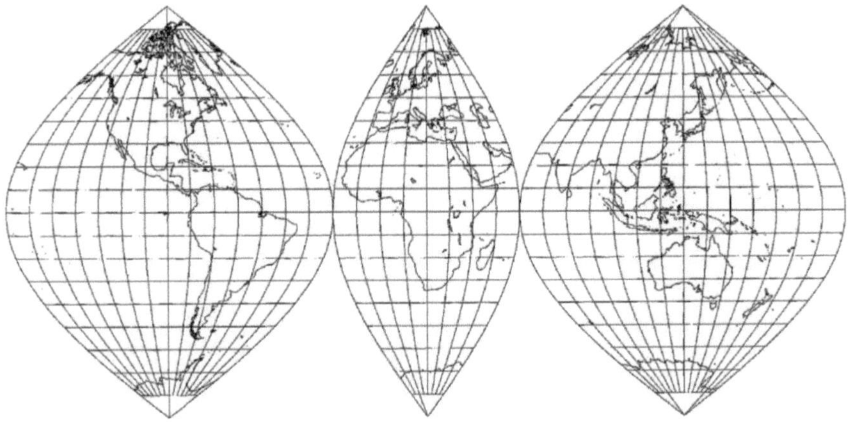

Figure 6.6
Interrupted sinusoidal projection. (Source: John P. Snyder and Philip M. Voxland, *An Album of Map Projections* [US Geological Survey Professional Paper 1453], Washington, DC: US Government Printing Office 1989, 43.)

What these examples show is that there are a number of distinct ways that a model such as a map can represent its subject matter, but each must do so in some particular way. No projection is *both* Mercator and sinusoidal. And what choices one makes will have consequences for what information is preserved, excluded, or distorted, and consequently what information can reliably be extracted from the model. The topological rail map excludes all information about surface geometry and magnitude. The Mercator and sinusoidal projections include such information, but in different forms. One needs to have a set of techniques for reading information off a particular type of map that is appropriate to that map. Many mapping techniques, when used for large-scale maps of small areas of the earth's surface, such as metropolitan Boston, will produce maps in which distance information is preserved well enough that one can calculate real-world distance by measuring distance on the map. (Greater distances can be calculated in a more complicated way if one knows what type of projection is used.) The same technique will not work with a topological map.

Relations between Models: Incommensurability and Combination

The differences in representational systems also have consequences for how information from different maps can be combined. If maps have the same projection system and scale, a transparency of one can be laid over the other to combine information visually. However, this will not work with, say, a geometric street map and a topological rail map. Indeed, a street map without indications of railway or station locations and a topological rail map without indications of street locations cannot be combined to plan a journey without additional information, such as an independent knowledge of where rail stations are located. Likewise, maps with different scales or projection schemes require additional knowledge of those variables and how to translate between them if one is to combine information from the maps successfully. Failure to observe such strictures can result in errors.

6.3 Blueprints

Architectural blueprints have a great deal in common with maps. They are two-dimensional representations of three-dimensional physical phenomena and are usually drawn at least roughly to scale. Geometric relations in the blueprint thus usually reflect similar relations in the building, at least along the horizontal axes. Of course, blueprints are normally constructed before the building is built, while maps are usually drawn to depict existing geographical features. But this need not be the case: one can construct a blueprint of an existing building, and planners create maps of cities and road systems that are not yet built.

Also like maps, architectural blueprints employ special conventional symbols to note standard elements of their domain: doors, windows, electrical wiring, lighting, plumbing, and the like (fig. 6.7). They may also indicate the locations of appliances and furnishings. Unlike the lines used for walls, these conventional symbols do not always mirror the geometry of the features they represent. A symbol for an electrical outlet, for example, is proportionally larger in comparison to the representation of the wall it is housed in than is the actual outlet to the wall, and the arc used in the symbol for a door indicates the direction in which it opens and not a curved object. Likewise, the different shading patterns used to indicate materials (wood, metal) are not spatially accurate representations of objects but conventions that indicate the materials used.

Blueprints, like maps, employ labels. But they also employ a special sort of label not always found in maps: labels indicating dimensions. Indeed, while blueprints are often scale drawings that preserve geometric proportions, they need not be drawn in this way. A hand sketch noting dimensions and features contains the information needed by the architect and the builder.

Much architectural design is now done using computer-aided design (CAD) software (figs. 6.8 and 6.9). The "deep" representational system used

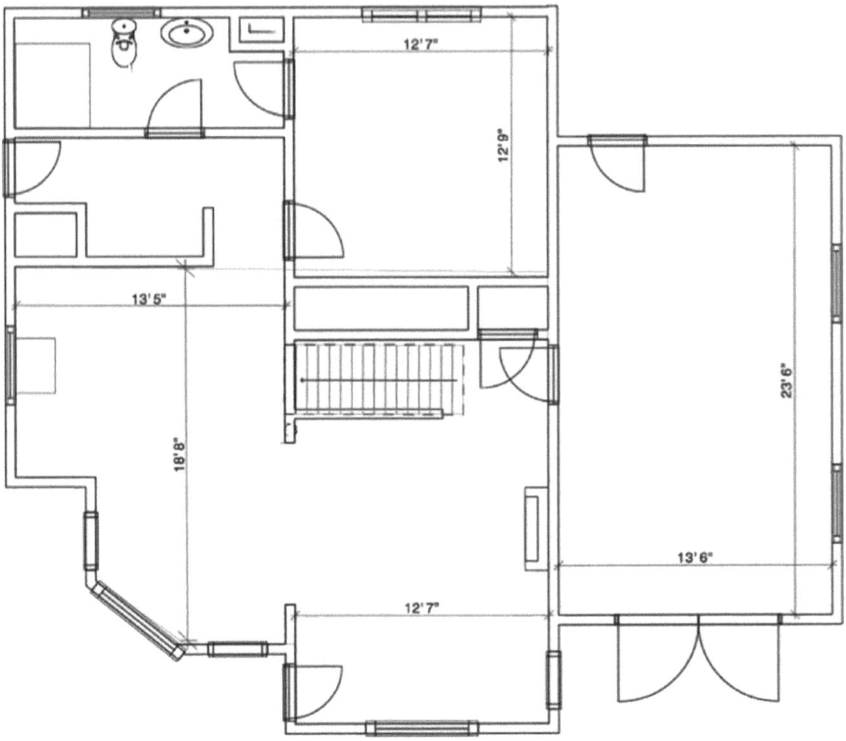

Figure 6.7
Blueprint of house.

by such programs is generally unavailable to the user, who interacts with the representation through a number of different "views": for example, a blueprint view (in which different features like wiring and dimensions can be made visible or invisible), a wireframe view (depicting three-dimensional features of contours without surface features), and a rendered view (in which surface features like texture, material, and color are added in, to simulate how the building looks from a particular vantage). Whereas the architect or builder working from paper blueprints must construct a mental model of the three-dimensional structure and its appearance based on the blueprints, CAD software generates three-dimensional representations and two-dimensional views of them, with different features included or abstracted away.

Any given production-model architectural software program employs some proprietary representational system, and a plan designed in one

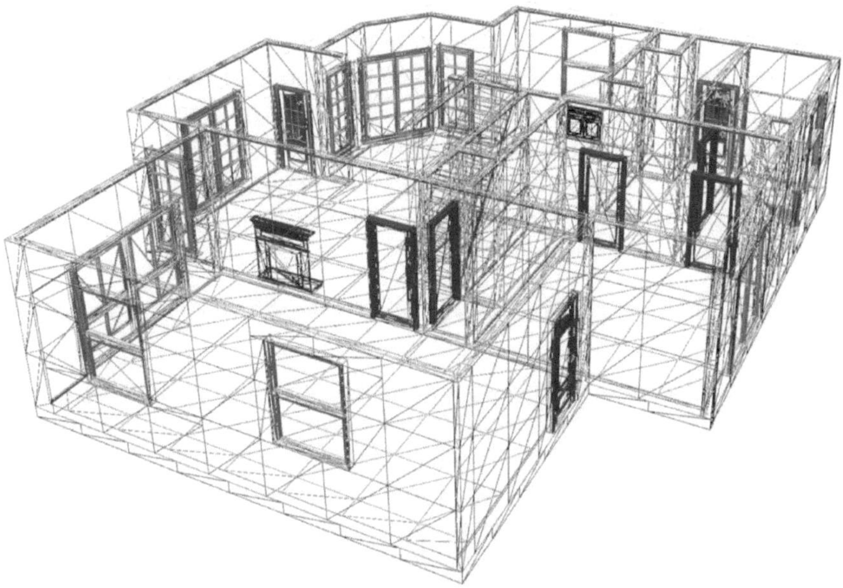

Figure 6.8
Wireframe view of house generated by the author using CAD software.

Figure 6.9
Rendered view of house generated by the author using CAD software.

program may not be readable by another. Such programs are, however, generally based on one or another more general set of CAD conventions, and their data types and algorithms are a superset of those of the more generic CAD conventions, much as the set of algorithms in other programs may be a superset of the algorithms contained in a packaged programming environment. The modeling language of a particular CAD program is thus a kind of special case of more generic CAD conventions, characterized by its particular data types and algorithms.

6.4 Program Code and Flowcharts

In programming and engineering, processes are often modeled by flowcharts (fig. 6.10). These are two-dimensional representations of functionally defined processes and the relations between them. These relations are both logical and dynamic. They are dynamic in that arrows in the flowchart represent a chronological sequence of processing. They are logical in that the flowchart includes branching points where the actual behavior of the program depends on particular conditions, understood in Boolean terms. For example, at a particular place in a program, if a variable *V* has a positive value, that value may be displayed on the screen, but if the value is negative, it may first be multiplied by –1 and then displayed.

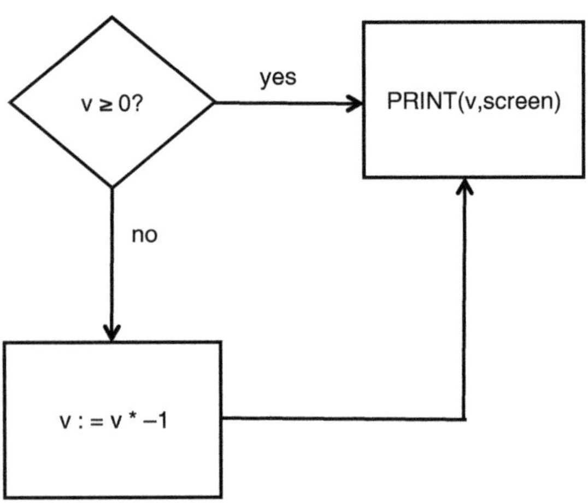

Figure 6.10
Flowchart of computer program.

Such operations can also be represented nondiagrammatically in program code. For example, the prior operations can be represented thus:

```
IF (v>0)
   PRINT(v,screen)
ELSE
   v:= v * -1
   PRINT (v,screen)
END IF
```

Format and Implementation

The same operations can be represented in a variety of general-purpose programming languages or in the machine language of a given computer. And the program can be *implemented* in an open-ended number of ways, as many ways as there are possible variations of hardware and software that will support the operations. The notion of a "program" is thus idealized away from its possible implementations. When a program is well designed and used wisely, these idealizations will not matter. However, implementation will sometimes matter. For example, in a programming language, the domain of values to which an operation like addition (or however we wish to think of its computer analogue) can be applied is open-ended. But in any actual implementation, there is a limit on the ordinality that can be represented (e.g., 2^{32} or 2^{64}), and different hardware and operating systems may behave differently if this ordinality is exceeded.

Generativity

Flowchart or coding-language representation of software dynamics is also idealized in a second way: the program itself is a set of potentialities for real events. For example, every time you use your word processor, what the word-processing program *actually does* is different from what it did before, because you are typing different symbols, performing different editing operations, and so on. But the *program itself* is unchanged.

Both flowcharts and program code can involve multiple levels of abstraction. For example, a high-level flowchart for a program might represent complicated algorithms with a single box. But one can also diagram the flow of operations within that box. Similarly, code is often "modularized." A repeatedly used algorithm is written once in source code and given a name. The name is then used in further code to "call" that algorithm without reproducing the lines of code. (Even the lowest-level operations of the programming language are generally translated into a series of instructions to the processor chip by an interpreter or compiler. The programmer may be quite ignorant of the machine language of the hardware she is programming for.)

6.5 Computer Models

Computer programs can also be used *as* models of noncomputational processes. The evening weather report, for example, might display weather maps for the next day or the next week. These are generated by programs that model meteorological phenomena and generate predictive simulations of how they might evolve: for example, how a storm will track, or how colliding fronts will produce high winds and precipitation. Researchers in computational neuroscience, artificial intelligence, and cognitive psychology use programs to test hypotheses about the mechanisms underlying neural and cognitive phenomena. And economists use computers to model the economic consequences of possibilities such as lowering the prime lending rate or a change in the oil supply.

To build such a program, one must begin with a set of hypotheses about what objects, properties, relations, and processes are relevant to the dynamics of the real-world phenomenon to be modeled. One then creates computational stand-ins for these variables, in the form of data structures and algorithms that are designed to mirror the relevant properties of the content domain. A model of billiard ball collisions, for example, would need to have some representation of the dimensions of the table, as well as the locations, dimensions, and momenta of the balls; and a sophisticated model might represent further factors such as the coefficients of elasticity for the balls and the sides of the table, and the coefficient of friction for the table surface. It would also require algorithms for computing trajectories over some quantized representation of time, for the effects of friction, and laws of collision. Such a model would be idealized in a number of ways. Real pool tables have surfaces that are not perfectly regular. A gust of wind might affect the real-world kinematics of a shot, or someone might interfere with a ball in play. The program may not have a way of modeling the effects of spin ("English") or the chaotic effects of many-body gravitational variables (which, I have heard it claimed, can produce macroscopic effects with a fairly small number of collisions). It may also not be prepared to handle the differences in high-impact collisions, where structural features of the balls become important, as they may significantly deform under impact or indeed lose structural integrity. These features are also relevant to the scientific models themselves on which such a computer model is ultimately based.

6.6 Features of Models

Models are cognitive tools that allow ready understanding of some set of features of the objects or systems they are models *of*. Models can be analyzed into parts or aspects that correspond to various features of the target system: objects, properties, relations, processes, and transformations. They thus provide a means of representing these features and reasoning about them. A model does not simply represent one fact about an object or system but provides a framework for representing various sorts of facts about them, and reasoning about these. A model is thus not itself simply a representation of an object or system but a kind of *system for representation* of things about them.

It is particularly important to note that the kind of understanding afforded by models is tied to the ways they are *systems*. And in this way models are unlike lists of propositions. If you have a list of propositions and delete one of them, you may lose some ability to infer things that depend on that proposition, but all the others are retained. But if you delete a road or rail connection from a map or remove the labels, the ability to use the map as a way of navigating from one point to another is more globally impaired. If you lose information about what kind of projection it is, or what the various sorts of symbols and lines stand for, the map cannot be used for all the same purposes and may become utterly useless as a map. If you delete a section of code from a program, it may cease to operate at all or result in vastly different behavior. If you alter the meaning of one theoretical term in a scientific theory or alter one of the laws, you generally must make a host of other alterations as well, or else the resulting theory will cease to have the same fit with the phenomena it is a theory *of*.

Models are useful for understanding in part because they provide a stand-in for the target object or system in which a certain set of its features is made available for inspection, reasoning, simulation, and manipulation in a fashion that is readily grasped. This almost invariably involves idealizing away from other features of the real-world system being modeled to reduce complexity and hence cognitive load. Of course, there will be problems for which the things that have been idealized away from will matter, and this will affect the aptness of the model for thinking and reasoning about such cases. The *way* features or aspects of a real-world system are modeled may also diverge from their real nature, and this can affect the aptness of a model as well. Often, such distorting idealizations make the system easier to grasp, and there can be a trade-off between ease of understanding, reasoning, and calculation on the one hand and accurate fit with the system modeled on the other. A ball-and-strut model of molecules is easier for human minds to grasp than one that models the behavior of charged particles, and whether it is more or less apt for a given application will depend on whether one is dealing with one of the cases where the differences matter. (And what matters is a pragmatic question, which in turn depends on practical, descriptive, predictive, and explanatory interests.)

Often, more than one model is available for what is, in some sense, the same phenomenon: maps with different projections, scales, and types of representations; scale models that use different scales and model different features. Different models make different sorts of information about a system available and are useful for different purposes, and we may *need* more than one model if we are to have ways of accessing all the information they jointly afford. But the very idealization of models that makes them useful aids to understanding can also result in their being difficult to combine, incommensurable, or inconsistent. Going between Mercator and sinusoidal maps requires careful calculations, and it is probably easiest to move from one to the other by going back to the spherical geometry from which they are both derived.

6.7 Models as Cognitive Tools

All the types of models we have surveyed can be viewed as *cognitive tools* for understanding, predicting, and interacting with phenomena in the world. They can afford both understanding and practical grip on the phenomena of which they are models. Some of them, like scale models, are physical objects that can be perceived and manipulated. In the case of computer models, our scientific understanding is encoded into a computer program and its data structures, and the computer does much of the computational work for us in tasks of simulation and prediction.

There is a sense in which one might say that *all* models are *mental* models already. That is, they are models only by dint of the fact that they are *used* in a particular way, by human subjects, in the course of various cognitive and practical tasks. Indeed, one might even go so far as to say that the very notion of a model carries within it the idea that something is a model only by relation to one or more minds. But this does not really help in getting us toward a clarification of the claim that cognition is, quite generally, accomplished through mental modeling, for that claim is not the claim that we always use *external* models (like scale models, computer models, or blueprints) when we think about things. And indeed, this use of external models is exactly what needs to be excised if we are to make sense of the idea of mental models.

There is, as well, a potential problem here, as the kinds of models we have examined have features that would be extremely problematic if we supposed that they must be found in mental models as well. First, using a scale model, blueprint, or map as a model requires us to make it an *object of attention*. But if we have understanding in everyday life through mental models of things like restaurant scenarios or core geometry, *these* "models" are *not* characteristically objects of our attention. They are not things we can perceive with our senses at all. We *can*, perhaps, *think about* them as theorists. But whatever the mechanisms through which we understand the world, we normally do *not* think *about* those mechanisms. It might be more appropriate to say that we think *through* them, in a sense analogous to how we see *through* a lens. And our ability to think through them does not depend on our having previously attended *to* them. Indeed, it is the other way around: it is only by reflecting on their *use* that we can attend to them as models at all.

Second, a model is a model *of* something. And with our examples of external models, using them as models requires us to *take them as* models of something. A child can play with the elements of a kit for modeling molecules; but for the child, this is not an exercise in modeling molecules but something more like playing with a Tinkertoy set. To use it *as* a model of molecules, one must also have an understanding of the target domain and understand that the thing one is interacting with is to be used as a model of that domain. But if this feature is taken to be a necessary feature of anything counting as a model, we cannot explain our understanding of the world just in terms of mental models. To use such a model—to use it *as* a model—we would need some independent form of understanding of the domain to be modeled. We *could* have mental models in the sense of, as it were, saying to oneself, "I shall use this as a model of X," in cases where we already had some other kind of understanding of X. But if understanding is supposed to be "models all the way down," then there is no way for this to get off the ground. In *Symbols, Computation, and Intentionality* (Horst 1996), I raised a similar concern about attempts to use the notion of "mental representation" to do any work in explaining the semantics of human thought. And so I, of all people, ought to be careful here not to make the same mistake that I accused others of making. So if we are to use the word 'model' here as a term for general features of the mind that are at work broadly in cognition, we need to explicate how the word is being used in a recognizable way that avoids this potential problem of incoherence.

6.8 Further Considerations

As a first step toward this goal, let us first ask whether *all* the types of models we have surveyed in fact need to be (a) objects of attention that are (b) understood to be models of some independently understood domain. In at least one case—that of computer models—I think we find an exception to this rule. It is true that the *designer* of a program needs to understand it as a model of some phenomenon—say, of weather systems. But it is not necessary that the *user* or *consumer* of the outputs of the program needs to do so. When I watch the weather forecast on television, I am aware of the diagrams that represent predictions of tomorrow's weather. But I am generally not actively aware that they are being produced by a meteorologist's model. And I am, in point of fact, completely ignorant of *how* the program does what it does, even at the level of what variables it takes into account. Another viewer might even be ignorant of the fact that these maps are produced by computer models at all.

The same holds true when I check the forecast using the weather widgets on my home computer. There is a modeling program running somewhere; but for me, the widget and the program from which it gets its information are more like a cognitive prosthesis. I have a skill at obtaining information about tomorrow's weather, just as I have a skill at looking out the window to see what the weather is right now. But I am unaware of *how* the forecast is produced, just as I am unaware of how my visual system allows me to see that it is clear and sunny. Indeed, one can imagine that a day might come in the not-too-distant future when people will be able to buy chips that will be directly connected to the brain to supply such information without the need to attend to some external object like a television or computer screen. If such prostheses were implanted early in life, a child might actually have to *learn* that his ability to check the weather forecast depended on a computer running a model in another location. The point of this is that there do seem to be ways that at least some types of external models can be used without attending to the model and without understanding that there is an external model operating somewhere that is the source of our information. Such models still need to be designed by people who understand that they are models, but this need not be a limit on their use. This is a useful analogy in thinking about mental models: *having* and *using* a mental model does not require that we understand the underlying processes of modeling or information extraction, or even that we recognize that there is a mental model at work.

7 Mental Models

As a transition from external models to mental models, consider what happens when you gain a proficient familiarity with a model or modeling system. At some point, you begin to be able to think about the kinds of structures found in the modeling system without actually having to perceive or manipulate elements of the external model. Perhaps you have been learning about molecular structures using the kind of kit for building model molecules I have described. Now you want to find the answer to a question—say, how many distinct types of organic molecules can be built on a chain of two carbon atoms. The modeling kit would be a useful aid in such a task; but suppose it is unavailable, or perhaps you are considering the question in taking a test in which you are not permitted to use it.

You may well find yourself able to *imagine* the kinds of structures you became familiar with through the modeling kit quite vividly. (Humans vary quite a bit in their capacities for imagery, and some seem to lack such a capacity altogether. If you fall within the latter group, you may find the ensuing discussion alien to your experience, though it is familiar to many other people.) Perhaps you imagine yourself *handling* a chain of two of the balls that represent carbon atoms in the kit, or perhaps you visualize such a structure without tactile or proprioceptive imagery. Perhaps you actually imagine yourself working with the kit, or perhaps you merely imagine structures (suspended before the mind's eye, as it were) that you think of as molecules. If you have learned to use the kit well, you do not simply imagine the pieces as solid geometric forms viewed in something like a mental snapshot. The affordances they offer as parts of the modeling system are available to you as well: the ways the balls and links can fit together in particular ways and hence be used to build larger configurations. Indeed, you may find that properties of the pieces from the external modeling kit that are irrelevant to modeling molecules—say, the material and color of the balls and struts—are not represented at all, but only the structural and relational properties. You can "build" different structures in your imagination, conforming to the principles learned from the modeling kit.

Again, there is a good deal of individual difference in how well different people can perform such an exercise. One person can do it successfully only if there is a single carbon atom, another can do it with chains of two or three, and so on. What the upper bounds might be is an empirical question. There are probably savants who can imagine much larger molecular structures than you or I, and each of us would likely be able to expand our upper bound through practice; but it might be that a DNA molecule would prove too large and complex for anyone to imagine in detail. But this is an observation about *performance limits* rather than *competence*, similar to the fact that there is some limit to the size of sentences you are able to parse, even though you have the necessary grammatical competence required. You can understand the sentences "Noah was the son of Lamech" and "Noah was the son of the son of Methuselah," and you understand the iterative nature of the "son of ..." construction, but at some point you lose track of the generations because you have exceeded some kind of limit on storage or parsing ability. Similarly, you might be able to concretely imagine one or two carbon atoms and what they are bonded with, but there is some upper limit to how many you can imagine concretely, though you are perfectly capable of imagining and thinking about the properties of any suitably small portion of a very large molecule.

How should we describe what you are doing when you perform a task such as imagining a chain of two carbon atoms with hydrogen atoms occupying the remaining positions? It seems natural to speak of this as "building (or using) a *mental* model" of such a molecule. You substitute an oxygen molecule or an OH group in one of the slots, and you have produced a mental model of a different type of molecule. But something deeper is going on as well. Your ability to imagine particular types of molecules and perform specific transformations on them—structural possibilities and transformations that correspond to those afforded by the modeling kit (and, assuming it has been well designed, to the ways molecules can actually be configured)—suggests that you have internalized the structural principles and possibilities that were engineered into the external model, and you now have a general and flexible capacity that you previously lacked: to *model molecular configurations in your mind*. With the external model, the mechanical properties of the pieces—the number and locations of the slots in the balls into which struts can be inserted, the lengths of the struts—enforce the "rules" for their combination and generate a possibility space of possible model molecules. If your purely mental modeling matches the external model in what possibilities it affords, there must be some *cognitive* principles by which you sort the elements into types, understand them to be combinable in a small and well-determined number of ways, imagine complex structures, and perform mental transformations on them.

Whatever the principles and mechanisms are by which you do this, they are almost certainly not consciously accessible to you. You simply *think* in categories like CARBON_ATOM when doing the exercise and *understand* that elements of that type have four bonding slots in a particular configuration. Operations like "add an OH group" are simply there in your mental repertoire, and you perform them as automatically as you would exercise the learned *manual* skill of attaching such a group using the external model.

A cognitive scientist studying the ability you have gained might propose theories about how you do it at various levels of abstraction: for example, in terms of rules that describe how pieces may be combined, or in terms of a neural architecture that could implement those rules. Some theories might hypothesize, for instance, that you are, at some unconscious or infraconscious level, *following* rules that must themselves be represented. I suspect that this hypothesis is false; but even if it is true, it is certainly not apparent to you from your experience of thinking about molecules. (Unless, of course, you are doing something *more* than the kind of modeling I have described, such as using a list of *verbal rules* like "carbon atoms can form four bonds" as a kind of mnemonic aid to supplement a modeling process that has not been fully internalized.) For purposes of developing the notion of a mental model intuitively, let us set aside for the moment questions about how it is achieved, and think of it as a real and interesting phenomenon for which there might be alternative explanations.

7.1 Two Observations

Now that we have reached this stage in an intuitive development of the notion of a mental model, it is necessary to take a brief detour to note two things about the kind of situation I have described.

Semantics

First, note that while the kind of capacity for mental modeling described in the previous section was produced (through some invisible learning process) as a result of working with external models, the thoughts we experience in mental modeling need not retain any reference to the external modeling kit. We may very well think of what we are modeling *as*, say, carbon atoms and methanol molecules, just as we may be thinking of carbon and methanol while using the external model. This is one of the peculiar features of models, that we use them to think about something else. When using them as models, we think *through* them rather than thinking *about* them. Of course, in many cases we can also think about the models as objects in their own right. We can think of the pieces of the molecular modeling kit as foam balls and struts of different sizes and colors that fit together in different ways, and in the process of learning how to use the kit, we may have to stop and think about what type of atoms the black balls are supposed to represent.

Models and Modeling Systems

The example we have been exploring suggests a distinction between *models* and *modeling systems*. A particular figure built from the modeling kit is a model of a water molecule, but the kit itself is not in the same sense a model of something in particular. Rather, it is a kind of system for forming models of more specific types of things. Moreover, at least in some cases, it seems most natural to say that "models" like that of an individual molecule are representations, but modeling systems are not representations but *representational systems*.

I think there is something deeply right and crucially important here: that modeling involves systems that are not themselves representations but provide the resources for generating representations. However, I argue in chapter 8 that there is not a hard-and-fast distinction between models and modeling systems: one and the same thing can be a model (and hence perhaps a representation) of something in one context and a generative system for modeling, say, states of that thing in other contexts. For example, an anatomical model of the sort used in medicine is in one sense a model of human anatomy; but it can also be used as a framework for modeling more specific things, like the effects of an injury or the possible ways of making a surgical intervention.

7.2 Beyond Internalization

I have introduced the notion of a mental model by way of an exploration, in the previous chapter, of external models and then, in this chapter, of how the use of an external model or modeling system can lead to the development of an ability to form purely mental models. But the acquisition of a mental model does not really depend on starting out with an external model. A child who inherits a chemical modeling kit and does not know that it was designed for modeling molecules can still play with the kit and learn the ways the pieces fit together. And, having done so, she can internalize what she has learned so as to be able to think about the same sorts of structures "off-line"—say, when the kit has been put away for the night and her busy mind is resisting bedtime. In fact, the process of learning to use the kit very likely involved the formation of a mental model of it as a system of balls and struts that can fit together in various ways as well, even though *that* model subsequently is little used or perhaps is used unconsciously to guide our hands as we construct models we *think of* in terms of the target system of molecules.

We learn about systematic regularities we find in a wide range of things we experience, use these regularities to organize our subsequent interactions with them, and think about them off-line in ways faithful to their actual behavior when they are not present. The formation of mental models on the basis of perception and interaction is actually a ubiquitous form of human learning. The sole difference in the internalization of an external model is that the property of being a model *of* something else (say, molecules)—perhaps something we have never directly perceived and interacted with—is transferred from the external model to the mental model. Starting with external models that are then internalized helps us to get a basic grasp of the idea of a mental model by analogy with the more familiar examples of external models; but not all mental models are formed in this way, and the relationship probably really works in the opposite direction: the ability to work with external models likely depends on the ability to form and use mental models. Indeed, many advocates of mental models believe that the capacity to form mental models originally arose as a way of internalizing regularities encountered in bodily engagement with the world.

Mental models can also be formed in other ways. As I have already suggested, some are formed through experience without an external model. Some are developmentally canalized. Some are formed through processes of verbal learning, social interaction, or explicit theory formation. In the ensuing sections, I describe examples of each.

7.3 A Mental Model of My House

Try the following exercise. Think of a physical environment you are intimately familiar with. For purposes of comparison, I will use the house I grew up in—a place I have not been inside for perhaps thirty years, though I lived there from ages two to seventeen. It has often struck me how vividly I can imagine that house. And when I say "imagine," I do not mean something like experiencing mental snapshots. What I have in mind is more like a kind of situated spatial imagination: I can vividly imagine myself being *in* the house, in a given location, and have visual, proprioceptive, and sometimes tactile and olfactory imagery of the environment around me. I can, in my imagination, turn and move, and things come into view. I can, as it were, *explore* the house I once lived in. I don't mean anything spooky here; I am not talking about some kind of astral projection. I am talking about an exploration of things that are somehow stored in memory, in a form that needs to be explored and exploited, often in a fashion not unlike what I would do if I were physically present, *only by interacting not with the house itself but with something in my own mind and brain.* So let us try to make the case for this in a way that, I hope, at least the readers with forms of imagining somewhat like mine can replicate.

Ask yourself: what color are (were) the living room curtains? For me, both a verbal answer ("red") and a visual image occur so quickly that I suspect I have more than one way in which I have committed this information to memory. But if I ask myself about their texture, or how they were hung, or how many panels there were, I find that the process of "recollection" definitely requires a good deal more imagination, and very likely some deal of *reconstruction.* For example, in thinking about the texture, I imagine myself *feeling* the curtains as well as looking at them. Perhaps some of this is based on stored episodic memories of times I touched them as a child. (Indeed, in some sense it must be based on that.) But I can also make myself imagine handling them in any fashion I can handle curtains—drawing them open or shut, pulling them outward, lifting them off their hooks (ah—in thinking about this, I recall that they were on those hooks that get pulled by drawcords!), and so on. To some extent, the parameters within which I can imagine those curtains reflect the actual ways I did interact, or could have interacted, with them.

Now, in one sense, this seems quite in line with the idea that cognition is based in capacities for bodily engagement with the world. But it is also in tension with the most extreme form of that view: that cognition takes place only through actual bodily engagement with the world, without features of the world being stored in the mind. Indeed, for me to be able to imagine interactions that supply some of the same sorts of information as analogous interactions through the body, there would seem to have to be *something* stored in my mind that echoes the affordances I once found in the things I am remembering.

I can also do other things in imagining my childhood home. Try this: ask yourself how many electrical outlets there were in your home (or environment of choice). What happens in your mind when you try to answer this question? For me it involves a kind of visual or spatial imagining in which I feel myself present at each moment in some particular position and orientation within the familiar room and move around it, as though focusing my gaze from place to place and noting the features I am looking for. What I am, in effect, imagining is walking around the living room, the dining room, the bedrooms, looking for outlets, and noting and counting as I go. Phenomenologically, it has a great deal in common with what I would do if I were really exploring a room I was present in. But there is, of course, a crucial difference: I am not bodily present in that room while I am doing it. I can do the same thing with any of a variety of other features of the room—the number and configurations of windows or radiators, the size of the doorways, the types of stones in the mantel, the flooring, and so on.

I should note that I fully expect the results of such an exercise to be quite fallible. Some might be fallible in the ways one would expect specifically for episodic memories. I might reconstruct the room as being larger than it actually was because many of my memories might have been from the perspective of the much smaller body I had as a child. But others might be fallible in ways that we expect more of a *reconstruction* of details from something more abstract. But a reconstruction from *what*? The answer that seems to suggest itself here is something like an abstract, schematic model, perhaps analogous to the information in the data used by an architectural design program. This answer particularly suggests itself when the imaginative task results in thoughts that are vague or blurry. My mind imposes *some kind* of casing around the windows; but if I am to be truly honest, I have little idea what the contours of that casing were. I remember a few of the details of the masonry around the fireplace in detail, the rough width of the grouting between stones, and some vague textures; but beyond that, my memory of the hearth is something like a blank template waiting to be filled in. I remember that several of the stairs to the second floor were visible from the living room, and the rest disappeared behind the wall and ceiling, but I could not tell you how many without reconstruction that goes beyond remembering or exploring what is implicit in memory.

Yet all of this still suggests that the basis of my knowledge of my childhood home comes not in the form of episodic or propositional memories alone but in the form of some kind of mental model (or models) of it, and of such a sort that I can explore them—and perhaps explore them most aptly—through an imaginary recapitulation of the operations I would have performed to collect the same information if I were bodily present in the actual room. And, as I assume that in so doing I am not interacting with anything *outside* mind, brain, and body, it would seem to follow that there is something *within* mind, brain, and body that somehow stores some of the relevant features and relations of the room itself. The accuracy and detail with which it does so are no doubt limited. But it is hard to see how, without this supposition, I can explain what happens when I think about my childhood home.

I can, additionally, do other things in imagination. I can imagine things that never happened taking place there. What would it have been like if Martin Luther King had visited us? I can certainly picture him there, imagine how many steps it would take to rise from the sofa to meet him at the entrance, and so on. What would the room have looked like had the walls been painted black? How would things have looked different if a window had been walled off, or if the porch that adjoined the room had been enclosed and the doorway to it been made into a doorless arch? What would the room have looked like from the perspective of a camera in a corner on the ceiling, or from above, if the ceiling were transparent? Imagining these things is *almost* as easy as imagining things from perspectives that I probably actually did occupy, which suggests that there are capacities at work that are quite different from episodic memory.

Now try a slightly different exercise: imagine a blueprint of the first floor of your house. For me, this takes a second or two to coalesce, but it happens quite readily. This is interesting, as what is present in a blueprint is a view we seldom if ever actually have of an actual physical house. We do not learn to understand blueprints by actually exploring houses bodily from above. But it is a form of understanding that comes easily to many of us. And, at least for some of us, it feels easy and natural to go from a blueprint perspective to the perspective of being situated in a room and vice versa. The information that is directly afforded in one is different from that of the other, but the mind seems quite adept at translating from one to the other. Give me a blueprint of a house, and I can fairly quickly form a rough image of what the view would look like from a given position and angle. The ability to transition back and forth is surely a form of know-how and is closely connected with ways of being embedded bodily in the world. But the "blueprint view" is not something we learn by flying above roofless buildings. Rather, we have some kind of competence for moving back and forth between an abstract schematic understanding of a physical layout and a sort of situated geometrical thinking that allows perspective taking.

I think the best explanation for this is something like the following. We encounter the world through situated, embodied, embedded cognition. But this cognition must have some mental *form*—one or more representational systems. The use of these systems is not limited to cognition through online perception but can take the form of various kinds of imagining as well. These are certainly not limited to perspectives that one has actually occupied in the world; for example, I can always ask, "What would that look like if I moved over there?" (And, perhaps more surprisingly, I can often quite routinely envision it.) Even more plainly, they are not limited to perspectives I am presently in. But they are also not limited to perspectives that I really could be in. Perhaps I can now be in a position to look down (say, from a crane or a hot air balloon) on a house under construction in ways most generations of my ancestors could not. But I cannot *really* be in a position to "see" what the organelles of a cell look like from a position in between them—though I can certainly imagine it and use that imagining to explore things about it.

Let us now try to be a bit more careful and precise. I am suggesting that I have mental models of things such as my childhood home. (Perhaps there is a single integrated model of it in my mind, perhaps a constellation of models of parts and aspects of it that I can move between and coordinate.) But by this I do not mean that "having a mental model" means that there "is something" in my mind that has the same geometric features as my house. That, indeed, would strike me as a category error. What, then, do I mean by "having a model" of my house in my mind?

One thing I mean is that there is something in my mind that allows me to store and explore features of the house itself. And in these examples, I am particularly concerned with forms of "exploring features" that bear a good deal of phenomenological resemblance to what I might do to extract information about the house by interacting with it if I were physically situated within it. Phenomenologically, I am imagining standing in a point in the room and looking in a direction, attending to the way the curtains are hung. This allows me to extract some of the same information about the room (assuming the imagining is accurate) that I would be able to extract were I actually situated in that fashion within the room. Phenomenologically, I imagine turning 45 degrees to the right, and other features of the room become available to my mind's eye. These correspond to some of the features that would become available for perceptual inspection were I really thus situated. Of course, the information extracted is accurate (and hence only "information" in the normative sense) only to the extent that my mental model of the room is accurate. It is surely finite in detail, and likely inaccurate in some respects. Mental models are abstract and can be erroneous. But for such things to take place at all there must be some kind of correspondence between something present in my mind that I use in imagination (a mental model) and the real-world system I might explore through bodily interaction. And the correspondence must be in some sense systematic: when I imagine turning to the left and walking into the dining room, the things I imagine encountering correspond to what I would actually see if I were walking through the house along a similar path. *Something* in my mind must still be keeping track of the information I am not imagining or attending to at the moment in order for it to be immediately and often seamlessly available to me when I call for it. The simplest explanation for this is that my mental model shares, at an abstract level, various features of what it is a model *of*, the house. Having already discussed external architectural models such as blueprints and CAD software, it is tempting to suppose that this must include something analogous to blueprints and three-dimensional structural models representing layout and features independent of perspective.

Let us now take it a step farther. Suppose I am looking into some question about my house by looking through the house itself—say, I want to know how many outlets there are in the bedrooms. Consider how different this task is if I am asking it with respect to my own house as compared with being deposited in the doorway of an unknown building. In the case of my own house, I know the layout, I know where I am situated with respect to that layout, and I know which rooms are the bedrooms—and hence, implicitly, the efficient routes to get there. In the case of an unknown house, there is a lot I have to explore and figure out to know how to carry out the task. Having a mental model that supplies this needed background information radically simplifies the task. Indeed, it seems to me that forming a mental model that includes information such as what rooms are bedrooms is a necessary part of the task of counting the number of outlets in the bedrooms of a previously unfamiliar house.

Let us now turn to another form of imagination: creative imagination. Suppose I am an architect, and my task is to build a house that answers to a particular set of parameters. I do not do this by first slapping together walls randomly and then exploring. Rather, I conceive the house imaginatively, imagine the consequences of my design, amend the design in light of these, draw up blueprints, and then turn it over to the builders. The architect may understand a house quite well and imagine it quite concretely before the foundation has been laid. Blueprints, cardboard models, and CAD models may indeed assist in this process. But an architect with a detailed imagination may well be able to conceive and design a home without these aids. The whole design process may take place with little or no bodily engagement with the thing being designed. Nor is this something confined to architecture: anything that is a product of design is imagined—and not just in the sense of there being images of it in the mind, but in the sense of it being understood in whatever ways will be relevant to its successful execution—before it actually exists. However much this may be developmentally dependent on skills of bodily interaction, it cannot depend on *actual* bodily interactions with what is being designed, because (more or less by definition) what is being designed *does not yet exist.*

This seems to lead to two conclusions about mental models:

1. Many types of skilled bodily interaction are "skilled" in the sense of using abstract mental models of what one is interacting with.
2. Human minds are capable, to varying degrees, of decoupling the use of these mental models from bodily interaction, affording capacities for imagination, information extraction, simulation, and reasoning.

7.4 Chess

Suppose I wish to learn chess. I might begin (though it is probably not the best way to begin) by reading a rulebook for the game. I then sit down with a chessboard and another player or with a computer chess program and begin to try to apply the rules in actual play. At first, I may consult the rulebook frequently. (Alternatively, I might learn to play by sitting across the board from someone who already understands the game and gradually introduces me to the rules, intermixed with practice of the perceptual and motor skills involved in playing chess on a board with a particular set of pieces.) But as time goes on, I begin to simply *see* pawns and knights as capable of particular types of movement, and of the state of play as affording particular types of strategic development. As this progresses, I am clearly developing some type of internalized understanding of chess.

It might be tempting to think of this process as though it were a matter of memorizing the rules. Of course, some people do memorize a set of rules from a book on chess, and it is a minimum threshold for competence in chess that one can determine which moves are permissible, what constitutes a capture, a checkmate, and so on. But a person can memorize the rulebook without understanding how to apply its principles properly, and one can learn to apply the principles without memorizing the rules as verbal formulas. One *does* end up with *some* form of understanding that mirrors the rules, but this understanding need not involve any reflection *on* the rules. Rather, we perceive the board *through* a kind of cognitive template that has been shaped by the rules. We might even say that this sort of no-longer-conscious understanding of chess amounts to having a kind of "chess-shaped understanding"—that is, one in which various types of mental representations and operations have the same abstract form as the principles of the game.

Understanding chess seems to require a mental model of the game, which in turn involves a number of kinds of abilities. At the least, these would include a way of representing the board as a geometrically and topologically describable space of possible piece positions, along with an understanding that at most one piece may occupy a square at any one time, representational types for kinds of pieces and permissive rules for their movement, definition of an initial state, and rules for turn taking, capture, and ending states (checkmate, draw, resignation). Together, these generate a rich (yet also tightly constrained) possibility space, including possible game evolutions and possible states of play. (Note that the possible legal states of play are generated by game evolutions. Not all permutations of pieces in spaces are legal—something that the expert recognizes but the novice does not. A chess game cannot, for example, result in a configuration in which a player's pawns occupy the row closest to him.)

The different components of understanding chess are tightly, even constituently, interwoven. My understanding of the piece type KNIGHT cannot be separated from my understanding of the rules of movement. Change the rules for how a knight moves, and you have redefined the type KNIGHT as well, and vice versa. Change the size or configuration of the board, and the move type CASTLING must either be altered or lost entirely. You can, to be sure, generate *alternative* games by starting with chess and changing things about the board, the pieces, or the rules. But what you have then is a new game, one that requires a distinct mental model, albeit one that might be *produced* by starting with a model of chess and making alterations.

As I move beyond the baseline understanding of chess—being able to play the game in accordance with the rules—I gradually begin to understand things about tactics and strategy as well. These are normative rather than permissive or constitutive in character, but they involve a subset of sequences of moves permitted by the rules and ultimately guided by the definition of the end state of having the opposing player in checkmate (or, if that becomes impossible, forcing a draw). Both the baseline competence and the increasing mastery of competitive play may start out guided by *explicit* rules and heuristics; but the more expert I become, the more automatically they occur to me, and the more intuitive and less calculated they seem.

While the typical way of learning chess probably still involves sitting across a physical chessboard from another player, understanding chess also involves a significant abstraction from this context. I may learn the game with a certain set of pieces, but I can recognize it as the same game if the pieces look much different. Just about anything can be used as a chess piece, so long as the players can remember what tokens are being used for which types of pieces. The *typology* of chess pieces *as* chess pieces is completely independent from their physical characteristics. Indeed, the pieces and the board need not even be physical particulars: there are many computer programs for playing chess, and good players can play mental chess, in which there is neither a real board nor real pieces, the players each keeping track of the moves and state of play in their minds.

7.5 Social Contexts

One of the better-known projects in AI knowledge representation was Roger Schank's script-based simulation of an understanding of dining in a restaurant (Schank and Abelson 1977). Like other frame-type models, this involved special types of representations for types of objects (menus, utensils, servingware, food and drink, checks, tables, chairs) and activities (arriving at the host or hostess station, seating the customer, taking the order, bringing the order, bringing the check, paying the check, tipping), as well as special social roles (customer, host or hostess, waiter or waitress, chef). The "script" part of the program consisted in a stereotyped sequence (or flowchart) of events in their typical order (see fig. 4.7 in chap. 4).

Schank's program modeled a *particular type* of social context (albeit one that allows further variations for different types of restaurants), but its general features seem to be required for understanding any type of social context. In our understanding of social contexts, people are characteristically understood through *roles*, often roles particular to that context. Knowing that someone is acting as a waiter, a judge, a coach, or a babysitter sets up expectations about the things he or she is likely to do. Some of these expectations may be mere associations or prejudices—I might expect waiters to have stage ambitions as a result of living in an area where out-of-work actors commonly take work waiting tables—but the more important type are expectations tied to the role. The expectation that a waiter will bring what I ordered is not a mere association or prejudice but part of my understanding of the role of being a waiter. These expectations are generally independent of who is serving in that role: *any* waiter should bring the check at the end of the meal. And our understanding of what that person will likely do *within* the role will generally offer little guidance in predicting how he might behave outside it. Indeed, within a particular context, it might be less confusing to have robots perform the roles "on script" than to have, say, one's waiter act like a judge or a babysitter instead.

The typology of objects within one context might likewise be orthogonal to typologies from other contexts. The judge's robe might be a piece of clothing also sold as an academic gown or a preacher's robe. The same production model of fork might be used as a salad fork for one course and a dessert fork for another, and my dessert fork might be made by Mikasa and my dining partner's by Lenox. In a pinch, I can use various chess pieces to play checkers; and if I do so, their different shapes and the ways they could legally be moved as chess pieces are irrelevant to their typing as checkers pieces. The same can be said for activities performed. A quarterback and a wide receiver might go through the same motions during a game that they go through just tossing the ball around at a barbeque, but only in the game do those motions count as a forward pass.

We understand many social contexts with special roles, local ontologies, and activities. Each requires a separate mental model, though some models may be related to, and variants on or special cases of, others. There are different types of dining contexts and different types of restaurants; and the experienced diner knows, say, that at *this* restaurant there is a special person called the sommelier who attends to the wine, and the check is not brought until requested. Social contexts vary greatly from culture to culture, and a great deal of cultural literacy involves understanding the particular social contexts of that culture. Moreover, cultures can vary greatly in how strictly ritualized their practices are, and in how much of their daily life is defined by roles and contexts. In medieval Europe or Confucian China, it would seem that a very great deal of how a person was expected to behave—and probably much of how people really did behave—was defined by roles of family relationship, guild, and social hierarchy.

Sometimes a person's roles can conflict and present contrary expectations. In the play *Antigone*, the title character is caught between the conflicting expectations of her role as a pious sister (who should see that her brothers' bodies are properly buried) and as a citizen (who should obey the Archon's decree that they should be left where they fell in their rebellion against the state), and this kind of situation forms the basis for much classical tragedy. To understand it *as* tragedy, one must understand the conflicting roles, which can make a story from one culture difficult for someone from another culture with different roles to understand. Sometimes such conflicts are themselves the material for further rules: a judge should recuse herself from a case in which she is an interested party, such as a trial involving a member of her family.

Like most models, social models provide a space for possible alternative events and sequences of events. People don't always order the same thing at restaurants. Sometimes they may just go for dessert or drinks. Different things happen at different trials, and with a different attorney, a given case might have resulted in a different outcome. Even a tightly constrained event like a play allows latitude. Lear must be played mad when he carries in Cordelia's body in act 5, scene 3 (though note that it is a dramatic sense of "must," not one suggesting physical or logical necessity), and the actors must say the lines that Shakespeare composed; but one actor might play Lear weeping, and another might play his madness as having taken him into a completely dissociative state, thinking that Cordelia merely sleeps and watching her happily for a time. Like other models, social models provide a *framework* for interpreting a particular type of situation, but there are various ways things might go within that framework. Some of these may be built into the model itself; others may be like improvisations on it. And, of course, in any real-world situation there will be many other things going on as well, some of which might ultimately impact on the aptness of the model. I might, for example, discover after several minutes in a restaurant that the seated people are not really diners but actors in a drama. Or I might walk in on an impromptu rehearsal of *King Lear* and mistakenly think I am really witnessing an old man mourning the death of his daughter.

Social models are *un*like many other models, however, in the ways they shape not only our interpretation of the world but also the social world itself. The behavior of chemical bonds does not change when our models of it change. But conduct in a courtroom changes when trial law changes and those changes are appropriated by judges and attorneys; and people in cultures that lack any understanding of the Japanese tea ceremony do not engage in that ceremony, however much tea they might consume.

7.6 Moral Models

One case in which the ability of models to influence behavior is particularly evident is in prescriptive models, such as moral models. The line between ritualized behavior (or even custom) and ethics is blurry from a sociological standpoint. Indeed, the Greek word *ethoi* means "customs," and it is only in the hands of philosophers that it came to mean what we mean by "ethics" or "morals" today. Professional philosophers generally distinguish between three broad approaches to ethical evaluation: consequentialist, deontological, and virtue theoretic. Each of these provides a general framework for evaluating persons or their actions. Consequentialism evaluates an action on the basis of its effects: the overall balance of good and evil, help and harm, pleasure and pain, it results in, compared to the alternative actions that were available. Deontology (from the Greek word for duty, *deon*) evaluates actions with respect to whether they are in accordance with one's duties. Virtue theory evaluates *persons* in terms of their virtues (beneficial character traits) and *actions* in terms of whether they were done out of a virtuous character trait. (And, of course, there are various cultural and theoretical models of what duties different classes of people have and of what character traits count as virtuous.)

While philosophers are concerned primarily with these accounts as frameworks for *evaluation*, they can also be used to guide conduct. A person can try to foresee the consequences of various alternative actions and evaluate their likely costs and benefits. She can take stock of what duties she believes herself to be under and reason out what they require of her. A person can practice the things needed to acquire a particular virtue like patience and can inspect her motives and veto prospective action plans that would be done out of vices like impatience.

The philosophical subdiscipline called metaethics contains a great deal of literature that attempts to adjudicate which of these frameworks is a better account of what is morally good. For our purposes, however, the more important point is that each framework provides unique tools for evaluating and shaping action and character. Each framework picks out different factors that are relevant for such evaluations—consequences, relation to duty, and character—and each idealizes away from certain things that another brings into focus. It seems almost certain to me that the ordinary person is better off having multiple models of human goodness than having just one of them. And I am inclined to suspect that even careful theoretical formulations are not enough to build all the insights of all three types of model into a single framework, unless perhaps it is a metaframework that uses each by moving back and forth between them. There are some things for which we have concepts—like *goodness*—that we need to approach from various angles, through a number of distinct (and sometimes conflicting) models.

7.7 Mental Models and Scientific Understanding

I have tried to explore several familiar examples of everyday understanding to make the basic idea that they are underwritten by mental models intuitively clear and plausible. I have chosen examples that readers would likely be familiar with and could think about "from the inside." Hopefully, the time taken to walk through these examples will help readers form a basic and nontheoretical idea of what a mental model is supposed to be, and to assess the intuitive plausibility of the idea for themselves.

These examples, however, share two features that might lead to some unnecessarily narrow assumptions about mental models. The first is that they are all taken from the middle ground of understanding—types of cognition that are acquired in mid- to late childhood or adulthood, as opposed to those found in infants and young children, yet not among the more rarefied forms of thinking that require very specialized theoretical learning, like higher mathematics or scientific theories. The second, related to the first, is that they are all cases in which the acquisition of a mental model is achieved through some invisible learning process on the basis of bodily and social engagement with the domain modeled, or with an external model of it, or with existing social and linguistic ways of thinking and talking about it. The types of models we have discussed may indeed be acquired in different ways from one another—for example, some may be formed largely as a result of individual bodily and perceptual engagement, and others from supervised learning—but none of the examples was taken from the forms of thinking found in very young children that seem to be present too early to be a result of learning at all, and none from the higher reaches of abstract theoretical understanding.

This might have suggested to the reader that mental modeling is something found in the broad swath of everyday commonsense understanding, but absent from infant cognition and higher theory. If so, this would be a misleading suggestion. Indeed, the reader may recall that a number of the main examples of domain-centered understanding described in chapter 4 included several forms of thinking that arise in early development and also scientific theories. In that context, the point of the examples was to show that domain specificity was by no means unique to Fodorian modules, and we would need to move beyond the bifurcation between "modular" and "central" cognition to find an appropriate way to characterize understanding. While it was this line of thinking that led up to an alternative model-based view, we have not yet talked about the natural conclusion: that these forms of understanding are based in mental models as well.

The idea that understanding comes in model-sized units clearly bears a strong resemblance to the idea, which became influential in philosophy of science beginning in the 1960s, that a scientific *theory* is an integral unit, which provides its own interconnected ways of conceiving and reasoning about a particular range of phenomena. Instead of thinking of a scientific theory as a set of propositions built out of a theory-neutral language, philosophers of this period began to think of the theoretical vocabulary of a theory as constitutively interdefined with its laws and other commitments. Scientific theories have their own proprietary ways of representing and reasoning about phenomena, tightly interconnected internally, but only more loosely connected to everything else. This is precisely what I have been saying about mental models more generally. I do not claim credit for this as a completely novel insight. A number of writers before me have suggested that we view understanding generally, even the infant's understanding (Gopnik 1996), on the model of scientific theory. I simply prefer to put it the other way around: scientific theories are a special case of how we routinely come to understand the world through mental models. Model-based views have become increasingly popular in philosophy of science over the past thirty years, and I will discuss the relation of some of these to my own view in chapter 9. Here, however, I will try to develop the idea that scientific understanding is based in mental models in the same sort of intuitive fashion I have employed with other mental models.

In most of the types of models we have already explored, a model is something that contains elements that are analogous to the objects, features, relations, and processes found in their target systems. A modeling system must have features that mirror key features of the target domain to be apt. This is also the case with scientific theories and laws. These postulate entities and their characteristic properties and relations, as well as invariants expressed by laws and physical constants. Like other types of models, theories and laws are a kind of surrogate that allow us to deal, in an idealized way, with features of the world in a fashion that affords both understanding and prediction. A theory of gravitation or a law like the inverse square law *is* in this sense a model of the dynamics of gravitational influence. Theories and laws about collision are models of how objects behave when they collide.

Like other models, scientific theories and laws are idealized in several ways. First, they employ *bracketing idealizations* in what they exclude from their scope. A gravitational model is concerned only with gravitational interactions, not with collisions or electromagnetism. This affords insight into one invariant in nature, but at the cost of resulting in a model that is increasingly inadequate for kinematic predictions (i.e., predictions of how objects will actually move) in cases where nongravitational forces play larger roles. Second, they employ *distorting idealizations*. A particular model of particle collisions, for example, may treat them as ideally elastic. We know this assumption to be false, but such a model may still be apt for low-energy collisions among some types of particles. Third, scientific laws often employ constants for the values of forces in a fashion that only approximates their real values. When the laws are considered in the abstract, this can be avoided by simply using a letter for the constant. But to put the law to work in a computation, some *finite approximation* of the value must be used. This is often innocuous but will cause inaccuracies of prediction in chaotic systems in particular.

As with other types of models, scientific theories can be thought of as models of real-world processes at two levels. First, the theory itself—say, the theory of special relativity—can be thought of as a model of very general invariants found in nature, and of space-time as a whole. But it can also be used as a framework for modeling more specific problems, such as how several bodies will behave with respect to one another.

But there are also several ways in which scientific models differ from some of the other types of models we have explored. Contrary to what some previous generations of philosophers supposed, scientific understanding is not simply a result of inductive associations from naive perception. Scientific concepts and principles often differ significantly from those of common sense and have to be constructed through difficult processes of experiment design and theory building. While the resulting models are ultimately the outcome of a long process of bootstrapping up from commonplace modes of conceiving and interacting with the world, the bootstrapping process requires the development of some highly specialized and regimented forms of thinking, including new forms of mathematics and regimens for experiment and confirmation. Many scientific models thus employ rarefied mathematical forms of representation that confer a form of exactitude not found in most other models, and scientific reasoning and experimentation are regimented in ways designed to minimize error far beyond those employed in everyday life.

We also frequently speak of science being "objective." This is a tricky word, one that upon examination has many different meanings. But one thing that it can mean, which really does seem to apply to scientific understanding, involves the minimization of individual and collective idiosyncracies and biases. If we assume that cognition arose first and foremost as a way of allowing organisms to interact more adaptively with their environments, it is natural to conclude that the baseline "default settings" for any mind are optimized for things that are biologically salient to the organism: paradigmatically, things relevant to survival, nutrition, maturation, and reproduction. From a biological standpoint, it is actually *good* that an organism's mind is attuned to things like what is dangerous, what is edible, and, in the case of social organisms, who is aggressive or cooperative. But features like these—what Gibson called *affordances*[1]—are not simply facts about the object but facts about its potential relation to the interests of the organism, and such facts vary from individual to individual and from species to species. What is food for the butterfly is not food for the lion, and what is dangerous for me may not be dangerous for you. Such facts are "subjective," not in the sense that they depend on opinion or taste, but in the sense that they are really *relational* facts that depend crucially on who and what type of organism you are and what your interests are. Animal minds (including those of human animals) have evolved so that certain interest-relative properties of the world stand out, and this extends even to how we conceptualize objects and the world itself. For example, we "naturally" understand *up* and *down* to be privileged directions—a useful and efficient way of framing spatial geometry for organisms that dwell on the surface of a planet, but one that long proved an impediment to a satisfactory cosmology. The privileged place of cohesive solids in our maturationally normal ways of thinking about a world of objects is likewise a useful framing, though one that makes contemporary physics counterintuitive. The color vision systems of various species no doubt attune them to affordances that are important to their mode of life, but color is not a good framework for understanding scientific optics.

One thing we mean in calling science "objective" is that, to achieve scientific understanding, we often have to peel away layers of "subjectivity" that involve understanding things in interest-relevant ways that are peculiar to us as individuals or as a species. This, however, does not mean that there is an end point of "complete objectivity." The explanatory interests of the sciences are still *interests*, and ones that occur to members of one particular species, and even so only within particular historical and cultural contexts. Our ways of conceiving things scientifically, even our mathematical understanding, are still drawn from the repertoire of ways that minds like ours are capable of thinking, and are also conditioned by the ways we have of investigating and acting on the world, even if these are greatly extended and transformed through technology.

The "objectivity" of science also does not imply a single theoretical perspective. Scientists use a number of distinct models: relativistic and quantum models, wave and particle models, evolutionary and molecular-genetic models. Like different types of maps, scientific models can be incommensurable with one another, have incompatible representational systems and background assumptions, and license contradictory predictions. This is one of the principal phenomena described as "disunity of science," and we will examine in later chapters what cognitive pluralism might have to say about the causes of such disunity and the prospects for unification.

7.8 Core and Folk Systems

At the other end of the spectrum of cognitive sophistication from scientific models, we find early appearing, developmentally canalized forms of thinking such as the core and folk systems described in chapter 4. These are each centered on domains that are constituted by their respective internal ontologies, which are represented and reasoned about in distinctive ways.[2] The core object system, for example, has an internal ontology with a single generic class of things ("Objects"), properties (location, shape, motion), relations (being in contact, colliding), and processes (causal launching, moving in a continuous path) and is characterized by constitutive "rules" that guide anticipations—for example, that Objects are cohesive and contiguous, move in continuous paths, and go into motion only when acted on. It seems natural to view at least the core object and agency systems (and perhaps the core geometric system) as models of broad and generic domains. They involve both bracketing and distorting idealizations and employ representational systems with different representational types and rules of inference.

There are *several* core systems and several folk systems. They latch onto different systemic features of the physical, mental, social, and biological world and provide systematic frameworks for representing facts involving those features. Two systems can often be applied to the selfsame object: Socrates can be constituted as an Object, an Agent, a thinker, an organism. But in spite of this, different systems can constitute their common referent in incompatible ways. A child's toy or doll might at one time be constituted as an Object and at another as an Agent, but nothing could really answer perfectly to the rules of both the Object and Agency systems, as they have incompatible rules about how motion is initiated.

Given the extended developmental sequence of the folk theories, it seems overwhelmingly likely that their acquisition, like those of several other types of mental models we have discussed, depends on extended engagement with the physical, biological, and social world, albeit with the important caveat that their species-typical features suggest that there may be strong developmental biases. Core systems would seem to need to be even more strongly developmentally canalized, as there is evidence that they are present so early in infancy that it seems unlikely that the child could have inferred principles like cohesion and local causal dynamics (for Objects) or goal-directed motion (for Agents) from observation and the severely limited capacities for controlled interaction available to human infants.

7.9 Conclusion

Mental models provide systematic ways of understanding particular content domains. A model functions as a unit, with interconnected components that track salient features of its target system. This allows us to use models as internal surrogates for their targets, and to think and reason about the targets in ways that track their invariant features. A model is more or less apt to the extent that it accurately reflects the features relevant to some particular set of theoretical and practical interests. In some ways, models might better be described as *modeling systems*, as a model supplies tools for representing and reasoning about particular situations within its target domain and generates a set of possible representations of those states, as well as supplying efficient forms of reasoning about them, such as simulating their dynamics.

8 Relations between Models

In this chapter, I explore five different sorts of relationships that can be found among different mental models:

1. Two models can be related by degree of abstractness.
2. One model can be a *variant* on another.
3. One model can be created from another through *metaphorical transposition*.
4. Two or more models can be used in tandem to *triangulate* objects, states of affairs, or entire domains.
5. Two models can be *dissonant* with each other in various ways, such as incommensurability, formal incompatibility, and generating contradictory interpretations or predictions when applied to particular cases.

These are five very different types of relations, and I have combined the discussions of them into a single chapter chiefly because together they come to a reasonable chapter length. The discussions of triangulation and of ways models can be dissonant with one another, while comparatively brief, are of particular importance, and both will be taken up again in part 3.

8.1 Abstractness

In one sense it should come as no surprise that two models can be related by degree of abstractness. Some paradigm cases of models, like those in the sciences, come in varying degrees of abstraction and can even be viewed purely in terms of their mathematical expressions in abstraction from their empirical content. What really concerns me, however, is something more specific. I have described mental models as providing spaces of possible representations of objects and states of affairs and have counted some general and abstract frameworks as mental models: for example, core and folk systems, systems used for spatial imagination, and a chess player's understanding of that game. Indeed, some of the examples I have given of external models, such as kits for modeling molecules and computer-aided design systems, are also quite abstract.

But one might object that this is a strained and unnatural usage of the word 'model': that the individual things I might build or imagine are models of molecules, but the modeling *kit* and the suite of mental abilities that underlie the construction of the mental models of molecules are not. The kit and the abilities might be specifically resources *for forming models*, but this does not make *them* models. We might call them modeling *systems* instead. And the same might be claimed for several of the other examples: the chess-related skills I have are a modeling *system* that can be used to build models *of* particular games or particular states of play; my abilities to imagine and rotate physical solids make up a modeling system that I use to form models of particular solids and their transformations; the folk biological notion ANIMAL is not itself a model of an animal but a kind of template that can be used to form mental models of individual animals or types of animals and thereby to reason about them; and so on.

The distinction between a model of X and a modeling system that provides resources for modeling many different possible Xs strikes me as a useful one. But it is not clear that we should see *models* and *modeling systems* as mutually exclusive categories. It is an important distinction of analysis, but not a classificatory system.

In one sense, a chemical modeling kit is a set of resources for building models *of particular types of molecules*. But it can also be viewed, at a more abstract level, as a general model of various types of atoms, bonds, and the ways they can be combined to form molecules. It is a model *of* the molecular bonding potentials of molecules. In this sense, a model of a particular molecule is (also) a *representation* within the space of possible representations generated by the representational system of the more abstract model.

Similarly, in one sense, a mental model of chess is a set of resources for playing games of chess. As such, it provides the framework for representing particular states of play, and for mental models of particular games as they evolve over successive turns. In this sense, it is a *modeling system*. But it can also be viewed as a model of "the game" in the broader sense of "the game called chess." The model of "the game" provides a modeling system that is a framework for representing particular states of play, moves, and games.

The system (or systems) used in imagination for imaginary transformations of solid objects is in one sense a modeling system, which provides the resources for forming models of particular solids and imagining transformations thereof. But it is also in a broader sense itself a model of the structural possibilities, affordances, transformations, and resulting views of objects in geometric space. It might also be a model *of physical space* in general, one that differs from other mathematically possible models that employ distinct geometric or topological principles. It is both an abstract model and a modeling system for providing representations and models of more particular states of affairs and processes.

A model of human anatomy, by contrast, is generally useful not as a kind of production system for various anatomical configurations but as a model of the standard human anatomy. It may be that some such models have parts that can be modified, removed, or rearranged to model nonstandard configurations, but to the best of my knowledge, this is not its standard or preferred use. It can, however, be used to model specific conditions and procedures, such as the anatomical consequences of removing a bone or organ or different approaches to surgical intervention.

Where we draw the boundaries around the term 'model' here is in part a matter of deciding what terminology will provide the most useful set of theoretical resources. I am inclined to use the term in fashion that treats "models" as offering a space of representational possibilities of states of affairs. But this usage is extremely broad, allowing models of varying degrees of generality and abstraction. Model A may provide the representational space for B, in which case B may be said to be a representation within the representational system afforded by A. But B may nevertheless itself be a model if it defines a space of representations of its own. Even a model of a particular object may do this: a scale model of a car or train may be capable of being manipulated to represent various states—say, doors open or doors closed, hooked to the engine or decoupled—and may allow other sorts of representations as well, such as one of the ratio between the length and width of the vehicle. Or a model of a 1961 Lincoln limo (the type of car in which President Kennedy was riding when he was assassinated) might be used in a *representation* of the assassination, or as a modeling system for conspiracy theorists attempting to prove that the shots could not have come from a single gun.

There can thus be a hierarchy of models, differing in degree of abstraction, much as concepts can come in hierarchies from the more generic to the more specific. This sort of idea is also familiar in knowledge representation, and more generally in object-oriented programming in computer science, as the notion of an *inheritance class*, where one kind of object is based on another. A computer simulation of a mental model of a particular kind of solid, for example, might be defined using the class features of a more general model for geometric objects and use more generally defined code for rotation of objects in space and production of perspectival views.

8.2 Variants

One model might be counted as a variant of another insofar as they share much of the same representational system and rules. As a simple observation about how we might compare features of two models, this is not in itself particularly interesting, though it does raise questions about how models should be individuated and differentiated from one another, analogous to questions about whether two people ever really speak "the same language" or possess "the same concept." I am interested in the notion of variant models, however, not so much as a classification by similarity as a classification by etiology: that is, I am interested in the psychological fact that we can alter models *for a purpose*.

There are at least two types of cases of creating a new model through alteration. In the first, we become dissatisfied with a model—say, a scientific theory—and make some minor change in it. If the change is successful, we likely no longer use the older version. I believe there is some reason to think of the "before" and "after" state of the mental model in question as two states of a psychological continuant that has undergone change over time. The model possessed at time t_1 is, from the perspective of its *features*, at least subtly different from the model possessed at t_2 and in that sense might be said to be a distinct model. But thinking of a model as a *psychological structure*, we might be more inclined to say that a person made adjustments *to* a structure for understanding some feature of the world.

The second case occurs when we see a use for a *second* model, very like one we already possess, and produced by using the first as a basis and making adjustments. The invention of non-Euclidean geometries might be seen in this way. A mathematician like Bernhard Riemann began with a model that he already understood well (Euclidean geometry) and then altered a part of it (the parallel postulate). The result was a new model with different properties. We also see such cases in scientific change. Whereas the former type of change is more or less what Kuhn called incremental change, the latter is revolutionary change.

The examples I have used are, of course, ones involving scientific and mathematical systems that have also been made explicit and brought into the public sphere. And it seems likely that having a model in a form that is available for conscious scrutiny, whether private or public, facilitates the process of altering it and producing a variant. I doubt, however, that this is a necessary condition for model alteration. Arguably, humans and many nonhuman animals routinely alter the models they use in everyday life through learning. And it seems plausible that when an animal uses very similar, but not identical, methods to access different food sources or avoid different predators, its learning history may have involved first having a model that was developed to deal with one situation, and then developing another model, based on the first, for dealing with another situation that shares many features but differs in some details.

8.3 Metaphorical Transposition

One of the more interesting and surprising lines of research in recent cognitive psychology has been about the role of metaphor in learning how to think about new content domains. George Lakoff and Mark Johnson (2003) claim that a great deal of how we think about the world is through the use of a comparatively few ways of understanding things used as metaphors for other domains of experience. They begin by introducing a metaphor that is particularly poignant for philosophers: that argument is war.

> To give some idea of what it could mean for a concept to be metaphorical and for such a concept to structure an everyday activity, let us start with the concept ARGUMENT and the conceptual metaphor ARGUMENT IS WAR. This metaphor is reflected in our everyday language by a wide variety of expressions:
>
> ARGUMENT IS WAR
> Your claims are *indefensible.*
> He *attacked every weak point in* my argument. His criticisms were *right on target.*
> I *demolished* his argument.
> I've never *won* an argument with him.
> You disagree? Okay, *shoot*!
> If *you* use that *strategy,* he'll *wipe you out.* He *shot down* all *of my* arguments.
>
> It is important to see that we don't just *talk* about arguments in terms of war. We can actually win or lose arguments. We see the person we are arguing with as an opponent. We attack his positions and we defend our own. We gain and lose ground. We plan and use strategies. If we find a position indefensible, we can abandon it and take a new line of attack. Many of the things we *do* in arguing are partially structured by the concept of war. Though there is no physical battle, there is a verbal battle, and the structure of an argument—attack, defense, counterattack, etc.—reflects this. It is in this sense that the ARGUMENT IS WAR metaphor is one that we live by in this culture; it structures the actions we perform in arguing. (Lakoff and Johnson 2003, 4)

That is, we use our way of understanding war to provide a way of understanding argument as well.

> *The essence of metaphor is understanding and experiencing one kind of thing in terms of another.* It is not that arguments are a subspecies of war. Arguments and wars are different kinds of things—verbal discourse and armed conflict—and the actions performed are different kinds of actions. But ARGUMENT is partially structured, understood, performed, and talked about in terms of WAR. The concept is metaphorically structured, the activity is metaphorically structured, and, consequently, the language is metaphorically structured. (Lakoff and Johnson 2003, 5)

In metaphor, the language and the manner of understanding the source domain (in this case, war) are taken over as ways of understanding the target domain (in this case, argument).

What Lakoff and Johnson say about metaphor has a great deal in common with what I have said about mental models: each metaphor works *systematically* and obscures (idealizes away from) some things so as to highlight others.

> The very systematicity that allows us to comprehend one aspect of a concept in terms of another (e.g., comprehending an aspect of arguing in terms of battle) will necessarily hide other aspects of the concept. In allowing us to focus on one aspect of a concept (e.g., the battling aspects of arguing), a metaphorical concept can keep us from focusing on other aspects of the concept that are inconsistent with that metaphor. For example, in the midst of a heated argument, when we are intent on attacking our opponent's position and defending our own, we may lose sight of the cooperative aspects of arguing. Someone who is arguing with you can be viewed as giving you his time, a valuable commodity, in an effort at mutual understanding. But when we are preoccupied with the battle aspects, we often lose sight of the cooperative aspects. (Lakoff and Johnson 2003, 10)

Moreover, we can use multiple metaphors to highlight different features of the target domain. Lakoff and Johnson note several different metaphorical frameworks for understanding ideas (as people, plants, products, commodities, resources, money, cutting instruments, fashions) and love (as a physical force, a [medical] patient, magic, war).

Lakoff and Johnson's book (which is both eminently readable and well worth reading) contains far more interesting observation and theory than can be explored here. From the perspective of model-based cognitive pluralism, the important lesson seems to be this: sometimes we can form a model of a domain by taking a model of some entirely different domain and transposing its structure to be used as the basis for a new model. The system of already-understood types, connections, and inference patterns from the source model provides a kind of ready-made scaffolding for a model of another domain. This is an exceptionally powerful learning technique, as it provides a way of acquiring the basis of a model of a new domain without having to construct it from scratch through laborious exploration of the target domain.

8.4 Triangulation

The thesis that we understand the world through using many domain-specific models that have proprietary rules, representational systems, and internal ontologies may explain important things about understanding, but it also presents problems. We may use many individual models to understand different things about the world, but we do not *only* use models one at a time and in isolation from one another. In bodily interaction with the world, we employ a number of distinct perceptual, motor, and cognitive systems, not merely simultaneously but cooperatively and in an integrated manner. Even in off-line thinking, such as imagination, reasoning, and theorizing, we can think about an object, phenomenon, event, or general type of problem from a variety of angles, framed in terms of different models. And this raises questions about (1) how we are able to coordinate them adaptively at all; (2) how we are able to recognize *one and the same* object, phenomenon, and so on, though we are thinking about it through multiple models; and (3) how we are able to combine insights from different models through reasoning, so that the resulting understanding we have as a whole is more than simply the sum of what we would get from several distinct model-based processes going on simultaneously.

If our *only* way of thinking about the world were through individual models, it would seem that at least the second and third questions might prove impossible, and even the first would require some explanation. If we assume that models (or at least those that constitute objects at all) each have their own implicit internal ontologies, positing special types of things, properties, relations, processes, and transformations, there is a danger that they would end up being separate gears spinning on their own axes without engaging one another. Suppose I am playing chess. I recognize a piece as a knight within the chess model. I also recognize it as a particular kind of physical object—a small wooden piece I can pick up, a large concrete statue that would need to be moved by heavy equipment, or perhaps a human being in costume if it is a game of human chess. But if model-based cognition were the whole story, I would not recognize *a single thing* as being both a knight and a concrete statue. Rather, I would have two separate representations: one of a chess knight, and another of a concrete statue. Either I would see them as *two different things* with no way to connect them, or perhaps I would not even be able to count across models at all, simply registering "chess knight present" and "equestrian statue present," in the way that a car's dashboard has separate indicator lights whose information the car is incapable of combining. And not recognizing them as the same thing, I would have no way of implementing my plans for the chess knight by reaching out and moving the little piece of wood or ordering the man in the costume to walk forward two spaces and one to the left. But plainly this is not what happens. I can think of *one* object, which I think about *through* two or more models, and I regard what I think about through the various models as *the selfsame object.*

I can also coordinate information from the different models: if the knight is also a heavy statue, I know I will need to use different techniques to move it than I would if the knight were a small piece of wood or a human in livery. If I know that I actually have the means to move some pieces but not others—say, I can pick up the fifty-pound pawn statue but not the three-thousand-pound knight, or I know that one wooden piece is glued to the board—it might affect my chess tactics. If I know that a particular actor in the human chess game has to leave soon, I might deliberately sacrifice the piece he is playing. I can even think thoughts that seem to cross the lines of models: I might, for example, wonder how the players in a game of human chess feel when, as pieces, they are captured.

I will speak of thinking about the same thing—the same object, event, process, phenomenon, and so on—through different models as *triangulation. That* we engage in triangulation through different models seems indisputable. *How* we do so is a much more difficult matter. I will in fact suggest that there must be *more than one way* of doing so, and I will outline several in the course of this section. This, however, will not go so far as to try to describe the *mechanisms* through which the various types of triangulation take place. That is a difficult empirical question. There are some candidate answers in some cases, but I think that in others what I am outlining are some important and fundamental open questions for further research in the cognitive sciences.

Three Great Problems

The problem of triangulation has been touched on several times, and in various forms, in philosophy and the cognitive sciences. In fact, there are at least three famous and significant problems that seem to be special cases of it: what Husserl called "Hume's problem," the binding problem, and what Dreyfus calls "the frame problem."

Hume's Problem

The great phenomenologist Edmund Husserl identified a central issue in philosophy as "Hume's problem." Hume was a British empiricist who viewed perception and thinking as a bundle of independent impressions and ideas, held together by association. In particular, in his early work *A Treatise of Human Nature* (1738), Hume denied that we have a notion of *substance* that can be used to knit together the various sensory impressions we have of the world—red, round, soft—as properties of a single object. Husserl noted that such an account of perception and thinking does not provide sufficient resources to explain why we do not simply think of various impressions, ideas, or properties but think of a world of *objects* that persist over time.

At least two distinct issues are at work here. One, which Husserl became interested in through his investigation of time consciousness, is how we are capable of reidentifying something as *the selfsame object* over successive perceptions or thoughts. Even if we are capable of constituting something as an object at time t_1, and something as an object at time t_2, how is it that we are able to additionally link these two thoughts as thoughts about the *same* object? To put it crudely, how is it that we do not simply think "an X" at t_1 and "an X" (or "*another* X") at t_2 but rather think at t_2 of "the selfsame X we thought about at t_1"? The second and more fundamental issue is how we are able to think, even in a single thought at a single moment, not merely of this property and that property being presented but of *a single thing* that *has* both properties. We are clearly able to do so, but a psychology like Hume's, whose resources are limited to the presentation of individual impressions and ideas, cannot explain how we are able to do so.

The Binding Problem

A related problem occurs in neuroscience. Philosophers since Aristotle traditionally assumed that, in addition to the various sensory modalities—sight, hearing, touch, smell, taste—there must be an additional faculty, the *sensus communis*, or "common sense," where the deliverances of these modalities are fused together into a polymodal representation of an object. Scientific study of the brain has revealed various neuroanatomical areas corresponding to the various sensory modalities, but not of a *sensus communis* where "it all comes together." Indeed, our contemporary understanding of the brain only makes things more complicated: not only is there no brain area where vision comes together with all the other modalities in a single representation, but there are distinct areas for different aspects of vision: color, form, and motion.

So the problem shapes up like this: given that the color of an object is represented in one part of the brain, and shape in another, and there is no downstream area for color plus shape, how is it that we know what shapes and colors go together? If we are presented with a red triangle and a blue circle, how do we know that RED goes with TRIANGLE (and not with CIRCLE or indeed with nothing at all) and BLUE with CIRCLE? Similarly, if we are inspecting things with eyes and hands and exploring a fuzzy triangle and a smooth circle, how do we know that FUZZY goes with TRIANGLE and SMOOTH goes with CIRCLE, and not the other way around, or indeed that representations in different modalities should be paired at all? One important hypothesis about how the brain accomplishes such "binding" of different properties was proposed by Crick and Koch (1993). The basic idea is that when two representations in different sensory areas are bound together, the bound representations are firing in phase with one another. The suggestion is thus that, at least in perception, binding of representations of the same thing is accomplished through a kind of *phase locking*.

The binding problem is a scientific variant on one aspect of Hume's problem: specifically, as a problem about the ability to unite different representations synchronously as representations of a single thing in perception. To the best of my knowledge, proposed solutions like the Crick-Koch hypothesis have not been extended to nonperceptual intentional states (say, the imagination of a red triangle) or to the integration of perceptual states over time as representations of the selfsame enduring object (the *same* triangle I saw yesterday).

The Frame Problem

As mentioned in chapter 4, one of the fundamental problems that second-generation AI encountered was what Dreyfus calls the frame problem.[1] Even if we assume that routine thinking is accomplished through the application of what Minsky calls "frames" and I call mental models, a further problem arises of *how to know which model to apply*. Dreyfus argues—plausibly, in my view—that this problem cannot be solved through standard computational methods, and this seems to be a significant problem for any theory that says that cognition is *nothing but* computation. It need not be a refutation of model- or frame-based approaches, however. On the one hand, a model-based approach to understanding need not be committed to the view that model-based cognition is purely computational. On the other hand, even if we assume that model-based cognition (in the sense of thinking *within* a model) is computational, this is compatible with the thesis that the processes that determine *which* model to apply are *non*computational.

However, Dreyfus's argument does seem to show that even if there is a need for a model-based story, it cannot be the *whole* story.

At first glance, the frame problem might not seem to have much to do with Hume's problem or the binding problem. It is generally presented as a problem not about the constitution of objects but about *which* frame or model (any one of which has its own internal ontology for thinking about objects) to apply. Yet there is an important connection: we do not merely use *one model or another* to think about an object, an event, a state of affairs, or a situation. We often use *several*, applying them to the same object or situation. And to do that, we need some way of coordinating them as applying to one and the same object or situation (and a way of individuating "situations" as well as objects that is model independent) in a fashion that does not depend wholly on any one model's way of framing objects and situations.

Relation to Triangulation

Each of these problems is about how we integrate different types of representations into representations of a single thing and coordinate the types of information and understanding afforded by different models and perceptual systems. It may be that some of the sensory information falling within the scope of Hume's problem and the binding problem is not based in models. If that is so, then there are issues about triangulation that are not about triangulation *through multiple models*. Alternatively, if we do end up having reason to regard even Humean concepts and early perceptual representations as embedded in models, the problem is more generally a problem of triangulation through models.

Cases

The ability to coordinate information derived from distinct perceptual areas and mental models is one of the chief arguments that has been given for the need of "central processing," either in language or in a language-like mental medium. To be sure, we do think in language and have language-like ways of thinking that may be distinct from the use of a public language in thinking. But this is not the *only* way of accomplishing such tasks and indeed is not a plausible hypothesis about how it is achieved in some of the cases. There do seem to be several distinct kinds of cases. I will survey those I am aware of here, noting proposals for explaining some of them.

Perceptual Binding

The perceptual binding problem is probably the case that has received the most scientific attention. A proposal like Crick and Koch's phase binding hypothesis provides an explanation of how information represented in distinct areas of the brain—say, the visual areas for color and form—gets coupled. Integration of information from different sensory modalities may additionally use association areas of the cortex that take inputs from multiple sources. However, these do not seem to be a *sensus communis* for panmodal representations. Rather, the integration of information from various modalities into a single percept seems to be a matter of how different areas are *coordinated and act in concert* rather than being accomplished by an additional area. There is no more an anatomical *sensus communis* than a string quartet ensemble is an instrument over and above the violins, viola, and cello. Moreover, the explanation invokes nothing like language-like representations, but neural systems in a kind of dynamic resonance pattern. And whereas a language-like representation supports the combined representation of two types of information in a single medium, phase binding treats the situation as involving the coordination of separate representations in different representational systems and representational media.

Sensorimotor Engagement

Perception is not normally a passive process like a camera recording an image. It usually takes place as part of engagement with the world, which is often purposeful and involves motor processes through which an organism acts on the world, even in the process of exploring it perceptually. Sensorimotor engagement involves more than merely binding together information from separate perceptual streams. It involves coordinating perceptual, motor, and orienting systems in a fashion that is propelled by the drives and interests of the organism and constrained by its physiological and cognitive abilities. Such processes, even when they are primarily "perceptual," also evolve as dynamic processes over time, adjusting to explore for new information in light of present states.

Advocates of "embodied and embedded cognition," "situated cognition," and the "enactive approach" have provided general frameworks for characterizing such processes, using resources such as dynamic systems theory. Silberstein and Chemero, for example, write:

> We propose that extended phenomenology-cognition be understood as a variety of niche construction, one in which the constructed niche is an animal's cognitive and phenomenological niche. In biological niche construction, the activity of some population of organisms alters, sometimes dramatically, its own ecological niche as

well as those of other organisms. ... These animal-caused alterations to niches have profound and wide-reaching effects over evolutionary time. Phenomenological-cognitive niche construction has its effects over shorter time scales—an animal's activities alter the world as the animal experiences it, and these alterations to the phenomenological-cognitive niche, in turn, affect the animal's behavior and development of its abilities to perceive and act, which further alters the phenomenological-cognitive niche, and on and on.

We depict a schematic extended phenomenological-cognitive system in Fig. [8.1]. Following enactive and ecological cognitive scientists, we take animals to be self-organizing systems. ... The animal's nervous system has an endogenous dynamics, which generates the neural assemblies that both compose the nervous system and constitute the animal's sensorimotor abilities. These sensorimotor abilities are the means by which the animal's niche couples with and modulates the dynamics of the animal's nervous system. These sensorimotor abilities are coupled with the niche, that is, the network of affordances available to the animal ..., and interact with it over multiple time scales. Over behavioral time, the sensorimotor abilities cause the animal to act, and this action alters the layout of the affordances available, and the layout of affordances perturbs the sensorimotor coupling with the environment (causing, of course, transient changes to the dynamics of the nervous system, which changes the sensorimotor coupling, and so on). Over developmental time, the sensorimotor abilities, that is, what the animal can do, selects the animal's niche. That is, from all of the information available in the physical environment, the animal learns to attend to only that which specifies affordances complementing

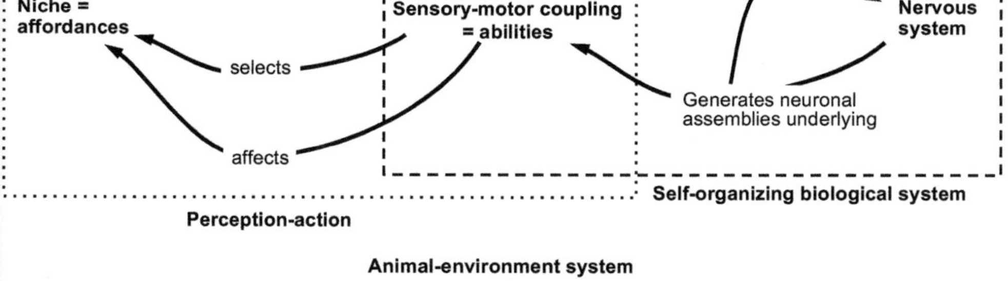

Figure 8.1
Schematic extended phenomenological-cognitive system. Reproduced from Silberstein and Chemero 2011, 8.

the animal's abilities. At the same time, the set of affordances available to the animal profoundly influence the development of the animal's sensorimotor abilities. So we have a three-part, coupled, nonlinear dynamical system in which the nervous system partly determines and is partly determined by the sensorimotor abilities which partly determine, and are partly determined by, the affordances available to the animal. (Silberstein and Chemero 2011, 7–9; figure number amended to correspond to the present text)

Depending on the state of interaction between organism and environment, control of the overall dynamics might be located in various places: in some neural subsystem, in processes adjusting the coupling, in features of the environment, or in a more distributed system including both an organism and objects and processes external to it. While such an explanation does not focus primarily on operation of subsystems like mental models, this is primarily because it is intended to be a general framework that is applicable to many types of relationships between organism and environment. The enactive approach need not *deny* that neural systems can be functionally decomposed into subsystems like mental models, or that these can play a predominant role in the dynamics of some kinds of tasks or be what is tightly coupled to more specific features of the environment (the system modeled) in the performance of those tasks. Coupling mechanisms specific to model-based interaction might be eminently relevant to the mastery and aptness of models, and the dynamics of neural change in learning an important tool for an account of model acquisition. Moreover, the capacity for off-line cognition may quite plausibly involve replacing an online coupling to external objects with a coupling to mental models that act as their surrogates. (In this respect, I think that a model-based approach to cognition provides a natural extension of dynamic coupling accounts of online cognition.)

The Selfsame Object

None of this so far really addresses the key issue Husserl identified as Hume's problem: how we are able to constitute something, perceived and conceived in various ways and over time, as *the selfsame object*, which endures through changes over time and is understood as the same thing even as our perceptions of and beliefs about it change. This phenomenon—which we may call *object-oriented cognition*—is really a special *type* of triangulation, distinct from the practical coordination of different models used in tandem.

I think there may be at least two significantly different sorts of cases here, which seem to require different sorts of accounts. The first involves object tracking within a perceptual episode. The brain has comparatively low-level mechanisms, particularly in the visual system, that seem to be involved in tracking an object over time. These seem to be capable of dealing with things like temporarily occluded objects, and perhaps even of constituting objects that are not yet directly perceived on the basis of cues such as scent or the vibrations in a spider's web. Such processes do not seem to require resources for the postulation of proposition-like hypotheses, though sometimes we find it easiest to describe them in terms of beliefs. They are found quite broadly in the animal kingdom. But they do not account for the abilities to reidentify something as the selfsame object in *distinct* perceptual episodes, or to engage in object-oriented triangulation in types of thinking other than perception.

On the other hand, it is not clear how many animal species are able to reidentify something as the selfsame individual on different occasions, or to think about it as an individual in off-line cognition. An ant can identify something as another ant, and things such as whether it is of the same colony, perhaps its "caste," and states such as carrying home food. But an ant probably cannot tell one individual worker or drone of its own colony from another. More likely, each encounter with another nest mate is for it simply a kind tokening—"worker ant," "worker ant again." Mammals and birds, on the other hand, are able to reidentify specific individuals, and this ability seems necessary for identification of mates and offspring, and to undergird social cognition for things like dominance hierarchies, at least among the more social species. It is unclear just where in the phylogenetic tree such abilities appear. Perhaps an alligator can see another alligator and remember it as the same alligator it saw a week ago. Or perhaps it just thinks, in effect, "male alligator bigger than me" on each occasion, not treating the two kind tokenings as presenting the selfsame individual.

At least some species seem able to reidentify not only particular conspecifics but particular inanimate objects as well. A dog seems able to recognize a *particular* toy, for example. And of course many animals are able to identify particular individuals of other species as well—a domestic dog can recognize its favorite human as well as it can recognize other dogs, and evidence suggests that even wild birds can distinguish between different humans they have encountered. I suspect that the ability to reidentify individuals may initially have been an adaptation for social cognition involving conspecifics within social species and is probably one of the key prerequisites for being a social animal. It may be that once this ability is in place, extending it to other things—individuals in other species, inanimate objects—comes for free or at least requires only the perceptual abilities to discriminate whatever cues are needed to distinguish between individuals.

(You and I have the requisite cognitive resources to think of separate individuals, but our perceptual abilities are not well suited to distinguishing between particular ants of the same species and caste.) Alternatively, it may be that these require further cognitive abilities that build on foundations originating in social condition.

The ability to reidentify specific individuals seems to require a particular kind of cognitive resource: concepts that are namelike rather than generic. Kind concepts (DOG) and property concepts (MALE, SPOTTED) are by nature things that can be applied to multiple individuals. Namelike concepts, by contrast, have the function of picking out and tracking just one thing. This strikes me as likely to be a *special* kind of concept, one that is not required for having ways of tracking kinds and variable properties, and one that has probably appeared comparatively recently in evolutionary history, though perhaps independently in several widely separated taxa.

Information Passing

While thinking within a model may often take place more or less autonomously, we are also often able to pass information between models. In this way, at least some model-based reasoning is not entirely encapsulated in the way that Fodor described modules.[2] My mental model of a social situation may supply general anticipations about what to expect, but I may need different types of information, not supplied by that model, to achieve a more fine-grained understanding of the situation and to make better predictions. Some aspects of how a waiter will behave—at least while "on script"—may be supplied by my model of restaurant etiquette. But others might require me to know something about his individual psychology: whether he is passive-aggressive, what his cultural background is, whether he is an experienced waiter or in training, what other roles he plays in the restaurant or outside it, what sort of mood he is in. It requires different sorts of models to understand these things, but if I do grasp them, they can affect how I run the restaurant model in this particular case.

The analogy that recommends itself to me here is that of parameter passing in computer programming. Contemporary programming is generally highly modularized, with distinct subprograms used for different operations. The different code modules, however, are not simply different operations that run in parallel; often one calls on another and passes and receives values of variables called parameters. One module M1 might, for example, need the area of a triangle but only have information about the length of the sides. There is a formula for computing area on the basis of the lengths of the sides, called Heron's formula.

$$A = \sqrt{p(p-a)(p-b)(p-c)}$$

where p is half the perimeter, or $\frac{a+b+c}{2}$.

It would be possible to enter code for computing the area within M1, but if it is an operation that will be called on by more than one module, it is more efficient to define it as a separate modular procedure. So we might define a separate module M_HERON for computing area from the lengths of sides. It would take three input parameters (one for the length of each side) and pass back a single output parameter for the computed area.

Of course, when we are doing explicit mathematical calculations, we can proceed in a fashion that fairly closely resembles the steps of such a program. (Turing's notion of "computation" was, after all, framed on the model of what human mathematicians do in calculation.) But in most cases that use mental models without such explicit calculation, the relation with parameter passing is probably only analogous. There is *some way* in which one model "requests" information that can be supplied by another, and some means of accessing and then using it.

Language and Language-Like Cognition

I have tried to make a case here that there are forms of intermodel triangulation that do not require a language or language-like thought processes. This, of course, does not mean that we do not *also* possess linguistic resources. At least *some* representations found in models can also be "exported" to be represented (or "re-represented") in a language or a language-like format, typified by a grammar and the ability to combine, through syntax, thoughts whose content may be derived from different models. This seems more or less to correspond to what Fodor calls "central cognition," albeit with two important caveats. First, the alternatives to this language-like thinking are not limited to Fodorian modules but include a wide variety of models as well. Second, if models form the basis of much intuitive inference, the *content* of nonsensory representations is not to be understood wholly within the context of central cognition (and so need not be "Quinean and isotropic") but is determined in large measure by their models of origin. But the other main assumption about a language-like mode of cognition—that it is a kind of ecumenical representational medium that is not domain specific—remains largely intact. (It remains only *largely* intact because there may be many model-bound representations that cannot be effectively integrated into central cognition. Think of all the things that are difficult to put into language.)

Language, or at least a language-like medium for thinking, also seems to be a prerequisite for particular forms of *reasoning*. Most obviously, it is needed for explicit argumentation in which the syntactic form of the reasoning is crucial. A language with names is a medium particularly well suited to representing and exploiting an open-ended wealth of information about individuals. But it would also seem to be needed for the ability to make our representations and reasoning objects of scrutiny that we can examine, test, and refine. I discuss these topics further in chapter 11.

How One Model Tests Another

Finally, the use of multiple models can in various ways help us refine the application and assess the aptness of each. In particular, different and especially contradictory results that emerge from the application of two models (discussed in the following section) may raise questions about the aptness of one or both in a particular context and may raise more abstract questions about the class of cases in which a model is aptly applied. In some cases, it may also lead to alterations in one or both models. These situations may often call for a kind of metacognition—for thinking *about* models rather than simply *through* them—that may not be available to nonhuman animals. However, we probably also have reason to think that there are ways, not involving metacognition, that a conflict between models can lead to changes in one or both models.

8.5 Dissonance

Using more than one model can help us to understand a more complex reality than either model could allow us to understand alone. But at the same time, sometimes the different models we use do not play well together, and generate puzzles and paradoxes. Sometimes they lead to contrary expectations and predictions. Sometimes they are actually *incompatible* with one another. We can, however, understand some of these problems by looking at the features of models as models.

First, consider a case in which more than one model of a complex situation is applicable: say, gravitational and aerodynamic models of the motion of a paper airplane; or a mother who is working at a corporation on an assignment that is running late, but also has to pick up her daughter from day care. If we use *only* the gravitational model, it would lead to a prediction that the paper airplane would fall in precisely the same fashion as a piece of paper of equal mass wadded into a ball. If we use *only* the aerodynamic model, it would lead to a very different prediction. In the second case, if we use only the social model of professional responsibility, it might lead to a prediction that the woman will stay at the office until the job is complete. If we use only the model of maternal responsibility, it might lead to a prediction that she will leave to pick up her daughter sometime between the originally expected hour and the time day care closes. In both cases, each model brackets the factors relevant to the other, and so neither model alone can be used as the basis for a realistic prediction of real-world behavior. Of course, in some ways the cases are different. In mechanics, we have the technique of summation of forces through vector algebra, whereas this approach is not available in the case of social duties, in which there might be other meta-techniques like assessing how one duty might trump another, or there might be methods of finding a satisficing solution. But in both kinds of cases, any kind of method that is used must go beyond the resources of a single model, as each model brackets factors relevant to the other, and both kinds of factors are relevant to the real-world situation. And there are probably both physical and social cases where we do not have an algorithmic method that ensures a way of integrating the factors.

Let us also consider a more serious sort of case, in which two models applicable to the same situation are either incommensurable or inconsistent. Wave and particle models of the same phenomenon—the propagation of light—lead to contrary predictions of behavior because they use different representational systems, inference rules, and processes for simulation. General relativity and quantum mechanics are inconsistent with each other, as are classical and relativistic gravitation. We might view these cases as analogous to the relationship between Euclidean and non-Euclidean geometries: the models in question have incompatible representational systems and rules. Each may be apt for a wide range of cases individually, but their formal structure ensures that when they are combined, problems will occur.

This may be puzzling from the standpoints of epistemology and truth theory—I will examine whether it should be deeply and abidingly puzzling in the final part of the book—but not from the perspective of psychology. Each model must use *some particular* representational system. And what representational system is employed will be constrained by (a) the demands of the real-world invariants to be modeled; (b) the general constraints of cognitive architecture (the specific set of types of representational systems that are available to a mind like our own); and (c) pragmatic demands such as utility for providing understanding, calculation, and prediction. As these constraints operate to shape models *individually*, we have no assurance that two models will turn out to be compatible with each other. Models that each work well individually and in particular ranges of cases may present puzzles and paradoxes when combined, and this would seem to be a natural consequence of a strategy of understanding the world through individual models that are suboptimized for a particular range of problems.

Dissonance between models will present a significant problem for the project of unification of understanding, and the reasons that model-based cognition can produce dissonance will help us understand cases of disunities of knowledge and of science, a topic to which I return in chapter 12.

9 Other Model-Based Approaches

In the preceding chapters, I have attempted to develop, in an intuitive an easily understandable way, a very broad notion of a mental model. While I have gone so far as to describe some of the features mental models might require, I have deliberately kept the discussion at a level of abstraction that avoids the kinds of technical details that we would need to do things such as create a computer simulation of model-based understanding and test its adequacy to account for various bodies of psychological data. Not only am I not in the right academic discipline to be pursuing that kind of project, but the differences between different model-based approaches to knowledge representation or AI are tangential to my purpose, which is to outline a general philosophical thesis about understanding and explore its further consequences for areas of philosophy such as epistemology and semantics. Numerous ongoing research programs in psychology, AI, and theoretical cognitive science employ notions of mental models as a central feature of their work, and this book is intended more to complement than to compete with them.

At the same time, the uses of the term 'model' in those academic areas vary considerably. Indeed, if the people writing about models agree on one thing, it is that the word is seldom characterized explicitly, and to the extent that it is, widely different uses of it are at work. Readers familiar with other model-based accounts of cognition may understandably desire some clarification of how my notion of a model accords with or differs from others. It is impossible to canvass all the uses in several fields exhaustively, but in this chapter I locate my account with respect to several influential projects in psychology, philosophy of science, and theoretical cognitive science. (I should, perhaps, warn that this chapter may seem more "academic" than most of the others and may be of interest primarily to readers already acquainted with some of the other work surveyed.)

9.1 Models in Psychology

The Scottish psychologist Kenneth Craik appears to deserve the credit for coining the expression "mental model" in the cognitive sciences. (And, even if someone else used it earlier, advocates of model-based approaches trace their usage of the expression back to Craik.) In a passage oft quoted in the mental-modeling literature, Craik wrote:

> If the organism carries a "small-scale model" of external reality and of its own possible actions within its head, it is able to try out various alternatives, conclude which is the best of them, react to future situations before they arise, utilize the knowledge of past events in dealing with the present and the future, and in every way react in a much fuller, safer, and more competent manner to the emergencies which face it. (Craik 1943, 61)

Craik published this in 1943, considerably before investigations in artificial intelligence. Indeed, at the time, computing machines were largely shrouded in the secrecy of the wartime code-breaking project at Bletchley Park, though Craik was familiar with and wrote about control systems in two posthumously published papers (Craik 1947, 1948) and sought a physiological and "mechanistic" explanation for mental processes. He was also writing long before evolutionary and adaptive approaches to the mind became popular (though, of course, this line of thinking had been pioneered by William James [1890]), but Craik framed the problem of thinking in terms of an organism coping with problems in its environment. Organisms are able to do so intelligently, according to Craik, through "explanations" that allow predictions of what will happen in the world and as a consequence of various things the organism might do to act on the world. The abilities to explain and predict require external processes to be "translated" into words, numbers, or other symbols, which are then used to "reason," and then "retranslated" into external processes, like building a bridge.

While Craik did to some extent explore the idea that such internal processes operate on linguistic or verbal representations, his central explanatory strategy was to liken it to the construction and use of external models. Indeed, as I have done in this book, he paved the way to developing the idea of a mental model by first discussing external models.

> By a model we thus mean any physical or chemical system which has a similar relation-structure to that of the process it imitates. By "relation-structure" I do not mean some obscure non-physical entity which attends to the model, but the fact that it is a physical working model which works in the same way as the process it parallels, in the aspects under consideration at any moment. (Craik 1943, 51)

A mental model, in turn, is something in the mind with a structure that "parallels" that of the external system we are thinking about:

> My hypothesis then is that thought models, or parallels, reality—that its essential feature is not "the mind," "the self," "sense-data," nor propositions but symbolism, and this symbolism is largely of the same kind as that which is familiar to us in mechanical devices which aid thought and calculation. (Craik 1943, 57)

Just what this nonpropositional symbolism might amount to is unclear, and we should probably resist the temptation to read later associations that the word 'symbol' has from within later computational approaches to thought. Johnson-Laird has suggested possible connections to Wittgenstein's pictorial or isomorphism view of thought and Peirce's iconic symbols, which represent by dint of structural similarity. Craik's interest, however, seems to have been not with semantics (more specifically, with what *makes* something represent something else) but with adaptive cognition: if structural similarities between a model and the system modeled are important, it is because this allows the model to function as a mental surrogate for the target system in off-line cognition through simulation or nonpropositional reasoning. Craik's suggestion that mental models include representations not only of features of the world but also of our own actions is also highly suggestive of the idea that models of this sort might bridge the gap between theoretical and practical reasoning and provide a basis for an account of understanding.

Craik never worked out this theory in greater detail, as he died in a bicycling accident on VE Day in 1945, two years after the publication of *The Nature of Explanation*. His general idea of a mental model, which seems altogether consonant with the views I have developed, received little attention at the time, as he was writing in the heyday of psychological behaviorism. The language of "mental models" lay dormant between Craik's death and the reintroduction of the terminology in books published in 1983 by Gentner and Stevens and by Johnson-Laird. Contemporary advocates of model-based views commonly cite both Gentner and Stevens and Johnson-Laird as pioneers of model-based approaches, but really they developed very different ideas of what sort of thing a mental model might be.

Although Gentner and Stevens's (1983) anthology was titled *Mental Models*, the word 'model' appears in Gentner and Gentner's own article in that volume, "Flowing Water or Teeming Crowds: Mental Models of Electricity," only in the title and the first sentence, which mentions "analogical models" (Gentner and Gentner 1983, 99). The preferred terminology in the article is "analogical comparisons." The article explores the idea that we think about electricity through "generative analogies," importing an understanding of a system of relations in one domain (such as water flowing through pipes) to another (electricity) through a process of "structure-mapping" (100).

> The analogical models used in science can be characterized as structure-mappings between complex systems. Such an analogy conveys that like relational systems hold within two different domains. The predicates of the base domain (the known domain)—particularly the relations that hold among the objects—can be applied in the target domain (the domain of inquiry). Thus, a structure-mapping analogy asserts that identical operations and relationships hold among nonidentical things. The relational structure is preserved, but not the objects. (102)

The system of relationships, and the structural isomorphism between them, can be represented diagrammatically (fig. 9.1).

If we may interpolate the language of "mental models" used in some subsequent publications (e.g., Forbus and Gentner 1997), the models are what are schematized in the two diagrams and consist of an understanding of the relations within a system, which are envisioned as residing in long-term memory and used in reasoning and simulation (Forbus and Gentner 1997). An "analogical model" would thus seem to be an understanding of a system of relations stored in long-term memory that is *formed* by the analogical transposition of the structure of another model to a new domain.

There are clear similarities between this view of mental models and the one that I have described. To possess a model is to posses a lasting ability to think about a domain in systematic ways. The model is an integral unit, defined by the systematic interrelations of its elements, and provides ways of framing, reasoning about, simulating, predicting, and operating on phenomena in the target domain. The postulation of *multiple* models of different content domains not only is implicit in the separate descriptions of different models but is made explicit in Collins and Gentner (1983) and in Forbus and Gentner's discussion of qualitative models in physics:

> The goal of most qualitative physics research is to build an idealized physical reasoner, a system that can reason with sophistication about the physical world in the way that the best human scientists and engineers do, without their frailties. This goal leads to a preference for systems that maximize generality and generativity. That is, the laws of qualitative physics are expressed in domain-independent terms, and knowledge of domains is expressed in situation-independent forms. It seems likely that people's mental models include laws and principles that are at least somewhat domain independent, as well as domain knowledge that is situation-independent. But there is ample evidence suggesting that much of what people know about the physical world and how they reason about it is more concrete than that. ...

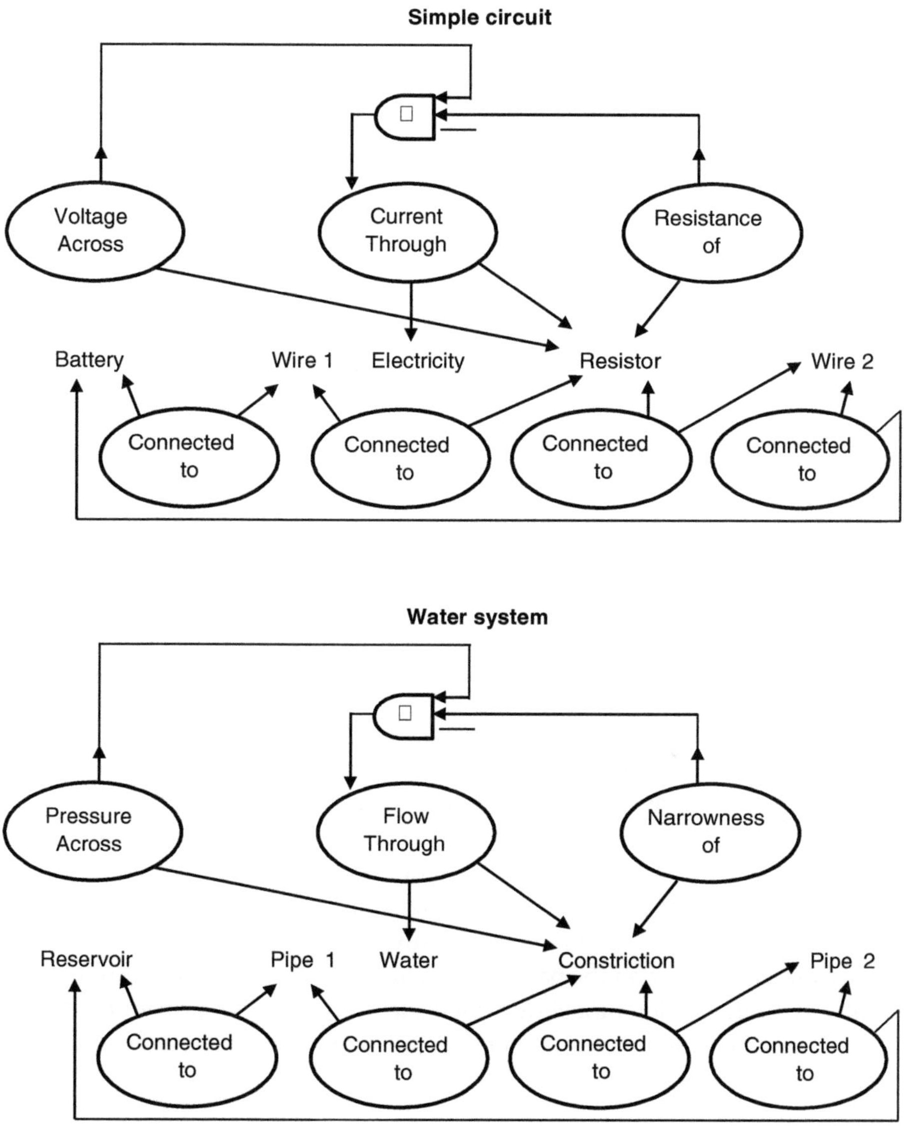

Figure 9.1

Representation of knowledge about (a) simple electric circuits and (b) simple hydraulic systems, showing overlap in relational structures. The relation stands for a higher-order qualitative division relation: the output (e.g., current) varies monotonically with the positive input (e.g., voltage) and negative-monotonically with the negative input (e.g., resistance). Reproduced from Gentner and Gentner 1983, p. 109, fig. 3.3.

The exclusive reliance of current qualitative simulation accounts on first-principles knowledge makes them implausible candidates for psychological models, except perhaps in very narrow ranges of high-expertise reasoning. (Forbus and Gentner 1997, 2)

Johnson-Laird's development of a notion of mental models also distinguishes models from both propositions and images and treats them as "analogical" representations of relational structures. His models, however, are not representations of systematic relationships stored in long-term memory but are constructed on the fly and held in working memory in processes of inference and discourse comprehension. Consider, for example, the problem of combining the information contained in the following sentences:

The spoon is to the left of the knife.
The plate is to the right of the knife.
The fork is in front of the spoon.
The cup is in front of the knife. (Johnson-Laird 1983, 160)

The information about the overall scene can be represented in a diagrammatic model of spatial arrangement.

spoon	knife	plate
fork	cup	

The diagrammatic model, unlike the sentences, encodes information in a format that employs relations analogous to the relations between the objects referred to—in this case, spatial relations. The processes of reasoning to further conclusions—say, about the relationship of the spoon to the plate or the cup, which is not explicitly mentioned in the sentences—could be approached as inferences over sentences (which would require further tacit knowledge, such as that the relationship "to the right of" is transitive, and of how to combine relationships along the left/right and front/back dimensions) or on the basis of applying capacities for spatial reasoning directly to the analog model (in which such principles are present by dint of the general structural features of the modeling system). Consider another similar set of sentences:

The spoon is to the left of the knife.
The plate is to the right of the spoon.
The fork is in front of the spoon.
The cup is in front of the knife. (Johnson-Laird 1983, 161)

These are compatible with two distinct arrangements, changing the nature of the inferential processes required on both the sentential and the model-based hypothesis about reasoning:

spoon	knife	plate		spoon	plate	knife
fork	cup			fork		cup

In Johnson-Laird's terminology, these would be two distinct "models," each of a particular possible configuration. Byrne and Johnson-Laird (1989) experimented with variations on such examples, arguing that the performance of test subjects indicated that they were reasoning using models rather than inferences on sentences.

Like Johnson-Laird, I view models as distinct from language-like representations and believe that understanding of discourse often involves the conversion of a linguistic representation into a model. We also agree that one thing that distinguishes models from language-like representations is their structure, which has a systematic correspondence to the target domain. However, what Johnson-Laird calls "models" are temporary structures in working memory, used in specific and transient tasks of comprehension and reasoning and forgotten sometime thereafter. The things that I have called "models," by contrast, are more like Gentner's: structures for understanding *kinds* of situations that are generally retained once acquired and reused as needed. Of course, there is no reason to suppose that the two hypotheses need be mutually exclusive. Indeed, it seems plausible that a "modeling engine" should be capable of producing models of systematic relations that are added to our permanent cognitive tool kit when they are useful for recurring kinds of situations, or of constructing them only temporarily to think on the fly about more particular situations.

9.2 Models in Philosophy of Science

The literature of philosophy of science has paid growing attention to the role of models in science, though this literature is complicated by the fact that the word 'model' is used in many different ways, some of which I discuss hereafter. In introductions of model-based approaches to science, it is standard practice to contrast a model-based approach with approaches that treat scientific laws and theories as sets of statements or propositions and scientific understanding as knowledge of, and inferences using, such propositions. Ronald Giere, for example, writes:

> The assumption that scientific theories are sets of statements goes along with the view that scientific representation is to be understood as a two-place relationship between statements and the world. A focus on the activity of representing fits more comfortably with a model-based understanding of scientific theories. (Giere 2004, 743–744)

Nancy Nersessian writes in a similar vein that "in standard philosophical accounts reasoning is employing deductive or inductive algorithms to sets of propositions. The understanding of deductive reasoning provided by classical logic stands as the model" (Nersessian 1999, 7). She notes a similar assumption in cognitive science and goes on to claim by contrast that subsequent studies "sparked numerous investigations in different domains that have led many cognitive scientists to conclude that much of human reasoning is by means of 'mental modeling' rather than through a process of applying a mental logic to propositional representations" (9–10).

In his introduction to the collection *Cognitive Models of Science*, Giere contextualizes several model-based accounts in this way in the course of his synopsis of an entry by Nersessian in the same volume:

> Most historically minded critics of logical empiricism took over the assumption that scientific theories are primarily linguistic entities. The main exception is Kuhn, who gave priority to concrete exemplars over linguistically formulated generalizations. Nersessian adopts a theory of "mental models" as elaborated, for example, by Johnson-Laird (1983). On this approach, language, in the form of propositions, may be used not to describe the world directly but to construct a "mental model," which is a "structural analog" of a real-world or imagined situation. Once constructed, the mental model may yield "images," which are mental models viewed from a particular perspective. This interplay of propositions, models, and images provides a richer account of the representational resources of scientists than that employed by either logical empiricists or most of their critics. It may be thought of as an extension of the model-theoretic approach to the nature of scientific theories as elaborated, for example, by Suppe (1989), van Fraassen (1980, 1989), and myself (Giere 1988). In any case, the cognitive theory of mental models provides the main resource for Nersessian's account of the dynamics of conceptual change in science. Some such account of representation seems sure to become standard within a cognitive approach to the philosophy of science. (Giere 1992, xvii–xviii)

Giere's and Nersessian's approaches, which are important articulations of cognitivist approaches to science, are particularly sympathetic to my view that *mental* models form the basis of scientific understanding. Other proponents of model-based approaches use the term 'model' in ways whose cognitivist commitments are perhaps less obvious. Some deal with the use of concrete external models, such as Watson and Crick's tin and cardboard model of the DNA molecule. Others use the term 'model' for analogical modes of understanding, including the use of diagrams, images, or simulations (all standardly referred to as "analog" representations), or models based in actual *analogies*, such as viewing particle collisions on the model of billiard balls. The use of such analogies is often viewed as playing primarily a heuristic role in theory construction and as being distinct from actual theories (Bailer-Jones and Bailer-Jones 2002; Hesse 1963, 1974; Holyoak and Thagard 1995; Kroes 1989; Psillos 1995; essays in Hellman 1988). Some writers, similarly, use the term 'model' for *simplified* or *idealized* characterizations of natural phenomena used in the process of formulating more adequate theories and perhaps retained thereafter for ease of calculation. There are, of course, also mathematical models in the sciences, including computer models.

While uses of the word 'model' are diverse, they reflect the common themes that scientific understanding involves the use of some kind of surrogate, be it mental or external, that has formal or structural similarities to its target domain. Moreover, while some particular uses of the word 'model' may differ—referring to external models, or to modes of thought involving analogy or idealization—the overall accounts in which they occur are more consonant with my cognitivist notion of modeling than at first it might appear.

The difference between using an external model and using a mental model is not really all that great. External models are already in some sense "mental," in that they are *models* only in light of how they are *used in thinking and reasoning* (though, of course, they are not "*purely* mental" in the sense that would imply that there is no external model). Moreover, often when one uses an external model, one is in fact thinking *about* the target system and not the model: one thinks about the target system *through* the model. Similar things can be said about analogies like the billiard ball model of particle collisions: while the model is *inspired* by the collisions of actual billiard balls, one uses the model to think about collisions of gas molecules or other particles, and in practice the source metaphor is forgotten or becomes transparent. The person using such a model is thinking about gas molecule collisions *in the same way* that she thinks about billiard ball collisions, but she is generally not thinking about billiard balls at all in so doing. Simplification through idealization is likewise already built into my characterization of models, and I do not regard this as a distinction between different modes of scientific cognition. Some models, theories, and laws are more simplified and reflect their targets less exactly than others, but *all* involve bracketing and distorting idealizations (sometimes called "Aristotelian" and "Galilean" idealizations, respectively, in the philosophy of science literature).

While some philosophers of science distinguish what they call models (e.g., visualizations, external models, or simplified heuristics) from laws and theories (sometimes distinguished as things that are framed as equations or propositions), others view scientific theories themselves as models or families of models (van Fraassen 1980; Giere 1988; Suppe 1989; Suppes 2002; though see the section later in this chapter on the semantic account of theories for clarifications about the notion of "model" used by some of these writers). Still others treat models as complementary to theories, filling in details absent from the abstract theory but needed for its application to a more particular context (Redhead 1980; Morgan and Morrison 1999). Again, for the most part, these verbal differences do not strike me as contrary to my claim that scientific understanding, including the understanding embodied in laws and theories, ultimately depends on mental models. Writers who distinguish models from theories often do so to emphasize ways in which scientific *reasoning* depends on models (Magnani and Nersessian 2002; Magnani, Nersessian, and Thagard 1999). If we need models as "mediators" (Morgan and Morrison 1999) between laws or theories (understood as "explicitly formulated principles" [Giere 2004, 744]) and specific contexts, then the *understanding* of the theory requires such models. I have no wish to claim that science does not *also* include equations, propositional representations, and symbolic operations over them. My claim is that the ability to work with equations and other symbolic representations in a purely formal way does not itself amount to understanding of a scientific domain, and such understanding is supplied by mental models.

Another use of the term 'model' in philosophy of science, however, is distinctly different from mine. This usage is taken over from logic and model theory. Model theory is used in the examination of formal systems in mathematics and logic. A mathematical or logical system is a set of axioms and their inferential consequences under the derivation rules of the system. Such systems are *formal* in the sense that they do not have interpretations of the terms assigned. A *model* of such a system is an assignment of interpretations to the terms of the system. In logic, this might be an assignment of a domain of objects to the referring terms and of relations to the two-value predicates. In mathematics, the integers under addition would count as a model of a group-theoretic structure with a particular type of operation and properties like transitivity. A "model" in this model-theoretic sense can serve a number of purposes. It can be used as a heuristic tool for understanding the abstract structure through a more familiar particular mathematical system. But it can also be used to test the structure for properties like consistency and closure.

The "semantic view" of theories in philosophy of science applies this notion of models to scientific theories. Different versions of the semantic view have been articulated by Suppe (1960), who takes models to be set-theoretic structures, and van Fraassen (1980), who views models as state-space structures (a survey of variations on the semantic view can be found in Suppe 1989, chap. 1). Proponents of the semantic view contrast models with linguistic formulations of theories, and some versions of the semantic view hold that a model and its target need to be isomorphic (van Fraassen 1980; Suppes 2002) or partially isomorphic (Da Costa 2003) to each other. But beyond this, the similarity with other notions of "model" ends. Model-theoretic models are *truth makers* for theories rather than nonpropositional cognitive structures through which we understand a scientific domain. Such models are characteristically *less* abstract than the things modeled. They are, in fact, special instances of the formal system for which they are models.

9.3 Models in Theoretical Cognitive Science

The claim that model-based understanding is a key feature of human cognitive architecture is found most clearly in artificial intelligence and theoretical cognitive science. Some writers use the term 'model'—or some variant of it, such as "scale model" (Waskan 2006) or "graphical model" (Danks 2014)—as their preferred technical term. But there is an equally large tradition that postulates systematic nonpropositional structures that form the basis of ordinary reasoning, but using other terminology, such as Minsky and Papert's (1972) "micro-worlds," Newell and Simon's (1972) "problem-spaces," and the semantic networks, frames, and scripts I described in chapter 4. Those projects had a great deal in common with my model-based account, particularly the ideas that understanding comes in domain-sized chunks and is stored in a nonpropositional form. Indeed, I count them as important pioneers of a model-based approach, and my acquaintance with them from the 1980s probably provided a substantial part of the inspiration for my own ideas. I will not, however, redescribe them here, as chapter 4 contains lengthy descriptions of some of them, particularly Minsky's notion of a frame.

At about the same time, David Marr (1982) published his influential book *Vision*, which provided an account of computational operations that might explain the processing of visual information from the raw perceptual array on the retina to a three-dimensional "model" of a scene. Marr uses the term 'model' only for the final stage of the process—a three-dimensional model that is object centered rather than viewer centered—though the prior processing stage, the "2½D sketch," might be seen as a viewer-centered model of a scene from a perspective. These are "models" in a sense more like Johnson-Laird's than mine, as they are transient structures that change with the perceiver's perception and orientation. However, the ability to construct such views requires the possession of visual modeling *systems* more in accordance with my usage. Marr's book is specifically about vision and does not propose that mental models are an important unit in understanding generally, and one could not extrapolate a more general model-based view from his theory, as three-dimensional visual perception is perhaps the aspect of cognition for which models that share formal features of their targets are most obviously suited.

Marr also introduced what has become an influential distinction between three levels of abstraction that are needed in understanding cognitive systems. The most abstract of these, which he calls the "computational" level, specifies what a system does and its function. The next level, the "representational" or "algorithmic" level, describes how it does them in terms of what kinds of representations it uses and the computational processes it uses to manipulate them. The third, the "physical" level, specifies how the system is physically realized: in the case of human vision, the neural structures; in the case of computers, the hardware implementation. This three-level system has been adopted widely by subsequent writers in cognitive science (though see Danks 2014 for some important clarifications of its limitations).

Discussions of cognitive architecture tend to be seen as involving Marr's computational or representational levels or the intersection between the two, addressing issues of the number and types of systems and at least a rough specification of the types of representations (e.g., models or propositional representations) they employ, independently of questions about their neural or physical implementation; and at least to the extent that one accepts Marr's classification of levels, my account is no exception. My account is committed to the existence of model-sized units that do a certain amount of their processing autonomously, and hence to at least some constraints on the kinds of representations and processes involved, and also to a separate system for language-like thinking, which requires different sorts of representations. I have not, however, proposed more specific features of the representations or representational systems or the algorithms that operate over them. This is in part because I wish to present a model-based cognitive pluralism as a plausible and broad thesis, which is compatible with various more specific hypotheses about the kinds of information-processing structures that might underlie the use of mental models. But it is also in part because I have a more guarded view about what "algorithmic-level" descriptions of such systems really provide: I doubt that cognition takes place through the *application* of algorithms, though cognitive processes might aptly be *described by* algorithms (much as I think that psychophysical effects are *described by* the Weber-Fechner laws, but very much doubt that this proceeds by the brain representing and applying a "rule" couched in terms of logarithmic or power function equations).

Some model-based accounts, however, are more specific in the type and nature of the models they propose, and I will briefly discuss two of these. The first is Jonathan Waskan's (2003, 2006) suggestion that "scale models" play an important role in cognition. Like most proponents of mental models, Waskan frames his view in contrast to the "mental logic hypothesis" that "thinking involves the application of syntax-sensitive inference rules to syntactically structured mental representations," proposing the "alternative hypothesis ... that thinking requires the construction and manipulation of the cognitive equivalent of scale models" (2003, 259).

Images and scale models fall under the more general heading of physically isomorphic models (PIMs), which are representations that possess some of the very same properties as that which they represent. Because forethought often requires the truth-preserving manipulation of representations of three-dimensional spatial and causal relationships, the PIMs that hold the most interest in the present context are scale models. (Waskan 2003, 261–262; internal references deleted)

Waskan begins, as I did, by considering external models such as constructing diagrams of situations on graph paper or building a model from Lego blocks.

Of particular interest is the fact that spatial representations can be used to generate predictions in a manner that obviates the need for rules (i.e., frame axioms) that specify the consequences of each possible alteration to a represented system. One might, for example, use a sheet of graph paper to represent the relative positions of Harry, Laura, and Carlene. Should one then wish to know what the relative locations of all of these individuals would be if Harry moved to a new position, one can simply delete the mark representing Harry and insert a new mark in the square corresponding to the new position.

Two-dimensional spatial media can also be used to represent the structure of objects, and collections of such representations can be used to predict the consequences of changes in their relative location and orientation. For instance, a cardboard cutout of my coffee table (as seen from above) can be conjoined with two-dimensional representations (of equal scale) of the rest of the items in my living room and a depiction of the room itself in order to predict the consequences of countless changes in the relative spatial locations and orientations of these items. A highly desirable feature of representations of this sort is that side effects of alterations to the representation mirror the side effects of alterations to the represented systems automatically—without, that is to say, requiring their explicit specification. As a consequence, such representations exhibit, at least with regard to a limited set of dimensions, immunity to the prediction problem. They can, moreover, easily be scaled up to include representations of further objects. (Waskan 2003, 262; internal references deleted)

As with other discussions of "analog" models, Waskan emphasizes that scale models reproduce systems of relationships in their target domains, and that as a result, transformations on the models track transformations in their targets. Waskan emphasizes the importance of such models in "forethought"—in anticipating how the world will behave (on its own or if intentionally manipulated) through simulation or other model-based reasoning processes.

Like Marr's account of vision, Waskan's theory is more limited in scope than mine. It is broader than Marr's, in that it includes rich possibilities of off-line cognition and anticipations of possibilities in the context of online cognition. But it is by its nature restricted to modeling spatial phenomena and hence is an account of only a subset of cognitive capacities. Like me, Waskan does not deny that there is also language-like cognition, but I also hold that there are also a variety of nonspatial models that preserve relations within, and allow nonpropositional reasoning about, other domains of understanding. Moreover, Waskan's scale models seem likely to be, for the most part, "models" in Johnson-Laird's sense, rather than abiding frameworks for understanding classes of phenomena. However, the capacity to form and manipulate scale models presupposes a more general modeling system (or a collection of such systems) suitable to this purpose.

In the most recent book on mental models in theoretical cognitive science, David Danks (2014) argues that understanding and reasoning (or at least a great deal of it; Danks seems to offer different claims about the scope of his view in different parts of the book) are underwritten by graphical models. The notion of a "graphical model" is considerably better defined than most notions of models, thus rendering his theory more directly available for testing against bodies of empirical data, an enterprise that occupies much of the book. But the notion of a graphical model was not originally developed as a notion of a mental model or as a thesis about cognitive architecture, and much of the point of Danks's book is to motivate the claim that it can nonetheless serve such ends.

In the original sense, graphical models are a type of mathematical model used to represent statistical relationships between variables.

A graphical model is a family of probability distributions defined in terms of a directed or undirected graph. The nodes in the graph are identified with random variables, and joint probability distributions are defined by taking products over functions defined on connected subsets of nodes. By exploiting the graph-theoretic representation, the formalism provides general algorithms for computing marginal and conditional probabilities of interest. Moreover, the formalism provides control over the computational complexity associated with these operations. (Jordan 2004, 140)

A distinctive feature of such models is their uses of a "graph" in a technical sense of a structure of "nodes" representing variables and "edges" representing probabilistic relations between them. (Diagrammatically, these look much like the structures used to illustrate semantic networks and frames.) The framework is flexible, in that it can be adapted to an open-ended number of contexts involving different variables and relations. Graphical models thus supply a kind of mathematical orthotic for, say, statisticians or economists reasoning under uncertain conditions, and Danks suggests that the same kind of structure already exists in the mind as a fundamental feature of our cognitive architecture.

At their core, graphical models can be understood as compact representations of relevance relations, where different types of graphical models represent different types of relevance (e.g., informational, causal, probabilistic, communicative). They thus address a key challenge for any cognitive agent: namely, determining what matters and, often more importantly, what can be ignored. The term "graphical model" encompasses many different types of mathematical models, including Bayesian networks (also called Bayes nets), structural equation models, random Markov fields, hidden Markov models, influence diagrams, social networks, command-and-control structures, and more. At their cores, these models all have a *graph*: a diagram of nodes and edges that encodes the qualitative relationships.

From a purely mathematical perspective, the nodes are simply objects, and the edges can be undirected (A—B), directed (A→B) or bidirectional (A↔B). (Danks 2014, 39–40)

Danks argues that the graphical-model hypothesis can be used to account for causal inference and reasoning, provide a useful reconstruction of major theories of concepts, and explain decision-making procedures that efficiently use the relevance information encoded in graphs. Graphical modeling, moreover, encompasses a broad class of more specific model types, differentiated by their features, such as whether they allow cyclic or only acyclic structures.

Danks's notion of a graphical model is broader than Waskan's scale models, as the latter are confined to spatial phenomena. It is both narrower and more exactingly developed than mine. In large measure this reflects the different agendas and intended audiences of our books. Danks's book speaks primarily to an audience of theoretical cognitive scientists and attempts to show that the specific features afforded by graphical models can be used to explain a broad range of experimental data and used to subsume other accounts of those data. I, by contrast, have attempted to write a book primarily for a philosophical audience—including philosophers working in areas like epistemology and semantics who have shown little interest in cognitive science—that would describe a very broad hypothesis about the role of mental models and explore its philosophical implications. I have thus intentionally left open the questions of how the hypothesis of a pluralistic model-based architecture for understanding might be developed (perhaps in a variety of different ways for different domains) at Marr's algorithmic and implementational levels. Further exploration of my view within theoretical cognitive science would require more concrete hypotheses about the types of models involved, and it is compatible with my views that some or all of the modeling might turn out to be accommodated within some more specific framework such as graphical modeling.

10 The Plausibility of Cognitive Pluralism

The most distinctive claim of cognitive pluralism is that we understand the world through many *mental models* of different content domains. The claim that cognition involves the use of a number of special-purpose systems that detect, track, and allow interaction with different parts and aspects of the world, taken by itself, may not be quite a consensus view; but many previous authors have claimed that the minds of nonhuman animals are composed largely of a grab bag of useful cognitive tricks, and argued that this is a good strategy—and perhaps the only biologically plausible strategy—for evolving intelligent animals. What is more controversial is the idea that a further proliferation of special-purpose ways of understanding various aspects of the world is *still* a good strategy once we have animals like human beings, who possess capacities for language and explicit reasoning. Really there are two separate sorts of issues here. The first arises from the peculiar disconnect we often see between discussions of human minds and the minds of nonhuman animals. Often, even people who think that animal minds are composed largely of a bundle of closed instincts and good tricks and acknowledge that human beings are products of evolution nevertheless talk about human minds as though we had been designed from scratch to be general-purpose reasoners. This strikes me as psychologically naive and scientifically unrealistic. Second, even if we acknowledge that our capacities for reasoning, language, and language-like thinking are things *added to* an older suite of more specialized abilities, there are quite reasonable questions about whether the newer and more sophisticated forms of thinking and reasoning also take the form of many separate areas of understanding about distinct domains, and whether this would still be a good design strategy for minds that are no longer restricted to developmentally canalized capacities that are products of natural selection.

10.1 A Good Design Strategy for Evolving Smarter Animals

The easiest question to address is why the minds of "simpler" organisms consist in—or at least rely heavily on—a kind of "grab bag of good tricks." This is well-explored territory. First, and most fundamentally, accrual of distinct traits that have individually proved adaptive is something one finds quite generally as an outcome of variation and selection. We have no reason to think that it should prove any different when the traits in question are *cognitive* traits. This does not make it a *good* strategy, but it does mean it is one we are *likely to get.*

Second, cognitive traits that are products of evolution and developmentally canalized also tend to have dedicated and efficient neural circuitry, and as a result, they tend to be *fast.* And when it comes to the things that count for selection, like avoiding predation, fast is good. In fact, speed is often more important than accuracy. In the case of predator detection, at least, false positives (i.e., thinking a predator is present when none really is) are less costly than false negatives (i.e., failing to detect a predator that is really there). An animal can afford to bolt for cover when a cloud passes overhead many times, but cannot afford to miss a predator that is stalking it even once. It also cannot afford to stand around thinking carefully about whether a stimulus was really caused by a predator. Careful rumination is good for some things, but avoiding a charging tiger or a swooping owl is not among them. So even if it were possible to start out with animals that represented the world through deliberative, general-purpose reasoning rather than reflexive good tricks, there is a good chance that they would all be eaten by tigers.

Third, with special-purpose mechanisms that are adaptive, more is better. Precisely because they are special purpose, each one is useful in only a limited range of situations. Consequently it is a sound evolutionary meta-strategy to accumulate individual adaptations that do different useful things.

Fourth, care of the young requires special cognitive mechanisms of its own, and these are lacking in most species. As a consequence, most animals (but particularly those that do *not* have parental protection) need to be able to behave adaptively very quickly. They thus need a basic tool kit of "good tricks" that are products of early development, even if they are capable of more sophisticated types of learning if they survive long enough.

Fifth, some of the mechanisms we find in animals are too specialized to be left to learning. This seems particularly evident in the types of perception and behavior related to mating. It is hard to see how animals of most species could *learn* to recognize possible partners or when those partners are ready for reproduction. Nature has solved this problem by making members of a species sensitive to extremely particular physiological, behavioral, and chemical cues that are specific to opposite-sex members of the same species, some of which are also indicative of preparedness for mating. It would probably be even more difficult for animals of most species to learn how to perform the mating act if they did not already have nativistic mechanisms to guide them. The courtship and mating routines of some species are quite complicated, and the animals who engage in those routines cannot learn them from sex education or pornography.

These considerations really boil down to two things. The *production* of separate special-purpose cognitive traits is something we should *expect* from evolutionary processes of variation. And the *accumulation* of such traits is something we should expect from evolutionary processes of selection.

10.2 *Still* a Good Design Strategy for Animals That Learn

It is not really all that controversial that multiplication of separate adaptive cognitive traits is an evolutionary strategy whose imprint is seen in *simpler* animals. Indeed, it was not uncommon for both philosophers and biologists of previous generations to think of human and animal minds in starkly contrasting terms: nonhuman animals have instincts but not reason, while humans have reason but few if any instincts. This bifurcation, of course, obscures the vast differences found among the minds of different nonhuman species. But it also misrepresents the human mind and implies a highly unrealistic evolutionary history.

The fact is that the human mind is by no means *devoid* of special-purpose mechanisms that are developmentally canalized and products of selection. Indeed, moral philosophy has often recognized this, though primarily in its castigation of those whose conduct is governed by "the appetites." But special-purpose mechanisms, ranging from the blink reflex to the contagion-avoidance reaction of disgust and incest avoidance, are not all appetitive, and many still serve useful functions. Now, if we are "rational animals," why is it that we still have all of these nonrational cognitive mechanisms?

Evolution Is a Hoarder

First, evolution is somewhat of a hoarder. Yes, ancestral traits can be lost. Humans possess neither the tails nor the sensitivity to pheromones that our ancestors had. Other traits lose their function but continue to exist in vestigial form, like the appendix and the coccyx. But generally there is a motley mix of the old and the new, as in the case of the three visual systems humans possess, or the ability to extract depth information from multiple cues. This is something that I think the evolutionary psychologists have gotten more or less right: it would be downright astonishing if whatever landmark mutations conferred our distinctively human traits *also* resulted in the loss of hundreds of millions of years of accretions of special-purpose instincts.

Another way of putting it is to say that evolution does not operate like an engineer, who can make large-scale changes between models and even go back to the drawing board and start from scratch. Evolution is limited by the range of mutations on genes that can produce organisms that are viable throughout their developmental sequence and capable of breeding. This evolutionary constraint also sets a limit on how drastic a change can be in species that reproduce sexually: a mutation that produces greater success for the individual organism in everything save breeding will not get passed on if the genetic changes make it reproductively incompatible with the rest of the population. We now know that mutations do not need to be as tiny and gradual as Darwin supposed. But compared with engineering, evolution is still a gradual and incremental process. When traits *are* lost because they no longer confer an adaptive advantage, this is likely to happen on the glacial timescale over which evolution operates on species. Perhaps our descendants will have progressively fewer special-purpose canalized mechanisms; but if so, the process will likely take thousands or tens of thousands of years. (This is, of course, something that could change if we take an active role in altering the human genome.)

"Heritage" Traits Are Still Useful

Moreover, it is by no means clear that *losing* some of these traits would be a good thing. Take the reaction of disgust. Scientists now generally believe that disgust evolved as a highly flexible mechanism for avoiding sources of contagion, such as excrement, blood, open wounds, skin diseases, vermin, and rotting tissue. The mechanism is triggered by many stimuli, operates quickly, and makes it difficult to force oneself to come into contact with whatever is regarded as disgusting. Of course, sometimes it is triggered by things that are not really dangerous. Most slithery things are not really dangerous. As I child I had a strong reaction of disgust to mushrooms, including the innocent and nutritious mushrooms used in cooking. And, as it turns out, leprosy—perhaps the paradigmatically disgusting disease afflicting human beings—turns out to be among the least contagious of diseases.

A more precise knowledge of fields like biology and medicine can provide a far more exacting understanding of what classes of things present biological hazards, the mechanisms through which they operate, and the protocols that would allow us to deal with them safely. To a limited extent, having such knowledge may allow us either to repress ill-founded reactions of disgust or perhaps even to edit the set of stimuli that provoke them. But such knowledge does not cause the disgust reaction to go away, and arguably this is a very good thing. The capacity for disgust cannot be unlearned because it is not part of a knowledge base that can be affected deeply by other things we learn. It is a deeply ingrained mechanism of its own. And it continues to be a *useful* mechanism, one that could potentially save us from sources of contagion—particularly novel ones—that we might not respond to sufficiently quickly by way of our scientific knowledge.

Or take the case of sex. Human sexual instincts are far more "open" than the hardwired programs for reproduction found in insects or in the species of birds that have characteristic mating rituals. But they are an important motivational component in human reproduction. It is true that we can understand human reproduction far more accurately in biological and medical terms. But a being that had *only* such scientific understanding of reproduction would have no intrinsic motivation to actually engage in reproductive acts. And it would probably be quite impaired at things like identifying suitable mates and their availability to breed, as our ability to identify these in large measure depends on the operation of special biological mechanisms sensitive to cues involving biological changes (pupil dilation, increased blood flow to the lips and genitals) and behavioral patterns indicative of gestational cycle and interest. (People *could* just ask each other, "Would you like to make babies with me?" but I will leave it to readers to decide for themselves whether that would be an effective sexual overture.) When it comes to the things that natural selection really cares about, like avoiding danger and producing offspring, evolution has provided a highly effective set of tools that it is difficult to improve on, the lack of which would leave us seriously impaired.

Or consider the ability to detect and respond to social cues, which is so important to members of a social species such as our own. There are human beings, such as autistics, who are impaired with respect to these abilities. Some high-functioning autistics can learn *about* them by way of a kind of theoretical reasoning and compensate to an impressive degree. But even they never fully compensate for the lack of elements of the maturationally normal cognitive tool kit that the majority of human beings possess. For the rest of us, learning *more* about social interaction by way of, say, research in social psychology *may* make us more adept at interacting with others, but it does not make for a viable substitute for specialized human capacities for social cognition.

10.3 The Advantages of Model Proliferation

What has been said so far has been an exploration of why minds like ours, in which learning plays such a weighty role, would still have a "grab bag of good tricks" as well. But another issue for cognitive pluralism is waiting in the wings. Why would the *more flexible* types of cognition we possess come in the form of *many distinct* models? The answers to this question basically fall into two categories: (1) it is a *good* strategy for increasing understanding and intelligence, and (2) there may not really be any alternatives, at least not for beings remotely like us.

Complex World, Many Patterns, Finite Minds

The world is a complex place. Indeed, even the little bit of the world we perceive at any one moment with our senses is quite complex. More is going on in the room you are in during the time it takes you to read this sentence than you could possibly think about in a lifetime. Perhaps God and the angels have minds that can take it all in, but you and I do not. Nor does any living organism on the planet. Our minds are not only finite but quite limited in their capacity when compared with all the facts and events that there are, or even those in our immediate vicinities.

To have any understanding at all, a finite, embodied mind has to represent the world in an abstract, idealized way, one that compresses much of the messy complexity into a finite number of concepts and disregards the rest (Simon 1977). And to be of any use to the organism, these concepts must be concepts of things that are salient to it and its interests and do a good-enough job of tracking real patterns (Dennett 1991b; Gibson 1966, 1977), either "out there" in the world or else at the interface between organism and environment. As minds become more powerful, the "interests" may become increasingly disconnected from biological needs—say, an interest in opera or in physics—and the tracking may well increase in fidelity. But even so, the form that understanding takes is bound to be heavily influenced by what interests the organism happens to have, by what regularities in the world are salient to those interests, and by which of those regularities the organism can perceive, conceive, and act on.

If the units of understanding are individuated by the regularities they track, or by the ways the organism can perceive and act on them, or by the interests that make them relevant, then understanding will come in "chunks" of corresponding size. Gravitation is a good candidate for a single theory because it is the uniting feature of certain regularities of the motions of bodies, and ones we encounter in ways that matter on a regular basis, as well as because it turns out to be a single and fundamental force. If you mix other things into a model along with gravitation, not only do you make the cognitive work unduly complex, but you also obscure understanding of real regularities. But we may have use for more than one model of gravitation, some theoretical and others not. It is a good thing that we have an intuitive understanding of falling bodies, which, while less accurate and projectable than general relativity or even classical mechanics, is better for catching falling vases or avoiding falling off cliffs. And even in scientific contexts, we may need multiple models for phenomena like collisions at different energy levels, partly because the particles themselves behave differently, and partly because the most useful ways we have of observing, conceiving, and acting on particle collisions differ with context.

So we have a first reason for a plurality of models: *the world demands it.* There are many real patterns in the world, and the most efficient way—perhaps the *only* way—to model them is one by one. (There are often ways of combining them, such as vector algebra, but this presupposes already possessing models of the individual forces.)

An Objection: Most Patterns Are Not Fundamental

Now, in some contexts, one might object that "many real patterns" can turn out to all be consequences of a single, more fundamental phenomenon. What we now call gravitation was once thought to consist of two distinct sets of principles: one for celestial bodies circling in the heavens, and another for falling bodies beneath the sphere of the moon. Electromagnetism and strong and weak forces were all accommodated within quantum theory, and many physicists expect there to be a unified field theory that will fold gravity into a single theory with them. This is a legitimate question in philosophy of science, one that I return to in part 3. But it will not do as an argument about what is *psychologically* possible.

Could God have made creatures *ab initio* that thought about the world solely in terms of the best theories we now possess, or a unified field theory, if there is such a thing? Perhaps. It's never a good gamble to bet against what God could do. Could an organism that thought in this way, and only in this way, be produced by natural selection? Probably not. Organisms that attained this sort of very general understanding would probably have to do so the way we have: by learning models of many more specific phenomena first, and then reasoning and experimenting their way to something more general. Without a plurality of such models, there would be no way to get to the more general theories. And even for purposes of the sciences, the more specific models—of electricity, magnetism, particle collisions, nuclear cohesion, planetary orbits, ballistics, and so on—are still used, and for good reason. A more basic theory may explain more particular theories as special cases but may still be useless for capturing the regularities captured by the more particular theories. Biological processes may all be a result of underlying physical processes, but neither evolution nor the Krebs cycle "pops out" of the physics equations, and attempting to reason about, say, the origin of species or even metabolism in quantum mechanical terms would be a very bad way to go about it.

Moreover, a being that thought *only* in general and basic physical terms would not be very good at interacting with its environment. The task of figuring out what is edible or dangerous would be impossibly cumbersome—and would not even make sense for an animal that did not *also* have models of feeding and predation. And since there are domains that are *not* special cases of physics—mathematics, ethics, aesthetics—the organism would not be able to think in those terms. Which is to say, it would not be able to think about all the things we can think about. In short, appeals to the *ideal* of unification are irrelevant to psychology: unification is an achievement that presupposes preexisting models, and a model that is "unifying" in the sense of capturing the causal or compositional basis for other phenomena is not necessarily a *substitute* for them in thinking or interacting with the world. What is important for an organism is the ability to efficiently track patterns that are *salient*, not those that are *fundamental*, and most of the salient patterns are not fundamental. Being able to understand increasingly fundamental theories allows us to think about new things, and to think about them in illuminating ways, but it does not diminish the importance of thinking about them in other ways.

Real Patterns and Affordances

I have spoken of mental models tracking "real patterns" in nature, such as the gravitational attraction between two bodies. It is tempting—perhaps irresistibly tempting—to think of such patterns as "objective" and "independent of us." And there are important ways in which there is something to this characterization. One of these is seen most clearly by comparing gravitation and the corresponding property of mass to a property like *edibility*. Edibility is always an organism-relative property. What is edible to one type of organism may not be edible by another. Meat is a potential meal for a lion, and a drop of nectar for a butterfly, but not vice versa. Likewise, what counts as dangerous, hospitable, or suitable for mating depends on what kind of organism you are. So some properties, including those that are most important biologically, are found neither in the organism, nor in objects in its environment, but in some sense at the interface between them.

J. J. Gibson, the founder of ecological psychology, called such properties *affordances* and would say that the nectar "affords eating" to the butterfly—that is, makes it available in a fashion the butterfly can exploit (Gibson 1966, 1977). Affordances *are* "real" in that they are not imaginary and in that detecting an affordance is not making a mistake but getting something right. They are "objective" in the sense that they do not depend on what anyone believes or how she feels about it. But they are not "objective" in the sense of *being located solely in the object to which they are attributed.* (In one philosophical terminology, they are not *intrinsic.*) Both metaphysically and scientifically, they are perfectly respectable on a par with gravitation, though of course they are not fundamental, and many of them may not appear among the principal theoretical terms of any science.

On the other hand, the *concepts* and *models* that we use to think about affordances can differ greatly, from the standpoint of psychology, from the models we use to think about a *world of objects and properties.* In particular, an organism can detect and respond to affordances without conceiving them *as things* or *as properties attached to things.* We can think of edibility or dangerousness as properties of things, to be sure. But even organisms that probably lack anything closely resembling human concepts or even object-oriented thought need to be able to detect and respond to affordances if they are to do things like eat and avoid danger. Mosquitoes are sensitive to differences in levels of carbon dioxide in the air and use these differences to navigate toward mammals to obtain a blood meal. They probably do not represent any of this to themselves in terms of objects and properties but merely have circuitlike mental models in their tiny brains that allow them to surf the affordance trail. Much of our nonconscious cognition consists in affordance detection as well: we simply register that a particular surface does or does not afford a stable foothold and adjust our path accordingly, for example. But much of our conceptual thought is oriented toward affordances, too: we see a chair as a comfortable spot to sit down or a water fountain as a potential source for slaking our thirst. When reliable connections exist between features of the environment and particular things we need or do, affordance detectors pick out real patterns as well. They are simply real patterns that involve *us*, with our particular set of needs, wants, and capacities, rather than patterns that do not depend on us (or at least depend far less on us), such as gravity.

Real Patterns, Idealized Models

Models may *aim at* tracking real patterns, but some of them fail. And those that succeed in biological terms—that is, track them *well enough* to be of use to the animal—may not track them perfectly. Which is to say, once again, that models are *idealized*. We have already addressed one type of idealization: the bracketing idealizations involved in isolating one set of features by ignoring others. But we have not addressed *distorting* idealizations, such as treating bodies as point masses or remapping wavelengths of light onto a subjective color space. Distorting idealizations can, of course, be a bad thing when we are confronted with situations in which the distortions *matter*. But in other cases, the very features that count as distortions can also enable easier comprehension and reasoning. And from a biological perspective, both evolved cognitive systems and ones produced by forms of learning that are not regimented like the sciences are shaped by processes that converge on good-enough models rather than ones that are optimal for purposes of exactitude.

Sometimes idealizations that distort by making simplifying assumptions are needed to make problems computationally tractable. And even when a more exacting model is computationally tractable in principle, it may not be so for every actual mind. Some people can grasp Newton's mechanics but not Einstein's; others cannot comprehend mathematical physics of any sort. It is a good thing that the vast majority of us have more intuitive models (learned or unlearned) that allow us to predict the trajectories of falling bodies and to throw footballs. And these are good things for the mathematical physicist to have, too. Mathematical physics is more apt for rocket science but is not very apt for avoiding falling objects. Because there are several epistemological and practical factors that go into the aptness of a model, different models of the same phenomenon can be apt for different purposes, and hence it is good to have more.

It is thus a wonderful thing that organisms like human beings are built to form models that are *useful* to them, and also that *some* of these models provide more exacting theoretical understanding. We have to begin with models that have a great deal of *us* built into them—affordance models, ways of knowing how to move about the world and how to manipulate objects with the particular types of bodies we possess, models that are centered on things like what is good (or bad) to eat, what is dangerous, who is friend or foe, and so on. Each of these provides a unique type of epistemic and practical grip on the world and is useful for a particular set of purposes. But we are also capable of refining our *interests* to things that are not of direct biological relevance, such as discovering increasingly exact and general models of patterns in nature, in ways that have increasingly less of us bound up in them—that is, models that are increasingly objective. In fact, we have acquired interests in an enormous number of things, and what we *could* form interests in appears to be remarkably open-ended. But to be able to incrementally expand a repertoire of interests in almost arbitrarily many ways requires us to be able to form models of arbitrarily many different domains.

Redundancy: Expected and Unexpected Benefits

One type of multiplication of models—redundancy—has particular advantages of its own. One thing it does, of course, is to make an organism more resilient in the face of injuries that compromise one of the redundant systems. But redundant systems do not function exactly alike. Redundant perceptual systems like those for extracting information about depth respond to different cues and hence are useful under different optical conditions. Redundant cognitive systems may represent their common target differently, thus conferring different inferences and breakdown patterns. And when two systems with the same target produce conflicting results, this provides impetus for deeper empirical and theoretical investigation, both of the phenomena in the world they aim to track and of the cognitive systems themselves. Redundant systems can be *productive* of greater understanding.

Multiple Idealized Models, Epistemic Triangulation, and Compensation for Forced Error

More generally, given that an organism is using idealized models of particular domains, it is naturally beneficial to have an abundance of models. On the one hand, the bracketing idealizations of models imply that multiple models are needed to cover different aspects of the world. On the other hand, as we have seen, even a single model of one target domain can be supplemented by another, idealized in a different way and employing a different representational system. Different models can compensate for one another's limitations. The result is a better understanding of reality, not in the form of a single broader or more accurate model (though sometimes that results as well) but through a kind of *epistemic triangulation* of the world through a number of different vantage points and lenses.

In the extreme case, any given model may be subject to *forced errors*: mistakes that are ensured by the way it processes and represents information.

The production of standard visual illusions in the human visual system would be an example of this. But conditions probably exist under which most or all models are forced into error. Consider our deeply ingrained assumption that every physical object has a determinate location at any one time. This assumption is encoded into classical physics, as well as the core object system, folk physics, and likely a lot of special-purpose models in between. And the assumption is safe unless we are faced with some of the problems of quantum mechanics. In those cases, any representational system that is forced to represent objects as having determinate positions is forced into error.

Note, however, that *we* are not forced into these errors, even if particular models we employ are. I can measure the lines in the illusory pair and confirm that they are of equal length. And with sufficient intelligence and training, I can understand quantum mechanical descriptions of the behavior of electrons and abstain from insisting on the assumptions of intuitive and classical models. What gives us the ability to go beyond the limitations of a given model is the very fact that we can have *multiple* models and use them to triangulate the world more precisely. If we could *only* think of physical objects through the core object system—or, for that matter, through classical physics or quantum mechanics—we would lack some of the cognitive flexibility that we in fact possess. So for anyone inclined to suppose that a single-model cognitive architecture would be preferable to a pluralistic one, ask yourself the question: *which* model would you choose? And what would be lost in abandoning all the others?

I wish to conclude this line of thought by considering a particular application of it. Kant held that the faculty of Sensibility always organizes sensation in terms of Euclidean geometry—so far, an empirical claim about the psychology of perception. But he went on to use this in a transcendental argument to the conclusion that space necessarily has a Euclidean geometry. I think the argument is a valid one, though I think the conclusion (and hence at least one of the premises) is false. One Kantian assumption is that "the world" we are speaking about in science is what he calls the *phenomenal* world: the world-as-interpreted-through-the-Forms-of-Sensibility-and-Categories-of-the-Understanding. So if the Forms and Categories force us to represent any possible object of thought or experience in a particular way—say, as located in a Euclidean space—then it is a necessary truth about the phenomenal world that any phenomenal object will have that property. And so *if* we are forced to think about physical objects in Euclidean terms and cannot think of them in any alternative terms, then the phenomenal world (of humans) is indeed necessarily Euclidean.

This argument was actually quite influential, so much so that it set back mathematics by about a generation. One of the first people to conceive of non-Euclidean geometries was Carl Friedrich Gauss, a nineteenth-century mathematician of the first order, who made many important discoveries, such as the field of differential geometry. Gauss, however, did not follow up on the idea of non-Euclidean geometries. Why? Because Gauss was a convinced Kantian and supposed that Kant had demonstrated the synthetic a priori truth of the claim that space is Euclidean. (Because of his deference to Kant, I like to think of Gauss as the father of *deferential* geometry.) But other mathematicians—the Bolyais, Riemann, Lobachevsky—*did* pursue the idea and discovered consistent non-Euclidean geometries. So even if Kant was correct that we can only *perceive* things through the lens of one particular geometry, we can nevertheless *think* in the terms of alternative geometries. And by early in the twentieth century, Einstein had proposed on empirical grounds that space is *not* Euclidean, a thesis that was to receive subsequent confirmation.

My point here is not to show that Kant was wrong about spatial geometry. (Plenty of previous writers have done that.) What I would like to point out is that *if* we *were* "Kantian cognizers"—that is, beings whose minds were really built in the ways Kant suggests—we would have been incapable of even *conceiving* of an alternative to Euclidean geometry. A Kantian mind could not concretely think or imagine possibilities contrary to the Forms of Sensibility and Categories of Understanding. And perhaps there *are* ways other beings could conceive of the world that our cognitive architecture does not permit us to conceive. But non-Euclidean geometry is not one of them.

A mind that can entertain and create multiple models with different representational systems has a far better chance of avoiding forced error than a mind that has only a single representational system. The errors that are forced by a particular model need not be forced on the *mind*, unless it is incapable of thinking through models that do not produce the same forced errors. *Without* recourse to alternative models, it would be impossible to distinguish forced error from synthetic a priori truth in one's own case. I will return to the somewhat disturbing question of whether there may be principled limits to what sorts of things we can understand in chapter 12. For now, the point is that a pluralistic cognitive architecture is a good strategy for avoiding such limits as far as they can be avoided.

11 The Complementarity of Models and Language

In previous chapters, I have explored the hypothesis that human understanding is grounded in domain-centered mental models. Model-based cognition, however, is clearly not the *only* component of human cognitive architecture. Nor is it something unique to human beings. As I have characterized models in a fashion that is independent of how they are acquired, there may be any number of comparatively simple animals with "nativistic" (i.e., species-typical, developmentally canalized) cognitive mechanisms that encode systematic features of domains in a fashion that would allow them to count as models. Even the capacity to acquire models through learning—the possession of a *modeling engine*—is clearly evidenced in a number of biological taxa. One sort of question this presents is how models and the modeling engine are related to other components of cognition and intelligence. Fully exploring such a question would require a very wide-ranging investigation, extending beyond philosophy and psychology into cognitive ethology and evolutionary psychology. I will approach it in a far more modest fashion, and principally to make a case that while there are distinct factors that contribute to the cognitive architectures of humans and nonhuman animals, some of which may have been historical precursors of others, they often play *complementary* roles in cognition and accomplish more in combination than the sum of what they contribute individually. This basic orientation—that distinct cognitive systems can be complementary—will then serve as the basis for a discussion of the relationship between model-based understanding and the most distinctive features of human minds: public language and language-like thinking.

Models and language allow us to think and reason in different ways, but they are not *simply* alternative ways of thinking. Model-based understanding provides the basis for much of what we often take to be the semantic properties of language. But language does far more than provide a way of communicating the types of thoughts we could have through model-based cognition alone. It provides a medium in which we can think different kinds of thoughts. It allows us to acquire models that nonlinguistic beings could not acquire. And it provides a platform for new types of reasoning, including not only syntactically based inference but also a variety of types of *critical* thinking that require us to think about thoughts (or their linguistic representations) as objects in their own right. Animals with modeling engines probably existed for hundreds of millions of years before language users appeared on the scene; but the addition of language vastly increased the scope and power of model-based understanding. Some of the forms of thinking that are built on the foundation of language are only hundreds or thousands of years old.

11.1 Cognitive Complementarity

Before turning to the distinctive features of human cognition, let us consider briefly some of the components of cognition and intelligence that we share with many other species. Evolved special-purpose mechanisms and the ability to learn through conditioning provide an ecumenical baseline for most animal cognition, and both can be found in some form even in animals as simple as flatworms. There are other strategies that are far from ubiquitous but are found in different forms in a number of species: incorporation of the environment into the extended phenotype (insect hives, spiderwebs, beaver dams), social distribution of cognitive function (whether in the different operations performed by different castes of ants, which presumably involve differential expression of a common genotype, or in social animals like vervet monkeys or prairie dogs taking turns as sentries), and the possession of redundant neural systems (such as our multiple systems for extracting depth cues) that not only increase resiliency but also extract information relevant to the same ends differently and under different conditions. Other advances have appeared more recently in our evolutionary lineage, such the increased curiosity and capacity for exploratory play characteristic of mammals, and tool use, which is found in rudimentary form in a number of species but is enormously expanded in human beings.

There are clearly interesting empirical and speculative questions in this vicinity for cognitive ethology and evolutionary psychology, such as which species have particular capacities, when they might have appeared in evolutionary history, what adaptive advantages each was selected for, and how some might have been precursors or enabling conditions for others. What I wish to emphasize, however, is a much more general point about how to think about the *relationships between* different components of intelligence that may initially have appeared as distinct watersheds in the history of cognition. And the point is that however they might have first arisen, they often exist today not only *alongside* one another but also in *complementary relationships*.

Two very different types of complementarity deserve mention. The first occurs when each type of system does something that the other cannot. "Nativistic good tricks" provide an organism with ways of doing things that could not be learned through "blank-slate conditioning." (Or even if they could be so learned in principle, the costs of having to learn them would be astronomically high.) For example, animals more or less *need* developmentally canalized ways of recognizing potential mates by responding to highly specific types of cues and likewise need to perform the mating acts typical of their species without having to learn them through observation or trial and error. The vast majority of species also are born or hatched with something much closer to full adult capacities for perception and locomotion than human infants and *need* to have such abilities immediately because they do not have the benefit of an extended period of parental care. Being able to learn through conditioning, by contrast, supplies precisely what "nativistic" abilities lack: the ability to increase the organism's adaptive repertoire beyond what has been selected for and in response to demands of its current environment, which is sometimes an environment that has changed significantly from that of its ancestors.

But there is also a second type of complementary relationship between canalized species-typical abilities and conditioning. There are some spectacularly "closed instincts," like the *Sphex* wasp's routine for depositing its eggs in a locust it has paralyzed. But many "instincts" are more "open," in the sense that they can be fine-tuned through conditioning. Some species of wasps, for example, can build respectable nests on the first try but build *better* nests on subsequent attempts. A kitten has an instinct to pounce on small moving things but becomes an efficient hunter only through practice. "Built-in" abilities and conditioning are in one sense distinct building blocks of intelligence, but much of an animal's adaptive competence is a result of their interaction.

Redundant systems—ones that provide parallel pathways to extracting what is at least roughly the same sort of information, such as the variety of ways the human visual system acquires information about depth—are also complementary with one another. Of course, one advantage redundancy confers is that it makes an animal's cognitive abilities more resilient if it loses one of the systems. A person who loses an eye can no longer use binocular depth cues, but there are a number of monocular cues left to draw on. But different systems also exploit different types of information, so that they are useful in different contexts. Monocular accommodation (adjustment of the focus of the lens of the eye) and convergence (the different angles of the eyes when viewing objects at different distances) are particularly useful at close distances, monocular movement parallax (provided by moving the head) at medium distances, and linear perspective (such as the sides of a road converging in the distance) and aerial perspective (the blue, hazy look of distant mountains) at long distances. In a single visual episode, a correct estimation of distance may require the use of multiple cues, and dissonance between the information provided by two systems prompts a search for more information.

While other species doubtless possess a modeling engine, and surely many can form models of things that we cannot because they are attuned to different affordances in the world through different perceptual organs and forms of embodiment, human beings vastly outstrip other species in the number and variety of things we can form models of. Perhaps this is due in part to some difference in the neural mechanisms that ultimately underwrite the formation of models or the greater sheer volume of such resources available in the human cortex. But it is just as plausibly related to other abilities we possess that are complementary to model formation. Capacities for tool use have often been singled out as prime factors leading to distinctively human intelligence, and the ability to use tools (and particularly to readily learn their uses, invent new ones, and pass them on through social instruction) seems to go hand in hand with the ability to learn new models. In one way, this is just obvious: one needs to learn a model for the use of a particular tool to employ it effectively. But beyond this, there are connections to broader systems of models. On the one hand, we don't just hammer or saw for the sake of hammering and sawing. These activities make sense, and the conditions for their optimization become manifest, only in the context of some further set of projects that require them. Hammering and sawing make sense only in the broader context of building things through carpentry and the requirements these impose on shaping and joining particular types of materials. More generally, having an expanded range of interests, and models related to those interests, creates a market for a capacity to invent and use tools. On the other hand, tools expand our ways of recognizing and exploiting the affordances the world offers, many of which would be not only unavailable but invisible to us without the requisite set of tools. Tools thus change what we can have models of. This is perhaps most obvious in the case of scientific instruments. Microscopes, telescopes, prisms, and centrifuges all allow us to explore new phenomena and partition them in new ways. Experimental setups reveal new systematic relationships between variables. But no sharp dividing line separates science from ordinary life here: simple machines like the lever and pulley supplied crucial transitions to modern mechanics but originated in more mundane ways, perhaps ultimately stretching back to whatever capacities for tool use emerged in early primate ancestors, and are found today in less-developed ways in other primate species. Likewise, carpentry tools opened up a whole new world of things that can be done with wood, which no one would have conceived of by interacting with trees and sticks using only their eyes and bare hands.

Active curiosity and exploratory play, which seem to be characteristic of mammals and are also found in at least some birds, are also highly complementary to the capacity to form and refine models. Indeed, being curious in ways that motivate systematic exploration of novel objects and environments seems to be so well paired with the capacity to form models of new content domains that it is tempting to suspect that there might have been a close historical relationship between them: that one was an enabling condition for the other, or they coevolved, or if they appeared independently, each made the other far more valuable than it would have been in isolation. Exploratory play likewise seems to be well paired with the formation of models, as it allows models of interaction with physical and social environments to be acquired and refined in a fashion insulated from selection pressures.

Identifying complementary relationships between cognitive systems or cognitive traits in present-day species can help to motivate further questions about their evolutionary history and how they interact in development. Two systems may have appeared independently, and their complementarity might be a happy coincidence. One system might be a necessary precursor for the other. Or they might each have appeared initially in different and more limited forms, and their interaction might have provided the conditions for each to be expanded and shaped in ways that exploit their combined benefit. While I am enthusiastic about the project of exploring such questions, I will not pursue them in any depth here. My principal concern here is with the relationship between models, language, and language-like thought.

In the case of the relationship between model-based cognition and natural language, it seems clear that model-based cognition and even the modeling engine appeared long before the capacity for natural language. The status of structurally language-like *thinking* and its relation to public language on the one hand and to mental models on the other are more complicated questions. It might be that there are (or were among our hominid ancestors) species that possessed some measure of language-like thinking but no capacity for syntactically articulated public language. Conversely, language-like thinking might first have appeared as a form of internalized speech and hence be found only among language users. It is also possible that the neural structures used for public language are the very structures used for language-like thinking, but the public and private expressions of these potentialities are in principle independent of one another, with the further possibility that just how fully each is developed may depend on the presence of, and interaction with, the other.

I do not have an optimistic assessment of the prospects for settling such questions decisively: assessing the *forms* of thinking in prelinguistic children or nonlinguistic adults is tricky business, and we have no way of testing the capacities of our long-dead hominid ancestors for either language or language-like thinking. However, even if we are unable to settle questions about origins and priority of appearance of baseline abilities, we will see that there is nonetheless good reason to think that some types of linguistically based skills that can be exercised without overt speech are likely to be acquired and mastered through the use of public language, often in ways that also involve linguistic instruction and the use of written symbols. Baseline competence in public language provides a scaffolding on which further language-based skills can be erected. But all of this also depends crucially on the complementary relationship between language and models.

In the remainder of this chapter, I will first stake out a position on the relationship between model-based cognition on the one hand and both language and language-like thinking on the other and then briefly explore a number of ways in which the addition of language and language-like thinking augments and transforms the forms of thinking available through model-based cognition alone.

11.2 Language and the Priority of Models

One clear feature of language and language-like thinking that distinguishes them from models is their *form*. Language is a medium characterized by a syntactic structure and a stock of lexical elements that can occupy syntactic slots. Models, of course, also have their own kinds of "formal shape," but it is not one characterized by sentence-like structures or transformational and inferential rules that operate on their syntax. In a model, the conceptual elements of the model are constitutively tied to the inferential rules, and also to ways of interacting with the world. In language, we can combine lexical items regardless of their origins or constitutive semantic inference patterns. In this sense, language and language-like thought are "domain general." The same linguistic structures and rules are applied to representations of various content domains and can be used to combine elements referring to different domains. On this point, advocates of central cognition and a language of thought had something right: human cognition involves more than the ability to think about particular domains in domain-specific ways, and this is true even if we recognize that domain-specific understanding can take the form of learned models and not just modules that are developmentally canalized and appear principally in perceptual preprocessing.

If you were to include within the scope of the word 'language' everything that might appear in a text on language or linguistics, it would cut a wide swath: not only syntax and semantics but pragmatics, phonetics, phonemics, prosody, dialect, and so on. Of course, much of what would thus be regarded as "parts of" natural language would not be parts of language-like *thought*, as the list includes acoustical and pragmatic features of public languages. Because I have characterized "language-like thought" as a form of thinking with structural units analogous to those of language, it would have to involve syntax, or at least something structurally analogous to syntax. And there are specialized uses of the word 'language' that require no more than this, as when we speak of "formalized languages." So there are "thin" notions of language that treat it simply as a system of syntactic structures, (uninterpreted) lexical types and tokens, and perhaps generative and inferential rules defined over these.

Philosophers have generally assumed that language-like thought would also have to include more or less the same semantic elements found in natural language as well. And in one sense this is clearly correct: by and large, when we think in a language-like form, we are not dealing with a purely formal language. But just how much of what we call "semantics," even with respect to public languages, is really *linguistic*? The question is not yet clear, as we have yet to specify the boundaries of what to count as "linguistic." But in the present context, where we are asking what the addition of language or language-like thinking adds to the cognitive resources of beings who already possess model-based cognition (or whose ancestors did), we have some reason to draw the boundaries of the "truly linguistic" more narrowly than we might in other contexts. If, as I have suggested, inferential semantics is grounded in models that are not themselves language-like, a good portion of what we call semantics is not "linguistic" in the sense of having its origins in distinctively linguistic capacities.

Consider an example. Suppose I am told (in language), "Napoleon was short and vain." Understanding this sentence allows me to make a variety of inferences: that Napoleon was short, that he was vain, that he was less than seven feet tall, that he would be likely to react badly to criticism, that there was at least one person who was both short and vain, and so on. How much of this understanding is a result of either linguistic competence or a capacity for language-like thinking? Of course, the mere ability to parse the sentence and make any sense of it at all requires competence in English. Moreover, inference from a sentence of the form "x is P and Q" to "x is P" or to "there is something that is P and Q" might well be based on understanding how the language works, and likewise for any syntactically based inferences I might make using it as a premise. But my understanding of what heights are counted as short when dealing with human beings, or what vanity is, or how a vain person is likely to respond to criticism is based not in truly linguistic competence but in having models of the relevant content domains. A person can have perfectly good linguistic competence and lack the knowledge of a particular content domain that he or she would need to make a semantically based inference.

The lexical items in a public language or a system of language-like thought do indeed somehow tap into the separate forms of understanding afforded by models: linguistic utterances and language-like thoughts can *express* what I *understand* through models. And if someone is learning a language, part of what she has to learn is what lexical items tap into which models. Someone learning English as a second language, for example, may already have a mental model of vanity but not know that it is expressed in English by the word 'vanity'. But if the problem is that she is a competent English speaker but has somehow never formed an understanding of vanity, this problem cannot be corrected by further learning of the *language*. Rather, she needs to become familiar with the phenomenon of vanity and form a mental model that tracks its salient characteristics. Someone whose social cognition was impaired in ways that rendered her incapable of understanding vanity might be able to learn what sorts of things people say about people labeled "vain" and mimic them, but this would not amount to an understanding of vanity. She would merely be faking it when the conversation strays onto that topic. Her deficit in social cognition, however, need not imply any impairment of her competence in language.

For purposes of exploring the relationship between model-based cognition, language, and language-like thinking, I will therefore construe "language" in a fairly "thin" way, excluding from it forms of thinking and inference whose basis lies in model-based cognition. Of course, there are other contexts in which it would make sense to adopt a "thicker" construal, one in which "language" includes all that can be expressed in language, what one can do with language, and so on. My aim is not to make a claim about the range of ordinary or technical uses of the word 'language' but to try to distinguish what contributions are made to cognition and reasoning by the capacities for language and language-like thinking over and above those grounded in model-based cognition. And in this sense, language and language-like thinking are representational media characterized by syntactic structure and lexical elements typed according to what syntactic slots they can occupy.

Once we have characterized language and language-like thought in this fashion, two implications that might otherwise be surprising seem clearly to emerge. The first is that much and perhaps all of what we think of as "semantics" does not really arise out of language at all. Many (perhaps all) truly semantic connections arise out of models. Lexical units can express and encode such semantic properties, but without models, language would be largely devoid of semantic content. And this leads to the second implication: without such a relationship with mental models, language and language-like thinking would be of little use and would confer little or no adaptive advantage. They would be something analogous to purely formal languages, in which the symbols can be rearranged syntactically but have no further connection to one another or to the world. And if this is true, it is no accident that a capacity for language arose only once model-based understanding was firmly in place. Language contributes a great deal as an addition to model-based cognition but confers little or no adaptive advantage without it.

11.3 Two Objections

Before passing on to further discussion of the relation between language, language-like thinking, and models, let us consider two possible objections. The first comes from the computationalists, who would likely concede that *natural* language is a recent and uniquely human trait, yet also claim that thinking in general must take place in some sort of language of thought, even if it must be in some ways very different from public language. The second is implicit in discussions of causal semantics: that what we need for a semantically pregnant language (whether public or private) is not a system of mental models but merely reference-fixing relations for lexical units in a language-like representational medium.

Computationalism

In the last section, I claimed that merely possessing an internal "language"—in the thin sense of a system with syntactic structure and some set of syntactically based techniques for producing symbol strings from other symbol strings—would not confer semantic understanding. To this, advocates of the computational theory of mind might point out that Turing showed us how syntactic operations can be made to track semantic relations, and this was the first (and perhaps the only) proposal for how there could be meaning-respecting mental processes that do not require a magical semantic interpreter. After all, some of the early explorations of model-based views of semantic understanding—Minsky's frames and other projects in AI that took place around the same time—were implemented in digital computers, and whatever semantic competence (or simulation of semantic competence) they demonstrated was ultimately based in syntactically driven computations.

However, note that the sort of "syntax" involved in such systems is *not* the syntax of the natural language in which the input and output sentences are couched. Such "syntactic" processes are machine language operations operating over binary digits that are not themselves representational at all. We must therefore be careful to distinguish two senses in which computational processes might be called "linguistic" or "language-like." One thing this might mean is that the computer has been programmed to support representations in a form something like a natural language and to perform operations on these—that it has representational types that are at least structurally analogous to sentence types or judgment forms. This is an important goal of artificial intelligence, but it is a feature absent from most of what the majority of computers do. We do speak of "the machine language" of a particular computer and call the rules of the operating system "syntactic." But what we mean by this is not that the symbolic units are structurally analogous to sentences or judgment types. The syntactic forms of machine languages have little in common with those of natural languages, and all that is really meant in calling them "syntactic" is that they are nonsemantic. So when we speak of computers essentially being devices that manipulate symbols according to syntactic rules, what we really mean is that programs have their own distinctive forms of symbols and syntax, and this ultimately drives the performance of the computer. This is quite generally true of computers—indeed, it is close to being a definition of digital computation—but it does not require the use of anything analogous to a natural language. Program languages, particularly those that operate at the most fundamental level of bitwise operations, bear little resemblance to natural languages.

As a consequence, advocates of the computational theory of mind tend to mix together two very different types of claims about the mind. (I do not mean to imply that they conflate them, though I am trying to caution the reader not to do so.) On the one hand, computationalists often claim that intentional states like beliefs and desires involve "mental representations" that are at least closely analogous to structures in natural languages—say, that intentional states are functional relations to representations with a compositional syntax that has a subject-predicate structure. On the other hand, they also claim that thinking in general is couched in some form of symbolic representation and driven by "syntactic" (i.e., nonsemantic) processes, presumably at some more fundamental level analogous to the machine code of a digital computer, with allowances made for the differences between the "implementing" systems of chips and neurons. The latter claim does not apply only to intentional states we humans experience, such as judgments. If human cognition is "computational" in this sense, presumably the cognitive processes of crocodiles and butterflies are "computational" as well, though the "computations" are simpler and perhaps couched in the different "machine languages" of different types of brains. But if this is true, the claim that thought is computational does not imply that it is structurally language-like. And if some language-like structures are built on such "machine-level" architectures, we have no reason to suppose that any *semantic* properties of the system supervene on their specifically language-like form. It might instead be a consequence of other structures that can be built on the machine-level architecture, such as frames or models. So even if we grant (1) that cognitive processes are accomplished through "computation" (in the machine-level sense), *and* (2) that structurally language-like systems are built out of machine-level computational resources, *and* (3) that some sorts of data structures and machine-level computational processes can, in the right configurations, confer (or at least mimic) semantic understanding,[1] it does not follow from this that the *kinds* of data structures and machine-level processes that confer understanding are the ones that are structurally language-like. Indeed, the very concerns that led to the development of structures like semantic networks and frames suggest that any semblance of understanding that can be programmed into computers requires model-like, rather than language-like, structures.

Causal Semantics

Since the influential publications of Putnam (1975) and Kripke (1980), a great deal of philosophical work in semantics has concentrated on how the reference of names and kind terms is fixed. In particular, there have been several variations on the general notion that reference is fixed through causal relations: Kripke's (1980) baptismal account, Fodor's (1987) causal covariation account, and Dretske's (1987, 1988) teleofunctional account in which the meaning (i.e., reference) of a term is what it has the function of indicating. For our purposes, what matters is the possibility that such accounts might supply a language-like system of mental representations with semantic content *without* recourse to non-language-like units such as mental models.

Let us assume, for purposes of argument, that some sort of causal or teleofunctional relation is sufficient for fixing references of names (or name-like mental representations) and kind terms (or concepts tracking kinds). (Indeed, I think such theories have at least some grain of truth, which I expand on in chapter 16, though I think such stories make sense only within a broader story about mind and world.) This supplies an explanation of semantic competence only to the extent that semantic competence is confined to reference—that is, to tracking individuals and kinds.

But clearly this is a much thinner and more limited notion of semantics than what motivated the need to posit things like mental models. Semantic *understanding* is not simply a matter of being able to name or track classes of things; it also involves being able to grasp their characteristic properties, relations, behaviors, and transformations. Even if we suppose that, in coining a term or concept for water, I thereby come to track a particular type of stuff, and even to pick out *that* kind of stuff in all possible worlds, it by no means follows that I thereby gain any *understanding* of that kind of stuff. Indeed, it implies neither (1) that I gain a grasp of the properties that kind of stuff really has (which is, in fact, an important part of the standard development of such accounts) nor even (2) that I have any particular "folk" understanding of its nature or how it behaves. Whatever forms of understanding I might have must be reflected in a systematic set of ways of representing and making inferences about it—the very sort of thing that I have argued requires a mental model. Not only is no such model ensured by the kind of causal relation that is supposed to secure reference, but according to such theories, it would have to be to some extent *independent* of reference, as I and my Twin Earth doppelgänger are supposed to have isomorphic models that differ only with respect to reference. Not only do causal accounts of reference not supply a way to explain such inferential-semantic competence, but their standard development ensures that they are independent of it. If there is any illusion that they obviate the need for something like mental models, it is only because discussions of them have sometimes redefined 'semantics' in narrower terms that exclude inferential dispositions from the domain of semantics.

Moreover, while causal accounts do not make reference a consequence of mental models, they do not make it a consequence of language either. Generally, what is supposed to secure the reference of a concept is some kind of causal relationship with things outside the mind, generally through perception. According to causal theorists, what makes a concept mean "cow" is not the fact that it can be embedded in particular syntactic structures or manipulated through formal inference techniques, but the fact that its tokening is caused in the right sorts of ways by cows. And what assures this is not the language-like structure of Fodor's central cognition, but the encapsulated processes of his perceptual modules.

11.4 What Language Adds

What sorts of cognitive benefits does possessing public language and language-like thinking confer that could not be had through model-based cognition alone? In this section, I discuss a number of these benefits, with attention to how they complement model-based cognition, both in the sense of adding things that cannot be had using models alone, and in the sense of how the two types of thinking interact to add up to something more than the sum of the parts.

Communication and the Acquisition of Models through Language

The most obvious advantage afforded by public language is that it allows for communication in ways that models do not. It is possible to build intersubjectively available external models for some things we think about, and sometimes these are essential for instruction, communication, and joint attention; but external model building alone would be an extremely cumbersome form of communication. But we should also make an important distinction here between linguistic communication of propositions about domains that speaker and hearer both already understand in similar ways and the use of language in the process of teaching one another the use of new models.

A great part of everyday linguistic communication involves the exchange of information already framed in terms of models that both speaker and hearer already understand. If I say, "There's another big storm coming in tomorrow," and you understand the local weather patterns this time of year, this may communicate something rich in specific implications. In a New England winter, a neighbor who heard and believed it might expect a substantial snowstorm, stock up on salt and gasoline for the snowblower, and so on. Someone uttering the same sentence in Florida in summer, or in the Arabian desert, would likely mean something quite different, and uptake would require models of different sorts of meteorological events. If I posted the sentence on a blog and followed it with "What should I do to prepare?" readers who did not know my location or did not understand the relevant type of thing called a storm would not know how to advise me (in one case, because they would not know which model to apply, in the other because they would lack such a model).

The purely communicative use of language generally succeeds only to the extent that speaker and hearer already share the kind of frame of reference provided by a mental model. When models are shared, a compact utterance can provide enough information for a rich set of inferences. But without such a shared understanding, there is unlikely to be uptake about the implications or even the sense and reference of the lexical items. If I read a random sentence from a book on a topic I know nothing about, I can probably recognize it as a grammatical sentence but will have no real understanding of what it is saying. It is the models that lie in the background that generate a space of possible *sensible* representations in language. To understand what is meant, I may have a great deal of work to do acquiring the requisite models.

But to do so, I need not engage in a process of rediscovery de novo. Language can also provide more efficient means for guiding me through the process of model acquisition. Indeed, the art of using language in this manner is precisely what is needed for a good explanation or the composition of a good textbook. In some cases, a really good verbal explanation may be all the external prompting a reader or hearer needs to form a model that at least roughly corresponds to the speaker's or writer's. In other cases, language may at least play important assisting roles, such as pointing out where the important variables lie or correcting or sharpening the learner's understanding. Learning a manual skill, a form of dance, or the structure of a new type of music probably cannot take place without exposure to, and practice with, the domain of study, and things like mimesis, pointed repetition, and reshaping of what the student is getting wrong may be indispensable. But even these can often be greatly aided by the use of verbal prompting. ("You are swinging the hammer from the side and bending the nail rather than driving it in." "In this style of playing the jig, the accents go on the first and third of six beats, not the first and fourth.") Communication using terminology based in other, *already-shared* models can thus help the learner quickly revise a new model she is building while learning.

And terminology based in the new model can adroitly be paired with carefully selected examples or demonstrations to help the learner attend to the units that the new model tracks. ("This passage is played *spiccato*—like this," followed by a demonstration of the bowing technique, and then perhaps examples and explanations differentiating *spiccato* bowing from other bouncing styles such as *ricochet*.)

Metaphor

In chapter 8, I briefly explored Lakoff and Johnson's (2003) claim that a great deal of human understanding is based in metaphor. I related this to model-based understanding by suggesting that metaphorical transposition involves taking a model of one domain and using its structure as the scaffolding for a model of a different domain, with subsequent revisions to accommodate the differences between domains. In the standard sense of the word, metaphor is a linguistic phenomenon, and so it might seem to follow that if Lakoff and Johnson are correct, a great deal of human understanding is possible only because we have the linguistic capacities that can form a basis for metaphor. We should, however, be cautious in drawing this conclusion, as nonlinguistic mechanisms might also be available for transposing the structural features of a model to a new domain. This strikes me as an important empirical question we should be careful not to prejudge. But even if nonlinguistic methods can achieve this result, mechanisms based in linguistic metaphor may be more efficient and have broader scope and are clearly powerful tools for the transmission of model-based forms of understanding. To take one of Lakoff and Johnson's examples, just *using* a conflict metaphor for argument can prompt the hearer to activate a conflict model and begin to use it as a way of thinking about argument. She might eventually have thought of this way of looking at things on her own, but language allows it to be transferred quickly and to proliferate through a community, enlarging the stock of shared models that can be used in routine communication.

Models That Require Language

Several of the examples I used in discussing the idea of a mental model involved forms of understanding that nonlinguistic beings could not acquire at all. It goes without saying that models of domains that themselves involve language—a model of the sonnet form, or even of restaurant interactions—would be unavailable to nonlinguistic beings. Scientific understanding requires more than a grasp of grammar and reference, but it is unlikely that our ancestors could have progressed very far beyond folk physics or folk biology without increasingly regimented forms of linguistic representation and communication. (And we might well wonder whether even maturationally normal forms of folk theories can be developed in the absence of normal linguistic interaction, which is after all part of the normal developmental environment of human children.) Likewise, the acquisition of forms of mathematical understanding that go beyond what is present in the core systems takes place through linguistic interaction and may directly exploit cognitive capacities originally used for language or language-like cognition.

It would at this point be a matter of speculation just what proportion of the mental models we use depend essentially on language. I am prepared to allow that the proportion may be quite large indeed. Language provides new forms of representation, communication, and interaction—not least in the ability to form concrete *questions*—and together these greatly expand the sphere of things we can form interests in, explore to find the relevant patterns, interact with, and think about. Mastery of a domain still requires us to form a mental model of it, but there are domains that are revealed to us only because we can think in language-like ways and communicate through language.

Representation and Reasoning across Models

One fundamental advance afforded by language is that it provides a medium in which we can think thoughts that combine elements from different models. The inferential patterns associated with lexical items in a language or concepts used in language-like thought may come from the life they lead within models; but from the standpoint of language and language-like thinking, a lexical or conceptual item's model of origin is irrelevant. It is simply a unit of a particular logical or grammatical type. Concepts and lexemes originating in different models can be combined grammatically, propositions involving them can be joined with logical connectives, and the resulting propositions can be assembled into forms of logical inference whose validity does not depend on their semantic values. This endows us with capacities for cross-domain representation and reasoning that go beyond simply using two models in tandem, such as triangulating a common referent.

While this is an extremely powerful cognitive advance—so much so that we might well regard it as one of the most important features of truly human cognition—the marriage between model-based understanding and language-like representation is imperfect. There are grammatically well-formed sentences that do not actually make sense once you understand the meanings and implications of the terms. A single lexical item may stand in for elements in more than one model, where it functions differently, producing ambiguity or even paralogistic reasoning. And we tend to assume in logical inference that the terms or concepts employed have sharply defined extensions, and forget that concepts as psychological entities tend to have fuzzy boundaries and originate in idealized models. Language, or at least logical reasoning based in language, seems to "want" things like bivalence and well-defined extensions, but this seems to be something of a mismatch with what model-based understanding provides. This, I suggest, is the psychological explanation of many apparent paradoxes and is a result of adding a language-like form of thinking optimized for one set of purposes to a preexisting model-based form of understanding optimized for others.

Linguistic Creativity

Language also affords its own forms of creativity. Precisely because grammatical sentence formation is relatively unconstrained by model-based semantics and new sentences can be constructed more or less at will, we can construct sentences that express things that we might never have thought of on the basis of model-based thinking. Of course, the vast majority of randomly constructed sentences have little cognitive value. Some are merely amusing (e.g., puns, at least if you consider puns amusing), others evocative (poetry). But sometimes they may spur new ways of thinking that prove to afford genuinely useful new ways of understanding the world. The initial creative act may not have been a *product* of insight but might nonetheless generate new insight.

Higher-Order Cognition

Language also seems to provide the basis for several forms of higher-order cognition, in which we think thoughts *about* thoughts or thought structures. Perhaps some basic forms of higher-order cognition are nonlinguistic, such as simply being aware that I am having a particular sensation, thought, or experience (though this might better correspond to what is sometimes called "higher-order *perception*"; cf. Armstrong 1968, 1984; Lycan 1996). The analytic power of higher-order cognition, however, seems to reside in the ability to attend to features *of* thoughts, and particularly to either linguistically formulated thoughts or linguistic representations of those thoughts. We can raise questions about the clarity of a term or concept, or about the meaning of a proposition, only when we can represent it as an object of thought, as opposed merely to thinking about other things *through* it. We can examine the validity and cogency of an argument or check the steps of an inference only when we can represent the argument and its constituent propositions as special kinds of objects. The capacity to treat thoughts (or their Fregean contents, or their linguistic or logical representations) as objects of thought is not required to be able to think at all, but it would seem to be a prerequisite for *critical* thinking.

This description, of course, leaves a number of questions open, such as whether various techniques of critical thinking are directed toward thought tokens, thought types, Fregean propositions, or linguistic representations such as sentences. It also leaves open the question of whether critical language-like *thinking* depends essentially on public language. I think there are two types of (potentially defeasible) reasons to assume that it does. First, critical thinking generally involves thinking about representations formulated in a public language, even if this is done without overt speech. Second, we generally learn to think critically through dialectical interaction with others. While I cannot rule out the possibility that a clever individual might develop forms of critical thinking on her own, it seems quite likely to me that, in general, the capacity for individual critical thought arises by way of a kind of internalization of interpersonal dialectic. (Of course, even a brilliant individual who pioneers critical thinking from the ground up on her own would do so using linguistically expressed thoughts, even if some of the vocabulary is of her own invention and has not been shared with others.)

External Symbol Storage

An important extension of natural language is provided by writing. While a good portion of the earliest extant examples of writing are accounting records, anything that can be said can also be represented in external symbols. Merlin Donald (1991) has argued that external symbol storage is itself one of the great watersheds in the history of human cognition, particularly because it allows the storage and retrieval of information to be freed from the vagaries of individual memory. Before writing, the stock of information available to a community was largely confined to what was known and remembered by its elders. Writing allows information to be preserved indefinitely and to be conveyed to broad audiences, widely separated in space and time.

But external symbol storage can also serve as an important cognitive prosthesis that expands and refines our ability to think and reason more clearly. A written sentence can be scrutinized and reexamined, allowing us to better understand it and to analyze it critically. As a consequence, it is also possible to set down in writing thoughts that would be difficult to express or comprehend orally because they stretch the storage limits of auditory parsing. In assessing a line of reasoning, it is an enormous benefit if one can use written formulations that *stay put*, which we can refer back to as needed. And certain forms of writing, such as the schematic representations used in logic, help to clarify the logical forms of propositions. It may be (as Plato claims in the Seventh Letter) that there are significant disadvantages to writing as opposed to the living dialectic of oral discourse, but writing also has signal advantages for clear analytic thinking.

Social Knowledge Production

Finally, while knowledge and understanding may be attributed in the strictest sense to individuals, their production is often inescapably social in character. Things like contemporary scientific theories and systems of laws are seldom the work of one person working in isolation. Not only does one person's work build on another's, but it is often the case that the formation of a model requires teams of individuals with different types of expertise and distinct duties working together to craft something that none of them could accomplish alone. Even if the final product is something that an individual is capable of understanding, it may well be that, in the process of developing it, none of the members of the team understood the whole product both comprehensively and in detail, and the process of coordinating the different components depends essentially on language. Mental models are indeed essential to the process, but their coordination takes place not via mechanisms like triangulation that operate within individual minds but through carefully developed special means of communication, crafted toward a particular end, but erected on the foundation of language.

11.5 Summary

Public language and language-like thinking are cognitive capacities distinct from model-based cognition. They are comparatively recent additions to the cognitive tool kit and would have been of little use to animals that did not have substantial capacities for model formation. They inherit much of their semantic content from mental models and do not themselves provide a basis for the same types of reasoning and understanding supplied by mental models. But neither are they simply separate additions to the cognitive tool kit. They provide distinct forms of thinking and reasoning that complement model-based processes and allow us to refine our mental models through more careful and explicit scrutiny. Moreover, most of the examples of mental models presented in earlier chapters probably could not have been formed without the assistance of language. Humans, like many other animals, understand the world through mental models. But the kinds of models we are capable of acquiring, and the ways we are able to work with them, are available to us only because we are also linguistic animals.

III Epistemology, Semantics, Disunity

12 Disunities of Knowledge, Science, and Understanding

We all find ourselves having some number of conflicting beliefs. We make different and conflicting judgments on different occasions or when primed in different ways. Sometimes, even upon deep reflection, we find ourselves committed to two opposing claims and do not know how to resolve the tension between them. It is not always easy to tell where the sources of conflicting commitment arise from, but they are often tied to the ways in which we frame situations through different models. And upon inspection, the different models may involve not only different ways of framing things but also different commitments and different patterns of inference whose internal logics lead to contrary conclusions. We might describe this by saying that our beliefs and our understanding are *disunified* in various ways. Moreover, we find such disunities not only in ordinary life but also in philosophical and scientific investigation. Indeed, recognition of the ways our beliefs and understanding *as we find them* are disunified is among the primary drivers of philosophical and scientific inquiry. Sometimes such inquiry leads to new ways of thinking that dissolve the old problems, but often it does not, and in some cases it presents us with new problems to replace the old.

This chapter seeks to do two things. First, it attempts to apply the resources of cognitive pluralism toward a diagnosis of why some types of disunity of belief and understanding arise as a consequence of a model-based architecture for understanding. Second, it examines the question of whether a model-based architecture might give us reason to suspect that *some* such disunities might be principled and abiding—not simply symptoms of our current ignorance but predictable and lasting artifacts of a particular type of cognitive architecture. First, however, we would do well to try to clarify some different notions of how beliefs and understanding might be "unified" or "disunified."

12.1 Visions of Unity and the Problems They Face

Let's start by reviewing what has been said so far about the units of knowledge and understanding. If knowledge is justified true belief, then the units of knowledge are sentence sized. I have used the term 'understanding' for a broader and more interconnected sort of comprehension of a domain—an understanding of chess, or of Newtonian mechanics, or of the Japanese tea ceremony—and argued that mental models are fundamental units of understanding. The implicit rules of a model are among the things that might be counted as dispositional beliefs (say, that bishops move only diagonally or that bodies exert a gravitational attraction described by the inverse square law). The conceptual resources of models are used to frame more specific judgments (say, that the white bishop can capture the black rook on the next turn or that this rock is accelerating in free fall toward the ground). The understanding associated with a model also includes abilities to recognize particular situations as ones to which the model may aptly be applied and to act in a fashion guided by the model.

To get clearer on ways that beliefs and models might be "unified" or "disunified," it is useful to start with science, if only because there has been so much prior discussion of "the unity of science." Science has often been characterized in terms of a set of beliefs about theories, laws, and empirical facts, but scientific theories are really paradigm examples of mental models. We can have beliefs *about* scientific models—say, that Newtonian mechanics provides an apt model for certain types of situations. Explicitly formulated laws express basic commitments of such models. And more specific types of scientific beliefs and judgments involve the use of a model to frame a situation and apply the rules of the model to reason about it. But scientific models also stand in various relationships to one another: they involve consonant or dissonant commitments, representational systems, and forms of reasoning and can be used to license judgments that are consistent or inconsistent with one another. I thus begin by discussing three visions of scientific unity, along with brief summaries of reasons that they have been difficult (and perhaps impossible) to achieve.

Three Visions of Scientific Unity

Some visions of the unity of science are cast in terms of scientific *knowledge*, *beliefs*, or *propositions* and based on two key ways that propositions can be related: consistency and implication. It would generally be assumed that two scientific beliefs—or two beliefs generally—that are *known* should be consistent with each other, because to be knowledge, they must be true, and two inconsistent beliefs cannot both be true, at least in standard logics. If two scientific theories imply conclusions that are not consistent with each other, we thus have reason to think that something must be amiss with at least one of them. This might seem to be a simple truth of logic and hence a fairly minimal type of "unity," but it is not at all trivial to make it work out the way we anticipate: the best scientific theories we have, general relativity and quantum theory, in fact seem to produce inconsistencies when combined.

A second and stronger vision of unity is based on the implicatory relations between propositions. Two such implicatory relations are particularly important. The first is the Laplacean determinist view that a comprehensive set of laws of nature (regarded as propositions) combined with a comprehensive description of the state of the universe at a time t should imply a comprehensive description of the state of the universe at any later time $t + \partial$. The second is the reductionist view that there is some lowest-level description of the world (say, in the vocabulary of fundamental physics) from which all the legitimate higher-level descriptions of the world (say, in the vocabularies of chemistry, biology, and psychology) can in principle be derived. Both determinism and reductionism have sometimes been regarded as views that *must* be true. No less a philosopher than Kant claimed that it is a synthetic a priori truth that we must regard all events as determined by a law of nature.[1] And some of the twentieth-century positivists and empiricists regarded intertheoretic reduction as a norm of the logic of science, eventually leading some philosophers who believed some supposed phenomena (such as intentional states and ethical facts) to be irreducible to become eliminativists about them. But the questions of whether all events are causally determined and whether all scientific theories are reducible to fundamental physics also seem quite plausibly to be higher-order empirical questions (a view taken, e.g., in Oppenheim and Putnam's [1958] "Unity of Science as a Working Hypothesis"). And as empirical hypotheses, both have fallen on hard times. Reductionism reached its zenith in the early 1960s, but since that time, philosophers of science who have examined the relationships between theories carefully have repeatedly found that they fall short of the standard of reduction as axiomatic reconstruction (Horst 2007). Quantum theory employs probabilistic equations and, in some of its philosophical interpretations, is regarded as truly indeterministic. And I argued in *Laws, Mind, and Free Will* that even nonprobabilistic classical laws do not imply a commitment to determinism, and a commitment to them is compatible with a commitment to anomic events, including free will and miracles (Horst 2011).

A third vision of unity is that an ideally completed science would employ a single comprehensive theory of everything. This would not mean that more particular theories of things like chemical reactions, thermodynamics, or neural dynamics would cease to be employed, only that they would turn out to be either derivable from the grand unified theory as special cases applicable (and perhaps much easier to use) in particular contexts, or else sufficiently close approximations of the implications of that theory to be useful in those contexts. As theories are, in my view, models and not propositions, we may regard this as a vision for replacing our many different models of distinct phenomena with a single comprehensive model without a loss of explanatory power or predictive accuracy. The search for a grand unified theory in physics is generally seen as an attempt to find a single theory to unite a relativistic theory of gravitation and a quantum theory of strong-weak-electromagnetic force. Many theoretical physicists assume (and we might well ask *why* they so readily assume it) that it is only a matter of time until we will hit upon such a theory, or at least that it is available in principle, even if no one in fact uncovers it. But this is by no means an obligatory view. Stephen Hawking, for example, has changed his mind on this question:

> Up to now, most people have implicitly assumed that there is an ultimate theory, that we will eventually discover. Indeed, I myself have suggested we might find it quite soon. However, M-theory has made me wonder if this is true. Maybe it is not possible to formulate the theory of the universe in a finite number of statements. ... Some people will be very disappointed if there is not an ultimate theory, that can be formulated as a finite number of principles. I used to belong to that camp, but I have changed my mind. (Hawking 2002)

I will return to Hawking's views shortly, as they are tied up with his "model-dependent realism"—a view that I discovered only late in the process of writing this book, but which has a great deal in common with cognitive pluralism.

Beyond the Sciences

Even if scientific knowledge and scientific models were to turn out to be unifiable in one or more of these ways, we would still be faced with the question of whether our other, *non*scientific beliefs and the models in which they are framed can similarly be unified. The most notable philosophical problems are about the relationship between normative claims (particularly ethical claims) and positive claims—that is, beliefs about natural laws and matters of fact. As Hume taught us, no set of claims about facts and laws entails any normative claim. And Kant's view that we are compelled to regard the world of nature as causally deterministic was in precarious balance with his view that we must regard human action as free in a sense that is incompatible with causal determination. Even *within* ethics we are led to paradoxes in which two opposing views seem compelling, and there are alternative metaethical models (consequentialist, deontological, virtue-theoretic) that seem both irreducible to one another and to often entail conflicting conclusions about cases.

There is also a set of problems about the relationship between mathematics and the world of nature: mathematical truths are not a consequence of how the world is, and have a kind of necessity that could not be underwritten by empirical facts and laws. Mathematics cannot be reduced to physics. But neither can physics be reduced to mathematics (despite the hopes of some thinkers of the seventeenth century), as there are purely physical properties like mass as well, and various mathematical systems provide descriptions of various alternative forms space might take, and waves, particles, and fields that behave in alternative ways. And the discovery of non-Euclidean geometries in the nineteenth century brought the startling realization that while individual mathematical systems might be internally consistent and derived from a small set of axioms, there are indefinitely many such systems whose axioms are inconsistent with one another.

We might, moreover, raise similar questions about many more ordinary domains of understanding. Chess pieces and chessboards are typically garden-variety physical objects—pieces of wood or plastic. (The ontology of the "pieces" in computer chess is more complicated, but still physical. The status of those in mental chess turn on whether the mind itself is physical.) But a complete understanding of the pieces and board in physical terms would not confer an understanding of the game of chess. Perhaps the diehard reductionist might insist that "the game of chess" is simply a set of conventions that could be reduced to the mental states of those who share the conventions, which could in turn be reduced to fundamental physics. And if such reductions could be carried out, they might well explain the *behavior* exhibited in chess matches. But it is unclear that they would explain the game as an abstract entity any more than an explanation of the mental states of mathematicians would explain mathematics.

We thus have three visions of how beliefs and models might be unified, each with corresponding possibilities for disunity. The first and most modest views "unity" in terms of global consistency: to be unified, our beliefs (including the commitments implied by our models, at least when aptly applied) must be mutually consistent. The second is framed in terms of implication: that there should be some relatively conservative set of beliefs and models from which the others can be derived. The third is that "unity" would consist of a single comprehensive and self-consistent model in which all our beliefs and judgments can be framed, and in such a fashion that nothing we know or understand on the basis of having multiple models is lost in the process.

12.2 Disunity as a Problem

At some level, most of us—not just philosophers and scientists—find disunities of belief and understanding troubling, and this leads us to seek ways to unify them. It would be useful to take a moment to examine the nature of this situation, and of our dissatisfaction with it.

First, we each find ourselves with some number of conflicting and even contradictory beliefs. Sometimes it is that we believe one thing on one occasion or when primed in a particular way, and another on another occasion or when primed in a different way. But sometimes we find ourselves, even upon deep reflection, equally committed to two contrary claims and do not know how to resolve the tension between them.

Second, we find this situation dissatisfying. It drives us to attempt to find a way of resolving the contradiction. It is no accident that paradoxes drive much of the histories of both philosophy and science. This seems to be a deep and productive feature of human psychology: that when we are pulled toward two contradictory beliefs, we are spurred to find some way to resolve the contradiction. Often we succeed, and some of these successes have been watershed moments in the history of human knowledge. A more complacent species would have learned far less about the world.

Third, we do not just have conflicting beliefs; we have multiple ways of *framing* things through models. (This is a slightly more contentious claim than the first two, because it involves a bit of theory.) Sometimes this is just a matter of having *different* ways of framing things. And sometimes it seems to amount to something more serious: the different ways of framing things are in various ways in tension with one another, such as being incommensurable or implying contradictory inferences.

Fourth, the disparities between models are also dissatisfying. They are probably not as immediately troubling as contradictions between beliefs, except perhaps when they generate contradictory beliefs. Yet something within us is puzzled and troubled even by having two models that frame the world in different terms, and seeks to find some way of bringing them into some deeper harmony.

We might summarize the second and fourth points by saying that the mind has a kind of drive toward finding ways to unify a system of beliefs and of understanding, respectively. This, of course, is a point famously made by Kant. Kant's attendant observation should be added: the mere fact that we have a drive to unify what we believe or the ways we understand things does not itself entail that we will in fact be able to do so. And to this we may add a fifth point, which is also a variant on Kant: we have notions of particular *ways* that knowledge and understanding might be brought into greater unity. Kant had his own hypotheses about what these are and their cognitive underpinnings. I would suggest that we see evidence of a distinct set of these in the types of unification that philosophers and scientists have in fact sought: global consistency of beliefs, reduction, and a single all-encompassing theory.

I will explore, in chapter 14, a way of developing a sixth point, also foreshadowed by Kant: the ways that features of cognitive architecture can generate particular types of *illusions*, some of which have to do with the mind's drives to unify knowledge and understanding in particular ways. For present purposes, it is perhaps enough to say that our dissatisfaction with disunities is clearly a feature of something about our minds, but a psychological drive to unify by no means ensures that the project can be carried out to its desired ideal conclusion.

Why might our beliefs and models, as we find them, be disunified? We have already explored one part of an answer to this question with respect to models: having various models of different aspects of the world is a good strategy for building a capacity for understanding many features of the world into finite minds, and each model must represent things in some particular way. What we have yet to explore is how the ways models are idealized can lead to disunities of understanding. If beliefs are constrained by the ways we can frame thoughts through the resources of models, some problems about disunities of belief may also be explained in the process. But in general, there need be no deep mystery about why a given finite mind, at any given time, might encounter conflicts among its beliefs and models. No one ever thought that everyone's system of beliefs was *in fact* comprehensive or consistent, and anyone who thought that about the models used to frame such beliefs was psychologically naive. Claims about unification of knowledge and understanding are claims not about our beliefs and models *as we find them* but about what it might be possible to do to bring them to greater unity.

The more trenchant question is this: is there a way of understanding why, in spite of millennia of philosophical inquiry and centuries of science, our beliefs and models might still present every indication of being *abidingly* disunified? And to this there would seem to be several distinct types of possible answers. The first is that the disunities of knowledge and understanding are only a symptom of the current state of our knowledge and understanding—our current state of ignorance, as it were. Given enough time and investigation, all disunities would all be resolved. A second is that there are indeed principled disunities of knowledge and understanding, and the reason for this lies in *something about the world*. A modest version of this would be that we need multiple fundamental theories because there are in fact multiple fundamental forces. A robust variant would be that understanding is disunified because the world was already, in some sense, disunified before it (Dupré 1993). A third possibility is that there are things about the world that minds like ours are simply unsuited to understanding—what Colin McGinn (1991) calls "cognitive closure." There are things about the world that we understand that dogs do not understand. Canine minds are cognitively closed to those principles. Similarly, once one thinks about it, it seems utter hubris to suppose that there is nothing about the world that human minds are cognitively closed to. I think this is a plausible idea, but what I wish to explore is a fourth alternative: that there may be features of our cognitive architecture—of the things we *do* understand in particular ways—that may themselves lead to disunities between the ways we understand them; and that minds like ours may be unable to produce a set of beliefs that is globally consistent or a single comprehensive model of everything without losing some of the epistemic grip on the world that we gain through many more localized, idealized models and the beliefs they license.

12.3 Model-Based Understanding as a Source of Disunity

The basic idea here is fairly straightforward. It is the idea that there are features of models that have implications for how two models are related to each other, and sometimes these involve incommensurability, irreducibility, and inconsistency. Not *all* combinations of models are fraught with these problems, but the fact that some are is simply a consequence of the fact that models are each idealized ways of thinking about particular content domains that involve particular systematic ways of representing and reasoning about their target domains. Viewed as a system for representing and reasoning, a model has a particular formal shape that can be described by rules. If two models do not represent their targets in the same ways, they can be incommensurable. Of course, if two models are sufficiently incommensurable that their domains do not overlap at all, they cannot produce contradictions. But if two models do have overlapping domains, and the formal features of the models are logically incompatible, they can jointly produce contradictions. And whether one model M_1 can be reduced to another model M_2 depends on whether M_2 possesses the right resources on which a construction of M_1 can be based.

While pure mathematics is special in several ways, we can understand some of the basic issues most clearly by considering it first. Consider two geometric systems: a two-dimensional Euclidean geometry and Riemann's geometry of the surface of a sphere. Each can be described by a system of axioms and theorems, and indeed they share a substantial number of axioms, differing chiefly with respect to Euclid's parallel postulate. In Euclidean space, a line stretches indefinitely in two directions, and for any point P not on a line L, there is exactly one line through P that is parallel to L—that is, the two lines do not intersect. On the surface of a sphere, the equivalent of a line is a geodesic, like the earth's equator. Start at any point and continue in any direction, and eventually you will return to the point of origin. There are no parallel lines on the surface of a sphere. (Cartographers do speak of "parallels," but they are not really lines as defined in the system, nor are they parallel in the relevant mathematical sense. To follow a cartographic "parallel" around a sphere, you would have to orient your path slightly to the left or the right. And two geodesics—say, the equator and the Greenwich meridian—intersect at two opposite points.) The parallel postulate is true in Euclid's geometry but false in Riemann's. And there are other differences as well: for example, the sum of the interior angles of a triangle on the surface of a sphere will always exceed 180 degrees; the Pythagorean theorem is false in a spherical geometry, as you can have a triangle with three right angles; and so on. If you reasoned from premises taken from the two geometries, it would be easy to generate a contradiction. Neither geometric system can be derived from the other, and there is no self-consistent third system from which all the axioms of both can be derived, because to produce the contrary axioms and theorems, the third system would have to already contain contradictory assumptions within it.

So here is the first crucial point: any particular model employs a particular representational system; it allows us to represent the world in some particular set of ways and not in other ways, and it employs a particular set of inference rules and not some other set. From this point, several others follow. First, one model may allow us to represent things or draw inferences that another cannot, and we may *need* multiple models if we need the different types of representations and inferences each supplies and we cannot find a single model that supplies them all together. Second, two models can turn out to be incommensurable or license incompatible inferences. As we shall see, this is not always as serious a problem as at first it might appear, because in some cases two models are not aptly applied to the same situations and hence should not really license contradictory judgments. But it is still an important source of potential contradictions and paradoxes. Moreover, models that are incommensurable or license contradictory inferences cannot be reduced to one another or to some third (internally consistent) model. There is also another reason that one model M_1 may be irreducible to another model M_2: if there are types of representational resources or inference rules in M_1 that are lacking from, and cannot be constructed from, the resources of M_2.

What we have arrived at here are the elements for explanations of various types of disunities of belief and understanding in terms of the features of our cognitive architecture. If understanding comes in the form of models, and each model employs a particular representational system that is optimized for particular ends, it is quite possible for two or more models to carve up the world in incompatible ways, to be formally incompatible, and to license incompatible inferences. Two models can each be eminently apt for their own purposes and yet not play well together. Given the assumption of a model-based cognitive architecture, we should *expect* to find such disunities, and the thesis about cognitive architecture supplies a kind of diagnosis of their source. In some ways, it is a deflationary diagnosis: conflicts in belief or understanding need not imply that "the world is disunified" (whatever that might mean) but may be artifacts of the cognitive architecture through which our minds understand the world.

Yet in another way, this is also a disconcerting insight. It may be that the disunities we find today are things we can bootstrap our way beyond, by finding a better set of models that do not conflict in the same ways, and perhaps even finding a single super model of everything that is self-consistent and allows all the legitimate insights provided by our existing models. But it could also turn out that this is impossible—that either features of our cognitive architecture itself or their combination with the particular set of things we encounter and care about in the world *requires* multiple models to provide maximal epistemic and practical traction on the world, and in some cases these models may, in combination, produce inconsistencies and paradoxes. If this is the case, certain disunities of belief and understanding may be *principled and abiding for minds like ours.*

The question of whether there *are* such principled and abiding disunities, I think, cannot be settled decisively in advance or by purely theoretical considerations. We cannot know a priori how many independent principles are at work in the world, and so any intuition that there should be just one must reflect a cognitive bias. Nor will it do to appeal to an intuition that "the world must be comprehensive and self-consistent," because the questions about unifying belief and understanding depend not solely on how the world is but on how we are able to understand it, and hence involve empirical questions about human cognition. On the other hand, the mere fact that our present ways of understanding the world are disunified in various ways does not entail that we cannot achieve better ways of understanding it that are better unified. There is no way of ruling out in advance a new insight that would change everything, because we could only evaluate its prospects by knowing what it was, and by definition future insights are unknown. Whether disunities of belief and understanding are principled and abiding is ultimately an open-ended empirical question.

So here is a second important point. A model-based architecture for understanding can (and does) result in de facto disunities of belief and understanding but does not itself answer the question of whether such disunities are a kind of principled limitation on knowledge or understanding. It *does* give us reason to be suspicious of any intuitions we may have to the effect that there should ultimately be a single, all-encompassing model of everything, or that we should be able to find ways to reconcile all our conflicting beliefs. There is no reason to assume that the evolutionary forces that shaped our cognitive architecture would have done so in a fashion that ensures such a capacity. (And even if one believes that there was a certain amount of special intervention by God in the design of human minds, there is little reason to think that the design would have included the capacity to have a fully unified understanding. Perhaps the sense of mystery engendered by paradoxes that cannot be resolved in this mortal life is precisely what God was after.)

The only kind of evidence we have on such a question would have to come from looking at cases. And here we find things that pull us in two contrary directions. The histories of philosophy of science are filled with examples of surprising unifications—say, that the things we see in the sky are not made of a different sort of matter operating on different principles, and the orbits of the planets and the behavior of falling bodies are explained by a single principle of gravitation. But they are also filled with discoveries of new phenomena requiring the postulation of new forces and other explanatory principles. And there are still deep disunities, both old and new, that we do not know how to resolve. To name just a few: the gulf between facts and norms, commitments to competing normative principles, the independence of mathematics from nature, wave-particle duality and other puzzles of quantum theory, and the tensions between it and relativistic theories of space-time and gravitation.

I began with examples of models of mathematical systems because they most clearly illustrate ways that the features of models can provide the basis for disunities of belief and understanding. But there are also important issues that we cannot illuminate using mathematical examples. *Pure* mathematics is special in that it is not accountable to anything outside the mathematical system. There are, of course, important questions about what physical systems each geometry can provide apt representations of, but a discovery that space is non-Euclidean does not amount to a falsification of Euclidean geometry. It is simply the system it is. And the fact that two geometries produce seemingly contradictory pronouncements on things like the parallel postulate and the Pythagorean theorem does not result in any deep paradoxes. Mathematically speaking, the parallel postulate and the Pythagorean theorem are not simply true-full-stop or false-full-stop; they are true-in-Euclidean-geometry and false-in-spherical-geometry. There is no further question about their truth, except for corresponding questions about whether they are true in additional types of geometric systems. Someone who mixes together incompatible commitments of two systems in an inference is simply making a blunder.

A different set of issues arises when we *apply* multiple models to real-world situations. It is here that issues about the ways the models are idealized and whether they are aptly applied arise. What makes a model good for its purpose is that it tracks patterns in the world sufficiently well for that purpose, and in a fashion that allows us to think, reason, and plan about situations in the world, and it does so by having a representational system and a set of implicit rules that do a good-enough job of tracking such patterns. The fact that any given model represents only *some* things, and in elegant and comprehensible ways, is what allows it to provide insight into particular real patterns. Two different models may track different patterns and may need to represent them in different ways to be elegant and comprehensible. But the very features that make each model apt for its given purpose may present problems when we try to use two models together to understand a real-world situation to which they both apply. This is a general and schematic story about a source of disunities of belief and understanding. In the ensuing sections, I explore its application to a number of cases.

12.4 Scientific Disunity

For approximately the past century, physics has had two highly successful, well-confirmed theories: a relativistic account of gravity and space-time and a quantum mechanical account of strong, weak, and electromagnetic forces. One of the major projects in theoretical physics is an attempt to unite the two into a single theory. It may be that there would be some physicists who would care about finding a grand unified theory simply because they found the postulation of two distinct fundamental forces that require different models to be inelegant or ontologically profligate. But this cannot be the entire reason that physicists care so much about a grand unified theory. After all, for a long time, physicists operating within classical mechanics were quite content with the idea that there are several independent fundamental forces—mechanical force, gravity, electromagnetism, chemical bonding. The difference was that scientists assumed that precisely because these forces were independent, *and all assumed a common Newtonian understanding of space and matter,* their contributions to a comprehensive and consistent mechanics could be computed through vector algebra. There were several models of what were thought to be independent fundamental forces, but they were not inconsistent, precisely because they were independent. The problem with relativity and quantum theory is that they are held to be incompatible with each other. Now, if you look into just *how* they are incompatible with each other, you will find an interesting variety of answers: there are differences in assumptions about renormalization, local causation, and an absolute or relative temporal frame of reference; and if you combine the mathematics in special cases like black holes, the results are either inconsistent or mathematically incoherent. Some of these concern incompatibilities between the formal features of the two models; others are about how, when combined, they lead to contradictory predictions or even mathematical nonsense.

It would be foolish to try to predict whether the search for a unified field theory will be successful. What my cognitivist account does is to supply at least a partial diagnosis of *why* such a puzzling situation might arise. We can gain some perspective on this conundrum if we assume that each supplies an *apt* but *idealized* model. Relativistic gravitation and quantum theory are each extremely well confirmed, and to an extraordinary degree of accuracy. For the phenomena to which they are applied, each is a highly apt model. But what does this mean? It means that particular mathematical descriptions do a very good job of tracking a certain set of empirically robust patterns. It is a marvelous thing that the human mind is capable of coming up with such models. But to be a concrete model at all, each model must use some particular representational system (and not some *other* system). What gives a model a simple and elegant fit with one set of phenomena may be an idealization that renders it inapt for others. And the representational systems that are simple, elegant, and apt for two different sets of phenomena may be formally inconsistent. It may or may not turn out that it is possible for us to find either a grand unified theory or two compatible successor theories that would resolve the problem.

Earlier I quoted Stephen Hawking expressing a growing pessimism about this prospect. Hawking (along with coauthor Leonard Mlodinow) seems to have reached a view similar to my own about *why* we might be faced with such a situation, in a discussion of what Hawking and Mlodinow (2010) call "model-dependent realism."

> [Model-dependent realism] is based on the idea that our brains interpret the input from our sensory organs by making a model of the world. When such a model is successful at explaining events, we tend to attribute to it, and to the elements and concepts that constitute it, the quality of reality or absolute truth. (7)
>
> Like the overlapping maps in a Mercator projection, where the ranges of different versions overlap, they predict the same phenomena. But just as there is no flat map that is a good representation of the earth's entire surface, there is no single theory that is a good representation of observations in all situations. There is no picture- or theory-independent concept of reality. (42)
>
> It is pointless to ask whether a model is real, only whether it agrees with observation. If there are two models that both agree with observation, like the goldfish's picture (of the world, through the distorted lens of a spherical fishbowl) and ours (without this lens distortion), then one cannot say that one is more real than the other. One can use whichever model is more convenient in the situation under consideration. … Model dependent realism applies not only to scientific models but also to the conscious and subconscious mental models we all create in order to interpret and understand the everyday world. There is no way to remove the observer—us—from our perception of the world, which is created though our sensory processing and through the way we think and reason. Our perception—and hence the observations upon which our theories are based—is not direct, but rather is shaped by a kind of lens, the interpretive structure of our human brains. (46)
>
> It might be that to describe the universe, we have to employ different theories in different situations. Each theory may have its own version of reality, but according to model-dependent realism, that is acceptable so long as the theories agree in their predictions whenever they overlap, that is, whenever they can both be applied. (117)

Hawking and Mlodinow echo my suggestion that the need for multiple models might be a consequence of our cognitive architecture. But an element in Hawking and Mlodinow's remarks helps us to see something else that is also important. Just *having* multiple models that frame the world in different ways may be a fairly innocuous consequence of how the human mind works. Where we really need to pay attention to potential problems is "whenever they overlap, that is, whenever they can both be applied." It is only in *those* cases that they can give rise to contradictory predictions or inferences. There are plenty of such cases, but in *some* of them, cognitive pluralism can help us see that, at least *sometimes*, we need not be committed to contradictory judgments licensed by the models if we attend properly to the ways the models are idealized and aptly applied.

First, consider an experiment in which a metal object is dropped in proximity to a magnet. If we apply a gravitational model, we will get a description of the resulting motion of the falling object. If we apply a model of electromagnetism, we will get a different description. They *will* be different descriptions; and if either model is used alone, the resulting prediction will diverge from the real-world behavior observed, because in fact both gravitation and magnetism are at work. There is nothing here that should be all that surprising. More than one force is at work in most mechanical situations, and the resultant behavior of objects is influenced by all the forces at work. There is good reason to have separate models of the forces, both because they are distinct and separable causal patterns in the world and because *understanding* them separately makes them comprehensible to us. The bracketing idealizations of the models are both psychologically necessary and empirically justified.

In *Laws, Mind, and Free Will*, I took some empiricist accounts of laws to task for treating laws (like those of gravity and electromagnetism) as though they directly license claims about real-world *behavior*, such as how bodies will actually move when dropped. If understood this way, the two laws would have to be seen as entailing conflicting predictions, neither of which corresponds to the real-world behavior in cases where multiple forces are at work. Interpreted as universally quantified claims about how bodies actually behave, the laws would be false. But the source of the error is not that the laws are false but that they should be interpreted not as universally quantified claims about actual behavior but rather as idealized claims about the different forces that are causal contributors to that behavior. The error in the interpretation of laws is a philosophical error—resulting, I believe, from an attempt to force scientific laws into a particular logical mold. It is not an error that any competent scientist or engineer would be likely to make, though of course the scientist or engineer might *overlook* a factor that was in fact causally relevant in a given case. On the other hand, the classical technique of combining the contributions of separate forces through vector algebra is applicable only when all the causal contributors in play have models that are quantified and independent, and this is not the case with all pairs of models. But really this is a case in which cognitive pluralism shows us how to *resolve* an apparent problem. Gravitational and electromagnetic models, applied individually and taken as licensing predictions, *do* generate inconsistent predictions. But understanding how the models are idealized should caution us not to take the models as licensing exact predictions individually, except in cases where only one force is at work, or the influence of one drowns out the influence of the other within the degree of accuracy employed.

The situation is different in the case of quantum complementarity. Wave and particle models represent phenomena differently, and wavelike behavior is different from particle-like behavior. In classical optics, particle theories (Newton) and wave theories (Huygens) were assumed to be *competitor* accounts, and experiments were designed to attempt to settle the question of which was correct. The underlying assumption was that there must be a single model that captures the "real nature" of light and can be applied to all the cases. Classical optics became absorbed into a broader theory of electromagnetism, and classical electrodynamics was superseded by quantum theories. But the availability—indeed, the *need for*—both wave and particle models persists in quantum theory. In some experimental setups, phenomena are aptly described by wave models, in others by particle models. This description, however, comes with an important caveat: the experimental setups are mutually exclusive, and so quantum duality is not understood in terms of two rival accounts of nature. But neither are the wave and particle models descriptions of separate forces, like gravity and magnetism.

Niels Bohr proposed the doctrine of complementarity to account for this counterintuitive situation. His description of the issues reflects both cognitive and pragmatic elements of his analysis.

> *However far the [quantum physical] phenomena transcend the scope of classical physical explanation, the account of all evidence must be expressed in classical terms.* The argument is simply that by the word "experiment" we refer to a situation where we can tell others what we have done and what we have learned and that, therefore, the account of the experimental arrangements and of the results of the observations must be expressed in unambiguous language with suitable application of the terminology of classical physics.
>
> This crucial point … implies the *impossibility of any sharp separation between the behaviour of atomic objects and the interaction with the measuring instruments which*

> *serve to define the conditions under which the phenomena appear.* ... Consequently, evidence obtained under different experimental conditions cannot be comprehended within a single picture, but must be regarded as *complementary* in the sense that only the totality of the phenomena exhausts the possible information about the objects. (Bohr 1949)

The first paragraph reflects a *cognitive* constraint, in this case on how evidence is conceived and expressed. The second reflects a *pragmatic* aspect: our epistemic access is mediated by measuring instruments, and there is no "sharp separation between the behavior of atomic objects and the interaction with the measuring instruments which serve to define the conditions under which the phenomena appear." This leads to a pluralist conclusion: Different experimental setups reveal different patterns of behavior, and we need different models to capture the patterns. If the models were taken as descriptions that were independent of experimental setups, they would generate conflicting descriptions and predictions. But in fact, the experimental setups are mutually exclusive, and the results one obtains with each are stable. I would interpret this as implying something about how each model is idealized and where it is aptly applied. Treating the models as unidealized would not only generate conflicting predictions but also result in an inability to see how we actually have ways of determining with some surety how matter behaves in the different cases. Conversely, in this example, properly understanding how the models are idealized, and more specifically the range of cases in which each can be aptly applied, prevents the formal incompatibilities of the models from entailing commitments to contradictory beliefs or predictions about more specific cases.

But there are other cases in the sciences in which multiple models are used to frame discussions that result in incompatible implications that cannot be resolved by restricting their apt use to particular experimental contexts. A useful example is the ongoing debate within biology about the proper way to understand species. If we are to believe the developmental psychologists, folk biology contains an intuitive concept of species that includes assumptions such as that animals and plants come in species kinds, individual organisms are produced by parents of the same species, and members of the same species share characteristic habitats, physiognomies, and diets. The intuitive essentialism of this notion was made explicit as a philosophical doctrine by Aristotle, who also initiated a project of understanding species through a taxonomy based on differences in traits, which culminated in the Linnaean taxonomic system in the late eighteenth century. Today many biologists and philosophers of biology would see this type of taxonomic classification as centered on criteria for identification and differentiation—pragmatic criteria employed by the observer that are at best contingently related to properties that are salient for a modern biological classification. Darwin repositioned the notion of species as the fundamental unit of evolution, and his theory of evolution provided two notions that have played a central role in more recent concepts of species: interbreedable populations and common descent. These two notions, however, have been emphasized and developed in different ways by theorists in biology, resulting in a number of distinct refinements of the concept of species. (Mayden [1997], for example, distinguishes no less than twenty-two different concepts of species in biology.) These may be grouped into two main traditions.

The first tradition emerged from the "evolutionary synthesis" of systematics, genetics, and evolution (Dobzhansky 1937; Mayr 1942). Mayr's formulation of the "biological species concept" was particularly influential:

> [A] species consists of a group of populations which replace each other geographically or ecologically and of which the neighbouring ones intergrade or interbreed wherever they are in contact or which are potentially capable of doing so (with one or more of the populations) in those cases where contact is prevented by geographical or ecological barriers. …
>
> Species are groups of actually or potentially interbreeding natural populations which are reproductively isolated from other such groups. (Mayr 1942, 120)

The notion of "reproductive isolation" is key to the biological species concept. One crucial question separating subsequent developments of the concept was whether merely extrinsic barriers to interbreeding, such as geographical separation or ecological barriers like mountain ranges, should count as differentiating species. Dobzhansky's definition of "isolating mechanisms" was sufficiently broad to imply such a conclusion, though the first quote from Mayr implies the opposite conclusion. The answer one gives to the question has implications for things such as whether populations that are temporarily isolated thereby count as separate species, and (more importantly) whether, if they are later reunited with related populations with which they interbreed, what were separate species can then remerge. (Perhaps most strikingly, if merely geographical isolation can be determinative for species separation, various isolated human populations were arguably until recently distinct species that have been reunited into a single species through the availability of long-distance transportation.)

But perhaps more interesting from the standpoint of biology was the discovery that in many species, populations that have overlapping habitats and are capable of interbreeding in fact do not do so. Isolating mechanisms (such as the different mating calls used by distinct populations of reproductively compatible birds) seem to be an important factor in speciation, at least among sexually reproducing species, and models that make such mechanisms central to the concept of species thus highlight and allow constitutive inferences about real patterns in nature. At the same time, this emphasis also involves a costly idealization: most organisms do not reproduce sexually—there is no such thing as an interbreeding population of bacteria or fungi—and in other taxa (particularly plants) hybridization also plays a significant role that is screened out by the biological species concept.

The second tradition, by contrast, emphasizes common descent. The most influential version of such an approach is the cladistic approach, based in the notion of a *clade*, which consists of an ancestor and all its descendants, living and extinct. (Depending on the details of the specific cladistic theory, an "ancestor" might be an individual organism, a population, or a species.) Cladistic classifications are treelike structures; but unlike the Linnaean system, they are intended to represent real and evolutionarily salient relations of lineage and divergence. (Linnaeus, like Aristotle, assumed species to be unchanging and eternal.) Cladistic classification produces representations of a hierarchy of taxonomic levels in a treelike structure, in which a species is a lineage between two branching points. Of course, most such branchings occurred in the remote past, and so the data for cladistic analysis must often be drawn from morphological "characters" reconstructible from fossils, though molecular features of DNA can also be used for modern specimens and for fossils from which reliable DNA information can be obtained. The use of characters as the basis for constructing cladograms is what allows the cladistic methodology to be extended to the remote past (something quite important for a classification based on lineage), but there is a great deal of dispute over the choice of relevant and reliable characters that are diagnostic of lineage and relatedness. (To take a toy example, using striped fur as a character would result in a biologically unrealistic analysis of the relatedness of lions, tigers, and zebras.) While the goal of cladistics analysis is a reconstruction of lineage, using characters as a basis for taxonomy runs the risk of the resulting system being a *classification* system that fails to represent a deeper evolutionary reality.

The distinction between classification systems and biologically significant units is often noted in the theoretical literature on species. But attitudes toward this, and implications drawn for the concept of species, differ. Mayden (1997), for example, excoriates the conflation of merely pragmatic means to differentiate species with the deeper goals of evolutionary biology and holds out hope for a species concept that is suitable for the latter:

Much of the turmoil embodied in the species problem ultimately derives from our packaging inappropriate criteria for species into a single concept. This results from a traditional conflation of function of concepts with their applications, definitions with concepts, taxonomic categories with groups, and ontological status of real species with teleological approaches to recover them. Analogous to classifications of supraspecific taxa, our forging inappropriate and ambiguous information relating to theoretical and operational discussions of species ultimately results in a trade-off between convenience, accuracy, precision, and the successful recovery of natural biological diversity. None of these expectations or intentions of species or classifications is attainable through composite, and possibly discordant, concepts of biological diversity or its descent.

With this in mind can one tease apart the theoretical concepts and operational definitions of species and develop a primary concept applicable to the various entities believed to be species? I think this is possible through a hierarchical view of species concepts and their definitions. (Mayden 1997, 383)

Dupré (2001b), by contrast, holds that we should view the concept of species in terms of units of classification rather than of evolution:

It is of course possible that the units of evolution could also serve as units of classification, and that this is in fact the case seems to be very widely assumed by writers on the topic. But, and this is the first thesis I shall defend in this paper, units of evolution are far too diverse a set of entities to provide a useable partition of organisms into classificatory units. If this is correct we need to decide whether the term species should be applied to units of evolution or units of classification. And here, if anywhere in biology, it seems to me we should honour conventions of priority. Species were understood to be basal units of a classificatory system millennia before anyone had thought to formulate a theory of evolution. And to this day almost everyone who is not professionally involved in theoretical systematics assumes that the species category is a classificatory concept. So I conclude that species should be treated as units of classification not units of evolution. (Dupré 2001b, 203–204)

Moreover, Dupré notes that classificatory interests, even within biology, are diverse, leading him to endorse a pluralistic view of classification and species.

We do, after all, generally classify the objects with which a scientific theory deals in relation to the demands of that theory. But this argument, even when more thoroughly fleshed out, is unconvincing. Classification in biology has a life of its own. Biologists in areas only tangentially connected to evolutionary theory, such as ecologists, ethnobotanists, or ethologists, need to classify organisms, as do foresters, conservationists, gamekeepers, and herbalists. As will be discussed below, for many, perhaps even most groups of organisms, evolutionary considerations are of little or no use for classificatory purposes. And finally birdwatchers, wildflower enthusiasts, or just biologically engaged members of the public, may choose to classify organisms, even if they do not need to do so. These diverse groups of people require workable classifications that enable them to communicate among themselves and to members of other such groups, record information about natural history, and so on. If, as I argue, units of evolution inadequately meet these needs they must be distinguished from units of classification. Furthermore, it follows that in biology classification and theory are more autonomous from one another than seems to be the case in most parts of science. The fact that classification cannot, at least, be closely tied to the central theory of biology leaves room for the thoroughly pragmatic and pluralistic approach to biological taxonomy that I shall advocate. (Dupré 2001b, 204)

While I am not in a position to take a stand on whether there is a viable unitary notion of species, or a variety of such notions, corresponding to the units of evolution, Dupré's analysis at the least sheds light on why we find such a variety of ways of constructing the notion of species, and of classifications of organisms into species. There are a variety of practical scientific (and nonscientific) standpoints from which the classification of organisms is called for, and these call for different ways of conceptualizing species. This is, moreover, not merely a matter of arbitrary convention or taste: the different approaches reveal different real patterns in the things studied, even if which patterns are regarded as salient depends on one's theoretical and practical interests. From the standpoint of cognitive pluralism, the fact that different pursuits—including various scientific pursuits—require us to adopt different models that cross-classify to bring the various salient patterns to light means that we have a need for a variety of models that are not fully aligned with one another, which produce different verdicts on the boundaries of species but are apt for different purposes. And this may well be a principled and abiding disunity, as the different purposes and salient patterns may well not prove to be reducible to a single common denominator. Dupré's other point—that there is more than one factor relevant to the "units of evolution"—likewise gives us reason to suspect that we may *require* multiple conceptual frameworks to fully understand them separately, and it may not be possible to compress these into a single model that can serve as an optimal successor to the intuitive notion of species.

12.5 Irreducibility

I will try, much more briefly, to summarize some potential barriers to intermodel reduction. One of the most straightforward of these is simply that the would-be reducing model lacks the formal resources needed for a construction of the model one is seeking to reduce. The most important and familiar family of cases of this type consists of normative models such as those involving ethical principles. The familiar adage "you cannot derive *ought* from *is*" may originally have been posed as a claim about deductive inferences in arguments, but it is equally applicable to the reconstruction of one model from another. Models suited only to talking about facts and regularities lack a kind of resource that is distinctive of those models that can be used to frame normative claims. The latter are not what mathematicians would call a "conservative extension" of the former. To get from a purely positive to a normative model, you need to *add something new*. We find a similar situation with the relationship between modal and nonmodal logics: whereas it is possible to, say, treat the various logical connectives in first-order logic as constructions from a single (albeit less intuitive) connective, it is not possible to construct the modal operators for necessity and possibility from nonmodal resources. The reasons for this sort of irreducibility are straightforward, but the implications are significant: whatever the state of our model(s) of purely positive claims, such as empirical facts and laws, they cannot provide a sufficient basis for reconstructing the content of our normative and modal models.

Two of the more familiar philosophical objections to reductionism are multiple realization and emergence. We speak of multiple realization principally when some functionally specified kind—like AND-gate or heart—is always (or at least often) built out of some physical components, but there are many physical configurations that could be used to make something with the same functional description. It may well be that, in each separate case, the functionally specified system has the functional properties it does as a result of its physical configuration. But this falls short of a reduction for two reasons. First, the models used for the realizing system (say, physics or cellular biology) do not contain the relevant functional concepts. *Given* a model containing those concepts *and* a sufficiently rich physical or biological description of an individual system, one might be able to deduce *that* it was an AND-gate or a heart, but only by way of comparing it with the standards supplied by the functional model. Second, the physical or biological model does not capture the relevant generalizations of the functional model: the functional laws do not "pop out" *as* laws, and thus, without the functional model, real and salient patterns would go unrecognized.

Dennett's case for the independence and need for the intentional stance seems to involve considerations of both of the aforementioned types. On the one hand, in his view, intentionality is a notion tied to rational *norms*, and those norms cannot be constructed from physical-stance or design-stance characterizations. On the other hand, when we adopt the intentional stance, we are able to discern real patterns that would go unnoticed without it. Those who take a functionalist view of intentionality agree with something like the second point—that the generalizations of intentional psychology do not "pop out" of physical or neuroscientific descriptions. I am inclined to agree with Dennett, however, that our understanding of intentional states is not purely functional, like our understanding of AND gates or even hearts. Rather, we have special ways of conceiving of agents and agency that have their own proprietary concepts, stemming originally from the core agency system, that cannot be reconstructed in physical *or* functional terms.

Arguments for the irreducibility of consciousness proceed along similar lines. Levine (1983), Jackson (1982), and Nagel (1974) argue that the frameworks provided by the physical sciences do not provide even candidate explainers from which we could derive the characteristic features of the first-person phenomenology of experience. Chalmers (1996) makes things more explicit: our understanding of the physical world is cast entirely in terms of structural and functional properties, but neither structural nor functional properties supply the right sorts of resources for deriving the qualitative character of experience.

These considerations might suggest that we encounter problems of irreducibility specifically in the case of human phenomena: intentionality, consciousness, normativity. And perhaps we find special kinds of problems for reduction here that we do not find elsewhere. However, as I argued in *Beyond Reduction* (2007), we find very few true reductions in the natural sciences, either. And at least in some cases, when we look at particular types of models and their would-be reducers, we can see principled reasons why this might be so.

The word 'emergence' is used to describe a number of supposed relationships in nature. Here I am concerned with a type of "emergence" that would consist in phenomena that occur only in systems of a certain size or certain degree of complexity, whose occurrence and principles cannot be derived from models of the behavior of their constituent parts. The level of complexity need not be a very *high* level of complexity or the size very great. Indeed, a useful example is found in the fact that atomic-level properties (e.g., the behavior of electrons in orbitals) and molecular-level properties (e.g., involving the sharing of electrons) cannot be fully derived from particle-level properties (e.g., of how free electrons behave). One way of looking at this, sometimes favored by emergentists, is to say that there are "new" properties that "emerge" only in complex systems. (And so we need separate models of all the relevant sorts of complex systems.) This would pose a problem analogous to the inability to derive the regularities of functional systems, though it is not a case of multiple realization. It seems to me that one could also view the problem from a different angle: that individual electrons *do* have (dispositional) properties to behave in particular ways in orbitals, to play roles in molecular bonding, and so on, but these properties will be screened out in experiments that are performed on free electrons. From a cognitive standpoint, to obtain a good model of free electron behavior, we need to bracket other (dispositional) properties of electrons, and hence the model of free electron behavior is idealized in a way that prevents derivation of the emergent phenomena. But there is no reason we could not read the dispositional properties they manifest in particular complex systems back onto the individual electrons. Yet note that in so doing, we are not reducing our model of atomic-level phenomena to models of how the constituent particles behave "on their own," but rather building dispositions for atomic-level behavior into our models of particles in the form of dispositions, capacities, and powers *that need to be specified in higher-level terms*.

This, in turn, leads to a broader issue. Some models are "outward looking" in that they assume something about a broader and more encompassing context. Take, for example, a model of an organism that treats it as embedded in an environment. Such a model may be designed primarily to bring out features of the organism but treats it as an open system. Perhaps some of the salient extrinsic variables will be specified but an open-ended number left unspecified. Moreover, some of the features of the organism that are thus highlighted will be inseparable from the broader context: respiration, nutrition, procreation, and being predator or prey cannot be understood as features of the organism in isolation. Indeed, if you were to try to reconstruct these notions in terms not centered on the organism as the canonical level of description, the natural way to do so would be by *moving outward*, to a description of a larger and more encompassing system, defined at a higher level of organization, rather than by microexplanation. Reductive microexplanation is thus a strategy ill suited to capturing the generalizations or explanatory power of such models, because what they depend on is a larger and more complex, rather than a smaller, system. Likewise, the biological species concept and cladistic concepts of species are defined in terms of actual and possible interbreeding relations in populations and of historical lineages, respectively. These concepts cannot be reconstructed from theories cast at the level of synchronic relations among the proper parts of animals.

Similar things could perhaps be said about most models of psychological and social phenomena. We can explain a great deal *about* perception and motor control through neuroscience, but perception and action are fundamentally relationships between an organism and things in its environment. Familial, social, and political classifications, though they may be strong influences on individual behavior, are fundamentally relations to larger social structures, as well as being tied to individual, cultural, and biological histories. You cannot reduce these features to synchronic relational properties of the proper parts of individuals, because the representational resources of an at-a-time model of the parts simply lack the requisite conceptual machinery.

12.6 Comprehensiveness and Consistency

Let us return in a more general way to the problem of unities of belief and understanding. Two things we might want in a system of understanding and the beliefs that can be framed in it are *comprehensiveness* and *consistency*. On the one hand, we want to be able to understand and have knowledge about as many things as possible. Perhaps there are some things that our minds are so utterly unsuited to understanding that they fall outside the scope of possible knowledge or even belief. It is impossible even to know what these might be, as we would have to have some level of understanding of them even to specify them that far. We can perhaps conceive of such limitations only by analogy: that just as we understand things that dogs cannot grasp, there might be some other beings that understand things we cannot grasp. There is little point in discussing them further, because they really cannot be discussed in any concrete way. So when I speak of the desire to have a system of understanding and knowledge be *comprehensive*, I mean this in the more limited sense of encompassing everything we *could in principle* understand, or at least all the things we *do presently* understand through the models we possess.

The cognitive pluralist thesis is that the basic strategy that a model-based cognitive architecture employs to gain more comprehensive understanding is through a proliferation of models that confer understanding of different things. A basic stock of such models is part of the developmentally normal cognitive tool kit, but a modeling engine supplies the resources for greatly expanding the number of models we possess, and hence the number of things we can understand and have beliefs about. Acquiring a new model is more powerful than acquiring individual beliefs because models, while confined to particular content domains, are nonetheless systematic. Models provide ways of representing and reasoning about any number of states of affairs within their domains. The range of models available to us is further expanded through their complementary relationship with language and language-like thinking, by social transmission, and by more regimented processes of inquiry such as those found in philosophical analysis and the sciences. These allow us not only to create new models, but also to refine existing models, to consider and use multiple models, and to test one model against another.

On the other hand, the very features that make this strategy so powerful—that models are used to frame and produce judgments, that model-based inference methods are well suited to their domains, and above all that models are idealized and employ particular types of representations and inference rules—can also result in disunities in understanding and belief. Most obviously, just having multiple models is itself a modest form of disunity. More troubling is that models with different representational systems and inference rules can produce inconsistent beliefs and judgments. And this offends a second epistemic sensibility: a desire that the things we understand and believe be *consistent* with one another.

A model-based architecture for understanding is characterized by a built-in tension between comprehensiveness and consistency. We acquire more comprehensive understanding first and foremost through having more models, but the addition of each new model carries the risk of generating more inconsistencies with the commitments of other models. Sometimes, of course, apparent inconsistencies turn out to be illusory: two models may have incompatible commitments, but these imply inconsistent judgments about the world only in cases where both models are aptly applied to the same situations. In cases like quantum complementarity, it turns out that the conflicting models are aptly applied to disjoint sets of cases and as a result do not commit us to inconsistent judgments about particular cases. And sometimes we are able to resolve the problem by replacing one or both models with one or more successors that do not jointly produce inconsistent judgments. Our drive to resolve conflicts and inconsistencies provides a kind of regulative norm for both philosophy and science. But there is no guarantee that it can be carried out to its ideal conclusion, whether in the form of a single comprehensive and self-consistent model or at least a set of distinct models that play well together.

What should we do when faced with a trade-off between comprehensiveness and consistency, beyond hoping for a better day? One way of eliminating inconsistency, of course, would be to simply *stop using* one of the models that plays a role in producing it. But in practice, both in science and in ordinary life, we in fact seem to care more about understanding as many things as possible than we do about having a single model or even eliminating paradoxes. Thus, to quote Hawking and Mlodinow, we countenance the possibility that "it might be that to describe the universe, we *have to* employ different theories in different situations" (Hawking and Mlodinow 2010, 117; italics mine). Outside of science, we are probably even more comfortable having different ways of understanding situations that work for us, are less likely to be aware of the discordances between them, and are troubled only when the discord is brought to our attention—and perhaps even then only to the extent that we are philosophically inclined. When push comes to shove, comprehensiveness generally trumps consistency: we care more about having good-enough ways of understanding and dealing with a variety of situations than about how well all that understanding and all those beliefs hang together as a whole. We *need* understanding, and hence the models that supply it. Consistency is something we *want*—and some care about it more than others—but it is not something we *need*.

But in what sense do we *need* understanding? In one sense, this might simply mean we have a strong internal *drive* for it. And it is easy to see why such a drive would be biologically adaptive: the more about our environment we understand, the more we can deal with in adaptive ways. What Hawking and Mlodinow seem to have in mind, however, is something slightly different: we "need" multiple scientific models if we are to explain all the data. Most human beings have gotten on quite well without any understanding of the phenomena that make up the subject matter of contemporary physics—we do not need it in a biological sense—but one of the effects of human cognitive evolution is that our potential range of interests has expanded well beyond "interests" in the biological sense. But the "need" here is still a kind of *pragmatic* need: we need a model M if we are to understand (and be able to intervene on) X.

Of course, the range of things we have interests in understanding and acting on goes well beyond science. And in many cases, the goal of a theoretical understanding that covers all the data, which looms large for a theoretical physicist like Hawking, is either far less important or lacking altogether. Philosophers working in metaethics care about the merits of various metaethical theories—consequentialist, deontological, virtue theoretic—and as a consequence they are interested at a theoretical level in the fact that these theories can produce contradictory evaluations of which actions are morally praiseworthy, and interested in questions about whether one can be reduced to another or whether one can decisively be shown superior to the others. But the metaethical theories are themselves likely grounded in ways of evaluating persons and actions that are part of the developmentally normal cognitive tool kit. We quite naturally evaluate actions in terms of their costs and benefits, regard people as having duties that they fulfill or shirk, and attribute character traits that we admire or despise. Sometimes we are perplexed when we find that different means of evaluation lead us to opposing conclusions about which people or actions are good. And a given individual might habitually use one mode of evaluation in preference to the others. But even when faced with conflicting intuitions, we probably seldom even consider the possibility of *giving up* a mode of moral evaluation. Each of them provides insights that the others cannot, and though these may be quite unlike the explanation of scientific data, a similar principle is at work: that, when push comes to shove, we are willing to pay the price of perplexity or paradox rather than lose a useful and powerful tool for insight.

This, moreover, may point to a different kind of "necessity": there may be some kinds of models that we are in some sense psychologically incapable of giving up. Modes of ethical evaluation and perhaps an assumption of freedom that is implicit in some of them might fall into this class, as might assumptions about a world of classical objects in space and time, and models of causation and teleology. It is not that people—or at least some people—cannot conceive of alternatives and even endorse them at a theoretical level. Some people can and do regard classical objects, causation, teleology, free will, and ethical properties as illusions or even as nonsense. But this does not mean that these theoretical commitments prevent them from forming judgments, or at least candidate judgments, on the basis of deeply entrenched models that assume them. At best, it may allow them to reflectively disavow the truth of the judgments or the aptness of the models. Note, though, that this too reflects a kind of disunity if, in making or being disposed toward such judgments, one treats them as true and the models as apt but reflectively disavows them.

And while the situation is most striking when it involves models that operate automatically and perhaps even ubiquitously, it also appears in a lesser form whenever we employ models about which we have reservations or we know to have limitations. Being an academician, for example, has provided me with no small number of occasions to become aware of the limitations of my understanding of particular topics, and the even greater limitations of the simpler models of those topics that I try to convey in my courses. In my course on early modern philosophy, for example, I present stories about figures like Descartes, Hume, and Kant. I try to avoid saying anything truly *false* but know full well that I leave out things that are ultimately important: for example, saying that Descartes equated matter with extension and hence matter has its properties necessarily leaves out other features of the Cartesian system such as the role of divine choice in deciding how matter would be divided into particles and even in deciding what the truths of logic and mathematics should be. I also know just enough about the scholarship on such issues to recognize that I do not fully understand how they fit together in Descartes's thought. (And, as my fellow philosophers have no doubt experienced for themselves, saying anything, particularly anything critical, about Kant in a professional context risks having any Kant scholars in the audience explaining many things I do not understand about Kant's philosophy.) I use mental models of the thought and works of other philosophers in my thinking, teaching, and writing, but at the same time I am aware that these models are not fully adequate. And, of course, this goes for any number of nonphilosophical topics as well.

This is, perhaps, a kind of expression of the Socratic virtue of knowing what you do not know (or understand). It strikes me that anyone who has developed this intellectual virtue is always in something like the position of the advocate of a mild form of negative theology. The *strong* form of negative theology says that our concepts are so unsuited to talking about God that we can say nothing true about God and perhaps ought not to try. The milder form says that all our theological claims must be hedged: "*This*, but not *quite* this, and much more as well." I think that a proper appreciation of the idealized and perspectival character of understanding based in mental models should lead us to adopt a similar attitude much more broadly. We make claims using the models we have—what alternative do we have beyond a principled silence? But we also know that the resources of those models are limited, some in ways we understand, and no doubt others in ways that we do not. To the extent that the model is apt, we count as true the best ways we have of expressing what can be said using its resources. But we should be reluctant to say that it is *the* truth, the *whole* truth, or *nothing but* the truth.

13 Models and Intuition

In ordinary speech, calling something an "intuition" is tantamount to calling it a hunch: an idea that pops into someone's head as though from nowhere and seems compellingly plausible even though they have no reasons to support it. The use of the label often carries pejorative implications: a judgment that we can give reasons for is regarded as better than a hunch, precisely because we can give reasons for it. And there are good grounds for such a suspicion: thoughts that occur to us spontaneously and without presenting their epistemic credentials are often the products of bias, prejudice, self-interest, fear, or wish fulfillment. At the same time, common sense also recognizes that some people have "good intuitions"—indeed, some people's intuitions are regarded as *brilliant*, latching onto insights that escape the rest of us and even the person who thought of them cannot explain.

In philosophy, "intuition" has traditionally been contrasted with reasoning. John Locke distinguished "intuitive knowledge" from "demonstrative knowledge," describing things that are immediately self-evident (e.g., the elementary propositions of Euclidean geometry) as "intuitive," and conclusions we reach only through explicit reasoning methods (e.g., working through a geometric proof) as "demonstrative." In more contemporary terminology, Lockean "intuitive knowledge" consists in judgments that are true, justified, and produced in a fashion that does not depend on a prior train of reasoning. Locke, however, did not regard intuitive knowledge as inferior to demonstrative knowledge. In fact, he regarded intuitive knowledge as the most certain form of knowledge: it is self-evident, and the other kind of knowledge (demonstrative knowledge) needs to be based on premises that are known intuitively.

The role of intuition in mathematical knowledge became problematic in the nineteenth century. Up until that time, it had been assumed that all of Euclid's axioms and postulates captured truths that were truly *necessary* and that denying any of them would result in self-contradiction. One of these intuitive claims was the parallel postulate: that for any line *L* and point *P* not on *L*, there is exactly one line passing through *P* that is parallel to *L*. The founders of non-Euclidean geometries, however, recognized that the parallel postulate could be denied without self-contradiction, and various alternatives to it would result in distinct alternative geometries. Euclid's parallel postulate may *seem* as though it must be true, but the reason for that is to be found not in the definitions and axioms of the system or the principle of noncontradiction but in a separate set of "spatial intuitions" to which the human mind is strongly biased. This discovery caused a crisis in mathematics, spawning the formalist movement, which sought to find ways of reconstructing mathematical systems in terms not dependent on mathematical intuitions.

In contemporary philosophy, appeals to "intuitions" are often used to justify important claims. In metaphysics, claims about what is necessary and what is possible—things that go beyond what could be settled by empirical experimentation—are often explicitly grounded in "modal intuitions": for example, that there could be a being that was a molecule-for-molecule duplicate of you but lacked conscious experience (Chalmers 1996). In ethics, not only claims that some particular action is good or evil, but also fundamental claims such as that human beings have natural rights, are justified by appeals to intuitions. However, unlike mathematical intuitions, which tend to be about things that everyone, or at least the experts, agree on, appeals to intuitions in contemporary metaphysics and ethics are often precisely about matters that people have *different* intuitions about. For some, what seems intuitively obvious is that a molecular duplicate of a person would have to have all the same mental states, and Jeremy Bentham famously quipped that the notion of natural and inalienable rights is "nonsense on stilts."

Intuitions, in the sense of judgments that are not reached on the basis of explicit reasoning from something else yet strike us as clearly true (and hence not requiring further justification), are by no means the sole province of mathematicians and philosophers. We think intuitively all the time in daily life. Indeed, if you were to monitor the judgments you make on an average day, you would probably find that most of them were not reached through any kind of argumentation or reasoning (or at least not any you could detect). Much of the time, intuitive judgments are nonetheless quite *dependable*. In fact, when you think about it, any reasoning you *do* engage in has to start with some set of assumptions that serve as the premises from which you derive the conclusion. The conclusion is derived through reasoning *from the premises*, but on what basis did you believe the premises?

To get an argument off the ground at all, you have to start by believing *something*. Moreover, the justification for the conclusion of an argument is no better than that of the premises you started with. So there seems to be something right in Locke's notion that, for deductive reasoning to accomplish anything, it has to be able to start out with judgments that are arrived at in some other, nondeductive fashion, which are justified (or at least *seem* justified) for reasons other than that we have reasoned them out on the basis of some prior argument.

But *how* are judgments produced when they are not products of explicit reasoning processes such as deductive arguments? And why do *some* of them have the ring of truth to them that we describe as seeming "intuitively true" or "intuitively obvious"? Not *every* thought that spontaneously occurs to us seems intuitively true, after all. Sometimes an idea pops into our minds, and our immediate reaction is "No, that doesn't seem right"—it seems *intuitively incorrect* or *counterintuitive*—and sometimes we have no intuitive sense about whether something is correct or not. So really there are two phenomena we must account for: (1) the types of psychological processes that produce judgments without employing explicit reasoning, and (2) the fact that some judgments thus produced seem self-evidently true (or at least plausible), others self-evidently false (or at least dubious), and still others lack any attendant sense of self-evident truth or falsehood.

The word 'intuitive' is used to mark both a type of process and a type of apparent self-evidence. We call a judgment-producing *process* "intuitive" when it does not involve explicit, conscious, stepwise reasoning. When we call a *judgment* "intuitive," we might mean either of two things: (1) it was *intuitively produced*, that is, produced by an intuitive process rather than by explicit reasoning; and (2) it is *intuitively recommended*, that is, it seems self-evidently correct. These two features of judgments need not go together. Not all judgments that are not produced by explicit reasoning come intuitively recommended; and explicit reasoning can produce conclusions that, once considered, are intuitively recommended. But they go together curiously often; and as our concern here is with the intuitive production of judgments, I will use the expression "intuitive judgment" for judgments that are both intuitively produced and intuitively recommended.

13.1 Discussions of "Intuition" in Psychology

A notion of "intuitive processes" has been explored in recent cognitive psychology as one of the two types of processes distinguished by "dual process" or "dual system theories." One of the main figures working in this area, Daniel Kahneman, writing with Shane Frederick, summarizes:

> The ancient idea that cognitive processes can be partitioned into two main families—traditionally called intuition and reason—is now widely embraced under the general label of dual-process theories. ... Dual-process models come in many flavors, but all distinguish cognitive operations that are quick and associative from others that are slow and rule-governed. ... We adopt the generic labels System 1 and System 2 from Stanovich and West. These terms may suggest the image of autonomous homunculi, but such a meaning is not intended. We use systems as a label for collections of processes that are distinguished by their speed, controllability, and the contents on which they operate. (Kahneman and Frederick 2002, 51)

The quote shows that the notion of "dual systems" really indicates two *classes* of processes, differentiated in terms of their speed and controllability. There might be numerous more specific processes sharing the System 1 profile, which presumably might operate through different mechanisms. In spite of our tendency to disparage intuition by contrast to deliberate reasoning, System 1 reasoning can be very good at what it does, and is recognized as an important feature of expertise, the acquisition of which involves the transition from System 2 to System 1 processes. Kahneman and Frederick continue:

> Although System 1 is more primitive than System 2, it is not necessarily less capable. On the contrary, complex cognitive operations eventually migrate from System 2 to System 1 as proficiency and skill are acquired. A striking demonstration of the intelligence of System 1 is the ability of chess masters to perceive the strength or weakness of chess positions instantly. For those experts, pattern matching has replaced effortful serial processing. Similarly, prolonged cultural exposure eventually produces a facility for social judgments—for example, an ability to recognize quickly that "a man whose dull writing is occasionally enlivened by corny puns" is more similar to a stereotypical computer programmer than to a stereotypical accountant. In the particular dual-process model we assume, System 1 quickly proposes intuitive answers to judgment problems as they arise, and System 2 monitors the quality of these proposals, which it may endorse, correct, or override. The judgments that are eventually expressed are called intuitive if they retain the hypothesized initial proposal without much modification. The roles of the two systems in determining stated judgments depend on features of the task and of the individual, including the time available for deliberation ..., the respondent's mood ..., intelligence ..., and exposure to statistical thinking. ... We assume that System 1 and System 2 can be active concurrently, that automatic and controlled cognitive operations compete for the control of overt responses, and that deliberate judgments are likely to remain anchored on initial impressions. (Kahneman and Frederick 2002, 51)

In the latter part of this quote, Kahneman and Frederick seem to assign to System 2 processes only a kind of monitoring function. I think this would be an overstatement, as we are also capable of using regimented reasoning processes like application of logical syllogisms or rules of column addition in a deliberately controlled way to produce judgments, and explicit reasoning allows us to derive conclusions we are unable to reach through System 1 thinking. But my main reason for introducing the foregoing quotes is to provide perspective on how some psychologists think about intuitive reasoning: it is a kind of thinking that occurs rapidly, automatically, and through processes that are often skilled yet "opaque"—that is, cognitively impenetrable.

Other psychologists speak of *concepts* and *judgments* as being intuitive or counterintuitive. In cognitive science of religion, Paul Bloom (2004, 2007) claims that both children and adults are "intuitive dualists." Pascal Boyer and Justin Barrett, by contrast, claim that supernatural concepts like those of immaterial souls and ghosts are "minimally counterintuitive" (Boyer 2001; Barrett 2004a, 2004b, 2009). Calling a judgment or a concept "intuitive" here seems to mean that it is "intuitively recommended"—that is, seemingly true, plausible, or, in the case of concepts, unproblematic. 'Counterintuitive' seems to mean the opposite of this: a counterintuitive judgment or concept is one that comes with an internal sense that there must be something peculiar about it.

These references to contemporary psychology, however, do not tell us much that we did not know before. Beyond the important claim that *expertise* is characteristically exercised through intuitive processes, they largely just confirm the distinctions we have already developed. The claim that intuitive processes are "automatic" amounts to little more than the claim that they are not the kind of processes found in explicit reasoning, in which we can control and monitor the steps. The fact that they are fast—that is, fast *by contrast with methodical explicit reasoning*—may be no more than a consequence of the fact that the latter sort of process is slow and laborious. Moreover, intuitive processes do not always *seem* to be fast: the development of a creative idea or an intellectual insight might require a great deal of time. When I reach an impasse in writing, I often do something else—take a walk or sleep on it—allowing unconscious parts of my mind to work on the problem. Sometimes it is a matter of hours, days, or months before an answer pops into my awareness. Of course, precisely because the process is unconscious, it is difficult to determine just when it is at work or how long it takes. It may be that the actual process takes milliseconds and happens just before I am aware of the outcome. Or it may be that some unconscious mechanism is slowly churning away at it all along. Moreover, since the notion "intuitive process" (or "System 1 process") is really a stand-in for a class of some unspecified number of actual cognitive mechanisms, they may operate in different ways and on different timescales.

What we would really like to know is this: what kind of cognitive process could produce judgments, including highly sophisticated expert judgments, quickly and "automatically," without a need for explicit reasoning, and why might the judgments thus produced come either intuitively recommended or flagged as problematic? My suggestion is that *one* way this can happen is when the judgment is produced as a consequence of features built into a mental model.

13.2 Intuitive and Counterintuitive Judgments

There may be more than one way a judgment might come to seem intuitive or counterintuitive. I will be offering hypotheses about one way this might happen, but I will speak of it, to avoid continual hedging, as *the* way it happens. The hypotheses, in their general forms, are the following:

1. A judgment is intuitive with respect to a model M if its truth can be "read off" the rules of M.
2. A judgment is counterintuitive with respect to a model M if it violates the rules of M in a fashion that is detectable—for example, because the rule violation triggers a cognitive dissonance.

There are several ways a model-based inference process can lead to a judgment being, and being recognized as being, intuitive or counterintuitive with respect to a model, and I will explore several of these by way of examples.

Consider a judgment such as "agents have goals" or "people act on the basis of their beliefs and desires." It is probably rare that anyone who is not involved in philosophy or the cognitive sciences would ever explicitly entertain judgments with these abstract contents at all. Even when philosophers or scientists do so, it is often in the context of debating specific theories (such as eliminative materialism) that *deny* these theses. To common sense, they are more like deep underlying assumptions that are so fundamental that it seems odd to even mention, much less to question, them. Why is this so? If the hypotheses that we possess core systems and folk theories are correct, the answer would seem to be something like this: From a very early age, we have models of agents and agency in the form of the core agency system and folk psychology, and the principles that agents act in a goal-directed fashion and that people act on the basis of their beliefs and desires are fundamental principles of these models. We use such models automatically and routinely; and to use the models at all is to constitute things as agents with goals or with beliefs and desires. You cannot constitute something as an *action* without imputing goals, beliefs, or desires. You can reject the aptness of the model for particular cases and assent to the use of another interpretive framework instead—it was a patellar reflex and not a deliberate kick, it was an animatronic robot following a program—but in so doing, you reject not only the beliefs and desires but the applicability of the whole framework of imputing *agency* at all. A candidate judgment that violates these rules triggers a particular type of cognitive dissonance with the model, and this is experienced as its being counterintuitive. A judgment that expresses the content of the rules is quickly and unconsciously confirmed as being an implication of the model, is experienced as being intuitively recommended, and can be produced by some mechanism that translates the implicit rules of the model into propositional form or confirmed by a mechanism that compares the propositional expression with the rules of the model. Judgments like "agents have goals" seem like *necessary truths*, perhaps even platitudes, because to think or talk about agency at all, one must be using a model that imputes goals, beliefs, or desires. Indeed, there are clear senses in which they *are* necessary truths. "Agents have goals" is something like an implicit axiom of models of agency, and hence is true-in-agency-models in much the way the parallel postulate is true-in-Euclidean-geometry. To use agency models at all, one must treat things constituted as agents as things that act to achieve their goals, and in order for the constitution of something as an agent to be apt, the imputation of goals to it must be apt as well.

Similar things can be said about the intuitive representation of objects in space. To represent something as being in space, one must represent it in a fashion that assumes that there is space on all sides of it. (Of course, "space" is said in a proprietary sense here, corresponding to a *geometrically* defined space, not as "*empty* space." If someone says, "There is no space to the left of the cabinet," because the left side of the cabinet is flush against a wall, they mean there is no *empty* space, not that the wall exists outside space.) This seems to be a basic feature of the model (or a set of models) we use to represent objects in space. As a consequence, the question "Is there (geometric) space to the left of it?" seems absurd to ask, as the answer is a straightforward consequence of basic principles of the model and hence, from the standpoint of someone working within the model, should go without saying. The assertion "There is nothing to the left of it, not even empty space," is highly counterintuitive, because you have to use a spatial model to make sense of it at all, yet what is asserted violates the implicit rules of the model.

In some cases, we may be unable to form a representation of the candidate content within the model at all. Our spatial intuitions may be unable to accommodate a representation of a three-dimensional space that curves back on itself like the two-dimensional surface of a sphere. We can conceive of such things through other means—say, mathematically—but they seem abidingly counterintuitive because they are dissonant with a deeply ingrained model of spatial geometry that is more or less Euclidean. Likewise there is a way in which ways of thinking about things for which I do not already have a model are bound to seem counterintuitive—that is, they seem counterintuitive at least until I acquire and master the model. The pronouncements of contemporary science often seem baffling, nonsensical, or even plainly impossible at first. They lose something of this air once we master the scientific model in which they are couched, and may even seem routine and obvious to the practicing scientist when she is using the requisite models.

The examples of agential and spatial representation both involve models that are deeply canalized and difficult to disengage. But similar points can be made about the intuitiveness of judgments that state basic principles of models whose use is more voluntary. Within the game of chess, it goes without saying that each piece in play must occupy a single square on the board. Indeed, within the model of the game, there is no such thing as a location halfway between spaces. Of course, we are perfectly capable, using *other* means, of seeing pieces as straddling two spaces. We can refuse to interpret them in terms of the chess model at all, or we can wonder which space an awkwardly positioned piece is "in" for purposes of the game, or we can interpret it as being in a particular space for purposes of the game. To not interpret it as being in a particular space is to not interpret it as being in play in a game of chess at all.

This strikes me as slightly different from the manner in which I intuitively see that the bishop can move forward one or two spaces on the right forward diagonal or can move a third space, capturing the opponent's pawn. If I am even a mildly proficient chess player, I simply *see* these possibilities and do not need to do any explicit reasoning to discover them. The question "How did you know that?" would be puzzling, unless perhaps asked by someone who does not know the movement rules for chess. The movement possibilities pop out for me because they are consequences of the implicit rules (in this case, permissive rules for movement) in my mental model of chess, and these permissive rules guide me in recognizing possible moves within the game. Conversely, if someone moves the bishop along a row or column, or moves it off the board entirely, this will immediately cause the type of cognitive dissonance associated with the counterintuitive. Presumably some unconscious mechanism linked to the model detects a violation of the model's rules, and the rule violation is flagged with this particular phenomenology. It is then impossible to keep operating smoothly within the model; and to make sense of what my opponent has done at all, I must reframe it outside the chess model: he is trying to annoy me, perhaps, or is a child who has lost patience with the game and is just horsing around.

Another type of intuitive process might involve a *simulation*—say, a simulation of what would happen in the game if I tried a particular strategy. This uses the internalized rules of the model to produce a series of representations of board states, or perhaps a representation *of* how a strategy is likely to unfold. Here the internalized rules guide the dynamics of the simulation process. Similarly, when I imagine the rotation of a solid, the rules of the model I employ for spatial reasoning generate a simulation, producing an intuition about the end result. I may well regard the result with confidence, as it is produced by the rules of the model.

Another type of intuitive process is what is often called "semantic inference." Many things seem intuitively obvious about dogs, even if we have not thought about them before: dogs are animals, carnivores, quadrupeds, and mortal; dogs have hair, hearts, kidneys, and emotions; dogs are offspring of other (pairs of) dogs; and so on. We would correspondingly find other claims counterintuitive: dogs are vegetables, vegetarians, bipeds, immortal, or robots; dogs have scales, gills, or opinions on philosophical issues; dogs are created on engineers' workbenches, are made of fire, or have parents that are fish; some dogs have masses greater than that of the earth; and so on. Some of the intuitively obvious claims might have been learned and stored as facts about dogs, perhaps even appearing in part of a definition or characterization of dogs you learned; but many of them are things you are unlikely ever to have thought of before, and yet they are intuitively obvious. Why might this be so? An answer that recommends itself is that the concept DOG either is itself, or is part of, a kind of richly structured semantic entity that affords a rich set of inferences that go beyond the judgments that have previously been formed.

Pascal Boyer (2001) suggests that species concepts like DOG are formed on the basis of a highest-order concept that is a constitutive part of folk biology, which he characterizes as the "ontological category" ANIMAL. Boyer likens concepts to note cards in an internal database and sees ontological categories as templates for such note cards that specify both category-general assumptions (e.g., animals are mortal and the offspring of same-species parents) and blank "slots" for other information to be filled in on a species-level basis, such as characteristic physiognomy, diet, habitat, and so on. When confronted with a new type of animal, we form a new database entry on the basis of the ANIMAL category. This leads us to assume that the species will have *some* characteristic physiognomy, diet, and so on, which guides us in a search for what information needs to be acquired to fill in the blanks. As we fill them in, we have an increasingly richly specified set of assumptions about that species, drawn from the "entry" for the species, the ontological category, and any levels (e.g., MAMMAL) lying in between.

The information thus "encoded" need not ever have been a part of an explicit judgment. At some point, I pieced together a set of assumptions about the internal organs—hearts, kidneys, and so on—I expect to typically find in some superordinate class of animals, and these are either "written into" my model of dogs or else linked with it in a fashion I can readily exploit. Others may be built into my mental model of dogs through experience: I observe what dogs eat, and encode the assumption that they eat meat into the model even if I never think a thought with the content "dogs are carnivores." Unless prompted, I may *never* form a judgment on questions such as whether dogs have kidneys, are carnivores, are mortal, or are the offspring of other dogs. But once I am prompted, the answer can be "read off" the mental model. The judgment is intuitively produced by being (somehow) read off what is already present in the model in a non-propositional form, and is intuitively recommended *because* it is a product of what is already present in the model.

Boyer develops his theory of concepts in *Religion Explained* to suggest that concepts of supernatural beings that figure in religious belief are produced by starting with a familiar ontological category (or one of its subordinate concepts) and then violating one of the rules—say, a concept of an immortal dog or a dog that can transform into a human being. These concepts are "counterintuitive" because they violate rules of the ontological category—say, that animals are mortal, are each members of a single species kind, or have a (constant) physiognomy. Boyer suggests that *supernatural* concepts are special because they (a) violate rules at the *category* (rather than the species) level and (b) violate *just one* rule. But ideas that violate rules at the species level (e.g., a six-legged dog or a dog without a heart) or more than one rule (a six-legged dog without a heart) would also be counterintuitive for the same general reason.

Saying that intuitive judgments (or a significant class of them) are ones that can be "read off" and "pop out" from the rules of mental models would seem to require three things:

1. The content of the judgments is implied by the rules of the model, either in themselves or as applied to particular cases.
2. There is some mechanism connected with the model that is capable of producing, on the basis of the model, judgments that track the implication.
3. There is some quick and unconscious process that compares judgments with the principles of the model and confers on the judgment the kind of confidence and sense of self-evidence that is the subjective mark of intuitively recommended judgments or the cognitive dissonance associated with counterintuitive judgments.

There are ways these three things can come apart. It is not clear that the *feeling* of self-evident correctness that comes with judgments arising from models one has mastered is different from the feeling that can come from wish fulfillment or bias, and even experts can make mistakes about which they feel a great deal of subjective certainty. Conversely, model-bound mechanisms may produce candidate judgments without a feeling of subjective certainty. The rules of a model may imply things that there may be no process for detecting, or which can only be discovered through laborious explicit reasoning. (For example, all the theorems of Euclidean geometry are implied by the axioms, but it takes laborious work to prove them, and hence some might initially strike us as unintuitive or even counterintuitive.) And there may be spurious judgment-producing processes that fail to track the rules of a model dependably, whether because they regularly produce intuitions that are not underwritten by the rules of the model or because they sometimes produce errors of execution.

13.3 Model Relativity of Intuitiveness

It is important to note that the intuitiveness and counterintuitiveness of judgments are characterized *relative to particular models*. Indeed, a judgment might be intuitive with respect to one model and counterintuitive with respect to another. Consider again the parallel postulate of Euclidean geometry: that with a line L and a point P not on L, there is exactly one line through P parallel to L. If one is using a Euclidean model, this seems intuitively necessary. Indeed, as Euclidean principles seem to be deeply ingrained in our folk-geometric thinking, it seems intuitively true, and even necessary, to common sense. It was not until the nineteenth century that alternative geometries were discovered. But once one knows, say, of Lobachevsky's hyperbolic geometry, which permits more than one parallel line through a point, the parallel postulate is no longer intuitive, at least when we are using a hyperbolic geometry as our model. Or rather, whether it strikes us as intuitive may depend on *which model we are using*, or perhaps whether we are dealing with the problem from the further level of abstraction of thinking *about* alternative geometries. If one is primed on hyperbolic geometry—say, if we are in the midst of a class on it—and someone suddenly asserts that there can be only one line through *P* parallel to *L*, the parallel postulate is likely to strike us as *counter*intuitive, because it is contrary to the rules of the model we are using.

Consider another case: wave and particle theories of light and two-slit experiments. We are quite capable of imagining both sets of dispersion patterns. But if we are thinking of light as made up of particles, one pattern may seem intuitively obvious, because that is what is produced when we run a mental simulation of the propagation of particles through the slits. If we think of light as made up of waves and simulate the results, the other pattern may seem intuitively obvious. If we are assuming the aptness of one model naively—that is, unaware that there are other models—the intuitiveness seems to have the force of *necessity*. Indeed, *if* the particle model is apt for this problem, light *must* behave in this way. But with the right sort of critical distance, we can see that the kind of "necessity" that is really implied by intuitive judgments is not really objective necessity. First and foremost, it is the kind of necessity involved with a rule: *in this model*, this principle is *treated* as necessary. Second, there is an additional kind of necessity imputed to the extent that this model is apt for a particular sort of problem: if (or to the extent that) the model is apt, the world must also behave thus. In classical optics, this generated trenchant paradoxes, because in some situations light seems to behave like a particle (the particle model is apt), and in others it seems to behave like a wave (the wave model is apt). The paradox is partially (though only partially) resolved by remembering that any given model might be apt for some cases but not for others. A good part of scientific competence consists in knowing which models to use in which situations, but puzzle cases like this one also drive us to seek a model that can accommodate cases that thus far need to be approached through different models.

13.4 Models, Intuitions, and Expertise

This model-based account of intuitions can also explain why discussions of intuition often include expert judgment as intuitive. One way that expert performance differs from novice performance on a task is that novices consciously think through problems in an explicit, stepwise fashion, whereas experts often report that "the right answer" simply pops out for them quickly and effortlessly. Dreyfus and Dreyfus (1992), for example, contrast the novice and the expert in the following way:

> The student automobile driver learns to recognize such interpretation-free features as speed (indicated by his speedometer). Timing of gear shifts is specified in terms of speed. The novice chess player learns a numerical value for each type of piece regardless of its position, and the rule: "Always exchange if the total value of pieces captured exceeds the value of pieces lost." But such rigid rules often fail to work. A loaded car stalls on a hill; a beginner in chess falls for every sacrifice. (Dreyfus and Dreyfus 1992, 114)

> The expert driver, generally without any attention, not only knows by feel and familiarity when an action such as slowing down is required; he knows how to perform the action without calculating and comparing alternatives. He shifts gears when appropriate with no awareness of his acts. On the off ramp his foot simply lifts off the accelerator. What must be done, simply is done.
>
> The expert chess player, classed as an international master or grandmaster, in most situations experiences a compelling sense of the issue and the best move. Excellent chess players can play at the rate of 5–10 seconds a move and even faster without any serious degradation in performance. At this speed they must depend almost entirely on intuition and hardly at all on analysis and comparison of alternatives. We recently performed an experiment in which an international master, Julio Kaplan, was required rapidly to add numbers presented to him audibly at the rate of about one number per second, while at the same time playing five-second-a-move chess against a slightly weaker, but master level, player. Even with his analytical mind completely occupied by adding numbers, Kaplan more than held his own against the master in a series of games. Deprived of the time necessary to solve problems or construct plans, Kaplan still produced fluid and strategic play.
>
> It seems that beginners make judgments using strict rules and features, but that with talent and a great deal of involved experience the beginner develops into an expert who sees intuitively what to do without applying rules and making judgments at all. The intellectualist tradition has given an accurate description of the beginner and of the expert facing an unfamiliar situation, but normally an expert does not deliberate. He does not reason. He does not even act deliberately. He simply spontaneously does what has normally worked and, naturally, it normally works. (116–117)

The Dreyfuses characterize expert performance as "intuitive" and contrast it with novice thinking that is based in explicitly identified features, propositional beliefs, and rules. Of course, saying someone "simply acts" does not supply an account of *how* she came to act as she did, and questions about the underlying causes are surely warranted, especially in a case like an acquired expertise. (The Dreyfuses have periodically explored the suitability of parallel distributed processing models of cognition as an alternative to classical computational approaches to answering such questions.)

The basic explanation suggested by a model-based approach is that the pathway from novice to expert involves the formation of mental models that allow for quick and apt responses that do not require conscious propositional thinking. The learning pathway often involves starting with explicit representation of propositions and rules that approximate the features of a model—perhaps in some cases, such as scientific theories, there is even a propositional articulation *of* a preexisting model available as an aid to learning—but becoming an expert requires the formation of something different, a mental model of the system. The expert's understanding is *better* than the novice's, because the characterizations the novice uses are approximations of an expert's model, and because the novice may not be fully competent even at using the approximations. The expert's performance is effortless and does not involve conscious stepwise reasoning because her answer is produced by the fast and nonpropositional processes of model-based cognition. The expert's judgment feels certain because it is driven by the implicit rules of the model, and because she has a history of using the model effectively.

13.5 Models and Dispositional Beliefs

A model-based account may also help to shed light on the nature of beliefs that are not judgments. These are often characterized in terms of "dispositions" or "normative commitments," but neither of these characterizations is helpful as a step toward a psychological theory. The ways that we have been talking about the intuitive production of judgments can help us to see *why* we are "disposed" to produce or assent to certain types of judgments and how such dispositions can be normative.

Let us suppose that as of yesterday, I had never considered the proposition "dogs have kidneys," and that we should nevertheless say that I *believed* it, in the sense that I was already disposed to readily assent to it if properly prompted. *Why* was I thus disposed? The hypothesis explored in this chapter would suggest that (1) the information was already available, albeit in nonpropositional form, in the structure of a model; and (2) there are fast-operating unconscious processes available that can (a) produce a judgment with that content and (b) if presented with the candidate judgment, confirm it as implied by the implicit rules of the model. If beliefs are dispositions that are grounded in stable features of models plus mechanisms that produce or confirm judgments in a noninferential manner based on their content when prompted, we have a characterization of beliefs that grounds a form of psychological realism but is not committed to the thesis that dispositional beliefs are sentence-like entities in the mind.

Moreover, we can also explain the fact that imputations of belief are also imputations of normative commitments: beliefs of the type I have described are implied by the implicit rules of their models of origin, and hence one is normatively committed to them when one is using the model. Of course, different models have different commitments, and those of one model may conflict with those of another. This would seem to imply that we have conflicting dispositional beliefs in such cases. But this is actually potentially a point in favor of the account. Numerous psychological studies have shown that the judgments we are disposed to make are highly subject to context and priming. So it will not do to simply characterize beliefs as dispositions to judgments without further qualification. Moreover, contexts do not simply magically produce changes in disposition. We need some way of connecting different contexts with the different judgments they evoke. One plausible way of doing this is to suppose that context and priming activate particular models, and as a result, one is biased toward producing the judgments that are commitments of *those* models and not others.

Not *all* dispositions to form judgments work in this way. The timid person is disposed to judge himself in danger. In a sixty-degree room, I am in some sense disposed to judge that it is a bit chilly. These dispositions are probably not based in mental models. But by the same token, it is not clear that we should think of them as involving beliefs, either. The timid person has a low trigger point for judgments of danger, as well as for various types of vigilance, but it strikes me as more appropriate to say that he is always on the lookout for danger and ready to impute it than that he always *believes* himself in danger. Nonetheless it may be that cases exist in which we would attribute beliefs that cannot be explained on the basis of the implicit commitments of models, and so the hypothesis should be taken as an explanation for at least *one class* of belief attributions.

13.6 Models, Intuition, and Cognitive Illusion

If expert understanding is often associated with intuition, there is also another and almost opposite set of cases equally associated with it: various sorts of *errors* that nonetheless seem intuitively correct. Much of the two-systems literature in psychology is in fact concerned with how people—including experts—make mistakes in reasoning, even when upon reflection they would understand the kinds of methods they employ to be fallacious. Sometimes such errors may be simple mental slips. But there also seem to be ways that we are systematically drawn to think about matters in inapt ways. In the next chapter, I make a case that the mind is subject to a variety of types of *cognitive illusions*, in which judgments present themselves as intuitively recommended, sometimes even as necessary, yet either they are erroneous, or at least we are inclined to greatly overestimate the degree of confidence we can place in them. (In Kantian terms, they are illusions of *knowledge*, whether or not they are accidentally true.)

14 Cognitive Illusion

Everyone knows about optical illusions like the Müller-Lyer illusion, in which two lines that are really the same length appear to be of different lengths because one is terminated in inward-pointing arrows and the other in outward-pointing arrows. In a perceptual illusion, things seem one way but are not really the way they seem. But not just any old mistake counts as an illusion. At least as psychologists use the term, perceptual illusions are persistent (e.g., the illusion does not go away just because you know the lines are really equal) and are products of properties of our perceptual systems.

We may also speak of *cognitive* illusions, in which it regularly and persistently seems (intellectually rather than perceptually) as though something is or must be the case due to properties of our minds. In philosophy, the idea that the mind has operations that can, under certain conditions, produce cognitive illusions was most influentially explored by Kant. But Kant meant something different by 'illusion' from what we would usually mean in ordinary language. In the ordinary sense, an illusion involves a misrepresentation: for a belief to be illusory, it would have to be *false*. Kant's notion of illusion, however, does not require falsity. What is illusory is the appearance of being *knowledge*, regardless of whether the belief is true or false. Kant's own discussion of what he calls "transcendental" or "dialectical illusions" has additional special characteristics that I will examine later, but first it is useful to distinguish between two notions of a cognitive illusion. Kant's notion—an illusion that a belief is *knowledge*—we may call an *epistemic illusion*. From this we may distinguish something closer to ordinary usage, an illusion in the sense that what is represented is *false*, as an *alethetic illusion* (using the Greek word for truth, *alethea*). A Kantian epistemic illusion can also be an alethetic illusion. But Kant was principally concerned with epistemic illusions of a sort where we in fact cannot know whether the belief in question is true or false.

What features of the mind might produce cognitive illusions? Kant explored one way they might be produced, which I discuss and expand on later in the chapter, but we have no reason to think that there should be *only* one way. An account of understanding based in idealized models suggests several ways that judgments can be generated from models and ways that such judgments can go awry when they lie outside the model's idealization conditions. And the suggestion, developed in chapter 13, that at least some kinds of model-based judgments carry a sense of intuitive correctness, sometimes even of necessity, can help account for their *seeming* to be knowledge, even a priori knowledge of necessary truths. There may be more types of cognitive illusion than I am able to explore here, but I will describe three types:

1. Illusions resulting from the inapt application of a model to a particular situation
2. Illusions resulting from treating the principles internal to a model as unrestricted truths not bounded by the idealization conditions of the model
3. Illusions resulting from attempts to unify understanding rooted in various models into a single system by unrestricted application of a second-order model of how more local knowledge fits together

Kant's dialectical illusions are of the third type, though I will argue that they are not the only examples of that type. It is this third type that is by far the most philosophically interesting, as arguably a number of familiar and controversial philosophical theses are in fact illusions of this type. In this chapter, I will speak more generally of cognitive illusion from an epistemological standpoint, and the chapter will conclude with considerations of a possible fourth type of cognitive illusion—what we might call *projective* illusions.

14.1 Illusions of Inapt Application

Part of what is required to master a model is to grasp its concepts and rules. But mastering a model also has a second aspect: understanding when it can be applied aptly and when it cannot. Newton no doubt understood his corpuscular theory of light well, and his experiments with reflection and refraction through prisms showed that the theory could aptly be applied to a number of problems in optics. Thomas Young did not devise the double-slit experiment, whose results cannot be accommodated by a particle theory, until long after Newton's death. Newton was a committed experimentalist and would surely have felt compelled to recognize the problems Young's experiment presented for his theory had it been developed within his lifetime. Yet we can well imagine that had he been posed with the question of how light would behave if shined through two slits in a partition, Newton would confidently have predicted a result based in his particle theory. What the particle model *would* predict if applied to the situation is fairly straightforward: particles, moving in straight lines, should result in two bands of light corresponding to the two slits. Indeed, between the evidence from other types of experiments and the fact that the idea that light travels in straight lines appeals to common sense (no doubt rooted in some folk model), Newton might well have assumed that light would *have to* go through the slits in straight lines and produce two bands of light on the opposite wall. But this is not what actually happens. Instead the setup produces a diffusion pattern of the sort one might expect if light behaved like a wave.

At the least, the experiment shows that Newton's corpuscular model cannot aptly be applied in cases like the slit experiment. Just what it shows beyond this is complicated by the fact that Newton had in fact devised experiments of his own whose results seem compatible with a particle model but not a wave model. And for this reason, classical optics was beset with a puzzle, as some experimental results suggested one model and others another, incompatible model. Without a model that made sense of *all* the experiments, no one was really in a good position to claim decisively that light is either a wave or a particle phenomenon. There were two models, each of which could aptly be applied to predict results in certain types of situations but not in others. Moreover, each model produced intuitive anticipations of what should happen in each context—indeed, what *must* happen if light behaves as the model characterizes it. Someone presented for the first time with experimental results contrary to the intuitive anticipations of the model he was using might well be astonished by them (as many of us are the first time we encounter Young's experiment) and feel a need to replicate it for himself. Indeed, even having known of the two-slit experiment for a long time, I still sometimes find the results counterintuitive, in spite of knowing that they are well confirmed.

To take another scientific example, in fluid dynamics, different models are applied depending on factors such as temperature, friction, viscosity, compression, and whether flow is laminar or turbulent. The Euler equations present a comparatively mathematically simple model that can elegantly be applied to many problems, and are viewed as useful simplifications of the Navier-Stokes equations. But as Mark Wilson writes, "These frictionless simplifications display a wide variety of counterintuitive consequences—viz., airplane wings should experience neither drag nor lift (leading many nineteenth century experts to glumly predict the impossibility of heavier than air flight)" (Wilson 2006, 186). The Euler equations are apt for predicting many types of fluid motion and capture important invariants. But they are also idealized, and in cases where the idealizations matter (such as the lift on an airplane wing), the model cannot aptly be applied, at least without being supplemented by additional models, in this case Ludwig Prandtl's boundary layer. Applying the model inaptly can lead to false predictions such as the impossibility of heavier-than-air flight. Before Prandtl's discovery of the boundary layer in 1904, we can well imagine that fluid dynamicists, applying the models they had, might confidently conclude that air flight was impossible, *because that conclusion was implied by the models they had to work with.* Of course, one can also apply a model inaptly even if one already has the wherewithal to know that it is not aptly applied there.

If we assume that (1) models can give rise to intuitions about what will or even must be the case, (2) models are idealized in ways that render them apt only under certain conditions, and (3) we do not always know where a model is aptly applied and may even be unaware that there are restrictions on its apt application, then we should expect that there would be cognitive illusions that arise when a model is applied inaptly but we are unaware of the inaptness. And such unawareness can arise in several ways: as a simple mistake (we know the conditions for apt application but forget that a particular case lies outside them), out of ignorance that a particular case is of a type that lies outside the conditions, or out of a deeper naïveté about the idealized character of models in general or about the particular model we are using.

Discussions of "cognitive illusions" in psychology are principally concerned with inapt applications of models. Consider Deborah Kelemen's notion of "promiscuous teleology." Teleological thinking is a common feature of maturationally normal human minds. It is also found in very young children, perhaps predating causal explanation. While scientifically tutored adults tend to reserve teleological explanations for artifacts and (perhaps with some qualms and hedging) for organisms, Kelemen's research points to a developmental pattern in which children at first apply teleological models much more broadly, and in preference to causal models.

> Despite similarities between teleological-functional intuitions in young children and adults, young Western-educated children reason in teleological-functional terms about a broader variety of phenomena than do Western-educated adults. In addition to biological parts and artifacts, they view non-biological natural kinds and their

properties as existing "for" purposes—a finding that has implications for proposals that teleological reasoning in young children is selective and is the earliest form of thought that is specifically biological. (Kelemen 1999, 461)

Frank Keil (1992) had proposed that humans have an innate system for teleological explanations linked to biological thinking. Kelemen's evidence that young children show a bias toward applying teleological reasoning to things they understand to be nonliving seems to belie this hypothesis, and so Kelemen hypothesizes that a system for teleological reasoning may be developed on the basis of an earlier-appearing system for understanding intentional agents.

Experiences that suggest objects exist in the world to fulfill the purposes of agents might subsequently contribute to the tendency to over-generate purpose-based teleological explanations when faced with explanatory gaps. In other words, later on, in the absence of other explanations, children might draw on their privileged knowledge of intentions and artifacts to conclude that, like artifacts, natural objects exist in the world because some agent put them there for a purpose. (Kelemen 1999, 466)

If we assume, with Kelemen, that the application of a teleological model to things other than organisms and artifacts is inapt, promiscuous teleology is a case of a bias to overgeneralize the applicability of a model in inapt ways, and subsequent learning about science that is ubiquitous in Western education narrows the range of cases in which teleological framing is applied. This range narrowing may well be a by-product of particular types of education and enculturation, without which we should expect to find animistic tendencies in adult populations. This seems to be the view taken by Guthrie (2002), who additionally ascribes "animistic" thinking to nonhuman animals and describes it as "illusion": "The attribution of characteristics of animacy to the nonliving and to plants appears widespread among complex organisms. Because illusion is an inevitable concomitant of perception, such attribution may be universal" (53).

A second example: In chapter 13, I mentioned research on heuristics and biases pioneered by Kahneman and Tversky (1973). Research in this tradition has identified a number of types of cases in which reasoning about statistical problems deviates from optimal Bayesian reasoning about probabilities. Scientists often regard these deviations as errors in reasoning and sometimes describe them as "cognitive illusions." In a critical reply, Gigerenzer describes Kahneman and Tversky's program in the following terms:

Kahneman and Tversky (1973) describe the gist of their view as follows: "In making predictions and judgments under uncertainty, people do not appear to follow the calculus of chance or the statistical theory of prediction. Instead, they rely on a limited number of heuristics which sometimes yield reasonable judgments and sometimes lead to severe and systematic errors" (p. 237). They see the study of systematic errors in probabilistic reasoning, also called "cognitive illusions," as similar to that of visual illusions. "The presence of an error of judgment is demonstrated by comparing people's responses either with an established fact (e.g., that the two lines are equal in length) or with an accepted rule of arithmetic, logic, or statistics" (Kahneman & Tversky, 1982, p. 493). Their distinction between "correct" and "erroneous" judgments under uncertainty has been echoed by many social psychologists: "We follow conventional practice by using the term 'normative' to describe the use of a rule when there is a consensus among formal scientists that the rule is appropriate for the particular problem" (Nisbett & Ross, 1980, p. 13). (Gigerenzer 1991, 2)

The word *illusion* is used here in a sense suggesting an *error*, either in the outcome of the judgment or in the choice of a heuristic unsuitable to the problem because it differs from a statistical norm. This is readily translated into the framework we have been developing, as Kahneman and Tversky's explanations of the errors in fact invoke mental models:

Many adults do not have generally valid intuitions corresponding to the law of large numbers, the role of base rates in Bayesian inference, or the principle of regressive prediction. But it is simply not the case that every problem to which these rules are relevant will be answered incorrectly, or that the rules cannot appear compelling in particular contexts. The properties that make formally equivalent problems easy or hard to solve appear to be related to the mental models, or schemas, that the problems evoke. (Kahneman and Tversky 1982, 129–130)

In our terminology, the diagnosis of the errors would seem to be that subjects frame the problem with which they are presented using a model that is not apt for that problem, and derive an answer using heuristics associated with the model. Gigerenzer likewise appeals to "probabilistic mental models" as a general framework for explaining people's confidence in their general knowledge (Gigerenzer, Hoffrage, and Kleinbölting 1999).

What seem to be primarily at issue between Gigerenzer and Kahneman and Tversky are (a) the latter's use of judgments of professional statisticians as a normative standard for determining when the results of heuristic methods are erroneous and (b) the extent to which the heuristics that people routinely use are truly inapt, and hence whether the results are truly "illusory." These differences are not relevant to our topic here, as we are concerned with epistemic rather than alethetic illusions, and so what is at issue is not whether the resulting judgments are correct but merely that we confidently assume them to be *known*. What is germane is a conception of how the mind—or more specifically the application of particular models to particular problems—can generate cognitive illusions. Reasoning under uncertainty—with incomplete or unreliable information, or information that is itself in the form of probability distributions—is a situation that we routinely face. Even formal techniques like the use of Bayesian methods do not ensure correct predictions, and such techniques require much learning and (being of recent invention) were unavailable throughout most of the history of the species. Using Bayesian analytic methods also requires possessing a model that incorporates them; but the use of what Gigerenzer calls "fast and frugal heuristics" is often necessary and, in many cases, produces as good or even better results (Gigerenzer, Todd, and the ABC Research Group 1999). But it is particularly clear in the case of reasoning under uncertainty that *which* models and heuristics are apt for a given case will depend crucially on details of the case and the information available to the reasoner. Using a model that is inapt for framing a problem is likely to produce an erroneous assessment.

Of course, this kind of case illustrates that it is often difficult or even impossible to know for certain whether a given model can aptly be applied to a particular situation, at least in part because the limited information involved may prevent us from knowing what kind of situation it is. The kind of "error" involved may often not be one where we "should have known better." The inaptness is ultimately determined by external standards, though we can also learn a great deal about different models and what kinds of situations they are aptly and inaptly applied to. When we do not actually possess an apt model and can only apply the models we do possess, we may be subject to a weak form of forced error. (It is a weak form in that it is forced only by the de facto lack of an apt model, as opposed to the stronger cases in which we are incapable of having an apt model or in which the use of an inapt model is automatic and cannot be interrupted or overridden.)

Note, however, that the mere fact that I apply an inapt model that results in an error is not itself sufficient to make the result count as an illusion. There are, after all, different ways of applying models. At one end of the spectrum, I might uncritically assume that the model frames the situation adequately, and completely trust the resulting judgment, as in the case of the intuitive judgments described in chapter 13. This would indeed seem to merit the label "illusion." At the other end of the spectrum, however, I might be reflectively uncertain about what would provide an apt model of a situation, and try out one or more models tentatively and experimentally. I might refrain from endorsing the output of the model in the form of a judgment at all, and if it is not a judgment, it cannot be an illusory judgment. Or I might endorse the judgment within the scope of a reflective awareness that the judgment is warranted only if the model is apt. Such a hedged judgment may indeed be *correct*, even if the model is not in fact apt, but it would not be an epistemic illusion because I do not endorse the model-based prediction in an unqualified manner.

14.2 Illusions of Unrestricted Assertion

The examples in the last section involved *applications* of models to particular cases. But when we move from practice to theory, we do not merely *use* models; we attempt to make their principles explicit and *assert* them. In science, it is well understood that any theory or model has *boundary conditions* governing where it may aptly be applied. Fully understanding an area of science requires understanding the boundary conditions as well as the theories. But particularly in the process of learning, understanding the model and understanding its boundary conditions are often distinct cognitive achievements. In science education, a model may be introduced using problems taken from its core area of applicability, and the student may learn only later (or not at all) what cases it can and cannot aptly be applied to. Reports of scientific models in the popular press often lack any discussion of boundary conditions. And indeed, someone who does not really understand the science may make remarkable mistakes of scope, such as thinking that everything about an organism must be an adaptation.

Outside science, the scope of a model's aptness and the ways it is idealized are seldom explored as carefully as the boundary conditions of a scientific model. The form of generalizations in a language—"dogs are four-legged," "people act on the basis of their beliefs and desires"—leaves the precise implications of such claims quite vague, and we often do not really know how broadly or exactly they apply. General claims are a blunt instrument from a logical standpoint. And this has its advantages: few generalizations could *accurately* be expressed as, say, universally quantified claims; and if we are often in the position of knowing that they are applicable in a broad range of cases, but not exactly *which* cases, it is useful to have a manner of expressing them that leaves such questions open. (We arguably could not do this if our forms of judgment were limited to the logical forms available in quantified first-order predicate calculus.) We may implicitly know *that* the generalizations are idealized, but not know exactly *how* they are idealized. The risk, of course, is that we *forget* that we are dealing with idealized claims and mistake them for perfectly universal claims. We can apparently do this in such a fashion that the intuitive force of the claims—that is, the intuitive sense of *correctness*—is not lost. But when this happens, we have drifted into epistemic and alethetic illusion: if you turn an idealized model into a perfectly universal claim, it may still seem like knowledge if you do not recognize that the idealization and the intuitive force are still present, and as a perfectly universal claim it is *erroneous*. Not all dogs do have four legs, and not everything people do is a result of their beliefs and desires.

While the production of cognitive illusions is in itself always lamentable, the mechanisms that produce them are arguably grounded in further features of our cognitive architecture, which confer distinct cognitive advantages. I will discuss two of these features: intuitive essentialism and the augmentation of understanding through explicit rules or principles.

Intuitive Essentialism

A good deal of psychological evidence indicates that, even as young children, human beings are "intuitive essentialists"—that is, in dividing the world into kinds, we tacitly assume those kinds to have hidden inner natures that are responsible for their characteristic observable properties and behavior. In an article reviewing research on essentialist thinking in early childhood (including preschool children), Susan Gelman writes:

> Essentialism is the view that certain categories have an underlying reality or true nature that one cannot observe directly but that gives an object its identity, and is responsible for other similarities that category members share. In the domain of biology, an essence would be whatever quality remains unchanging as an organism grows, reproduces, and undergoes morphological transformations (baby to man; caterpillar to butterfly). In the domain of chemistry, an essence would be whatever quality remains unchanging as a substance changes shape, size, or state (from solid to liquid to gas).
>
> Essentialist accounts have been proposed and discussed for thousands of years, extending back at least to Plato's cave allegory in *The Republic*. Numerous fields, including biology, philosophy, linguistics, literary criticism, and psychology, stake claims about essentialism. Here we are concerned with essentialism as a psychological claim. Although there are serious problems with essentialism as a metaphysical doctrine, recent psychological studies converge to suggest that essentialism is a reasoning heuristic that is readily available to both children and adults. This article reviews such evidence, and discusses the implications for human concepts. (Gelman 2004, 404)

Psychological essentialism is arguably a deep principle of how children think and learn about the world. It is not simply that children impute *particular* features to kinds of things—say, to species—as essential properties; rather, they *expect* kinds (or at least some types of kinds) to have essences, even if they do not know what those essences might be.

> Medin and Ortony (1989) suggest that essentialism is a "placeholder" notion: one can believe that a category possesses an essence without knowing what the essence is. For example, a child might believe that there exist deep, non-visible differences between males and females, but have no idea just what those differences are. The essence placeholder would imply that categories permit rich inductive inferences, capture underlying structure (in the form of causal and other non-obvious properties), have innate potential, and have sharp and immutable boundaries. (Gelman 2004, 404)

Essentialism is an *implicit* feature of how children think about the world. (The question of how psychological essentialism is affected by further learning into adulthood is a topic of greater debate, though it seems clear enough that essentialist biases continue to operate in some fashion.) But it is also expressed in generalizations: "lions are carnivores," "women are nurturers," "Germans are strict." The surface form of the generalization admits many fine-grained interpretations; but when it expresses an essentialist assumption, it imputes a trait as something internal to the nature of the kind. As the examples show, the "kinds" to which such thinking might be applied are not necessarily restricted to scientific kinds like species.

Essentialist generalizations can be produced as a result of experience but can also be transmitted socially through language. Essentialist thinking is a powerful inductive tool. Making essentialist generalizations explicit can also be a powerful tool (see the next section) and also makes them available for social transmission. Of course, essentialist thinking about subcategories of human beings is notoriously present in social bias and prejudice, and so the benefits of a powerful tool for learning and categorization clearly come with serious cost and risk. This, however, helps to illustrate some of the types of errors to which essentialist thinking is prone. Even if some kinds of things *are* aptly thought about in some contexts as having essences (or essential properties), we may think about *other* kinds of things in essentialist terms *inaptly*. And even if there are some kinds that can aptly be imputed essential properties, we may mistake the *wrong* properties as essential or mistake whatever "essentialness" amounts to for *universality*. If species have essences, being quadrapedal may be essential to dogs, but not all dogs have four legs. Female mammals may possess biological features that make them more likely to manifest nurturing traits, but the expression of these traits depends on other variables, and there are men and women at both ends of the range of nurturing behavior.

Various kinds of essentialism have, of course, played significant roles in the history of philosophy. Aristotelian philosophy may be the most comprehensive essentialist philosophy, but contemporary talk of "natural kinds" as privileged ontological kinds, and "new semantics" accounts of reference pioneered by Kripke (1980) and Putnam (1975) seem to posit that words like 'water' are able to refer to an inner essence of the stuff being referred to (e.g., being of the molecular kind H_2O) even if that essence is as yet unknown. As explicit theory, however, such philosophical views are better discussed in the next section.

Explicit Rules and Principles

One can *use* a model with its principles remaining implicit and unarticulated. Indeed, the many nonhuman species that can use model-based cognition are probably capable of using a model *only* in an unarticulated way. Language, however, gives humans the means to make principles of models explicit. Zebras can systematically *treat* lions as dangerous predators, and so can we; but we can also think and say, "Lions are dangerous predators." The ability to think it as a judgment affords certain aids to learning through memorization, and the ability to say it enables social transmission of knowledge without the need for each individual to learn through experience. It would be difficult to overestimate the role that learning through the adoption of explicit principles plays in human life. (Certain familiar styles of pedagogy are based almost solely on it.) Of course, making a principle explicit also enables us to examine, test, and refine it, though an explicit principle can also be used and even believed uncritically.

Principles often take the form of grammatically general claims:

Xs are Ys
Xs do A
Xs have property P

Such a form can of course be an informal way of expressing universally quantified claims like "All Xs are Ys." But probably the vast majority of even the useful general claims would be false if interpreted as universal claims. And it is not clear that there is some *other* logically exact meaning that they "really have," unless perhaps the form "Xs are Ys" is itself counted as a logical form. To spell out the *apt* use of such generalizations, we would probably have to do something equivalent to defining boundary conditions for scientific theories and would have to do so separately for each such generalization. But we are seldom in a position to do this; we may think a particular generalization is a useful rule of thumb, and also realize that there are exceptions, but not know what the full range of exceptions is or how to characterize (or anticipate) them. And in many cases there may not be a way of exactly spelling out where the generalization is and is not aptly applied.

Again, in one way all of this is both intelligible and useful from a psychological standpoint. To the extent that we are ever to learn when a principle is and is not aptly applied, we first have to learn the principle and try to apply it. And this requires that learning the principle be dissociable from mastering the conditions for its aptness. This opens the gates to inapt application to particular cases, as discussed in the previous section. But the unqualified form of the principle also allows it to be interpreted as a perfectly universal and unidealized claim.

Mistaking a generalization for a universal claim can, of course, result in epistemic illusion: the intuitive force of the general claim may still be present, but it has been covertly turned into a different sort of claim. (Whether it is also an *alethetic* illusion depends on the nonpsychological facts of whether the content of the judgment is true.) That "Xs are Ys" is a useful rule of thumb in some unspecified range of contexts by no means ensures the truth of "*All* Xs are Ys." And when we (perhaps unwittingly) transform a useful generalization grounded in a model that is often apt into a universal claim, we may still feel the intuitive force of the generalization, not realizing that it has illicitly been transferred to an unsubstantiated universal claim, and hence fall prey to a cognitive illusion.

But the mind's tendency to elide from general to universal claims can also be the beginning of a fruitful kind of dialectical process of inductive testing. "Xs are Ys" is probably too vague to test at all. "All Xs are Ys" can be tested, often by first being *con*tested, and most claims of this form turn out to be false. But the dissonance between the intuitive appeal of the general claim and the proven falsity of the universal claim drives the mind to look for a more adequate exact formulation. This is, in fact, one of the predominant methods we see at work in philosophical discussion. A thesis may first be proposed on the basis of seeming to be an apt way of viewing things based on a small set of paradigm examples. The thesis is proposed as a universal claim (often in the form of putatively necessary and sufficient conditions), and either the theorist herself or other philosophers then test it by considering possible counterexamples. Often the account is modified to accommodate counterexamples and subjected to a further round of criticism, and so on.

From this perspective, I think we can see some of the familiar weaknesses and strengths of this approach to philosophy. On the one hand, it is hardly a new observation that many philosophical theories were motivated by what in retrospect might seem a woefully small set of paradigm cases.

From the outside, it might seem puzzling how anyone could seriously have believed theories such as that *all* thoughts are images or that *everything* people do is done on the basis of rational calculation involving their beliefs and desires. The corresponding general claims, like "thoughts are images" or "people decide what to do on the basis of their beliefs and desires," might be plausible and useful within some more restricted scope and might be tied to models that are indeed apt for understanding *something*. But the universal claims do not have good inductive support, and one would be led to very different conclusions if one started from different paradigm cases. That anyone could have thought the universal claims plausible suggests that they were victims of a cognitive illusion that involved losing sight of the restricted scope of the general claim. This type of interpretation is often offered as a *criticism* of a theorist and his reasoning process. But it only really works as a criticism if we assume that the process of finding good inductive generalizations works by first assembling all the evidence and then drawing a conclusion from it. Viewed as a *dialectical* process, often involving multiple participants playing the roles of defender and critic, the process of proposing a universal claim based on a restricted range of cases and then subjecting it to critique by counterexamples can in principle work quite well, even if in other cases it ends up being unproductive, falsifying the universal claim but offering nothing constructive in its place. What is perhaps more abidingly peculiar from an epistemic standpoint is that professional philosophers who adopt the role of advocating a particular theory in such dialectic often seem to also feel committed to the theory being *true*, even when they are aware of poignant counterexamples. I suspect that the roots of the commitment are really to be found in a commitment to an underlying model being apt for understanding a particular range of cases.

One important question for further empirical investigation is whether the formulation of generalizations within the resources of ordinary language facilitates the production of illusions. It is not that we *cannot* devise careful uses of language that include explicit hedges indicating the idealization of the models underlying generalizations or employ more formal logical tools. Indeed, whole schools of philosophy have been devoted to "correcting" the vagueness and ambiguity of ordinary language. The interesting question, rather, is whether some of the structures of ordinary language, like "Xs are Ys," *lose information* that was present in purely model-based understanding before it was transmuted into language—for example, whether we are more prone to forgetting the idealization conditions of a model we are reasoning with when we express it in language in the form of a general claim.

14.3 Illusions of Unification

One of the many fine things about Kant's epistemology is his attention to the fact that the mind does not rest content with "knowledge," in the sense of knowing individual things, but has a natural drive to *unify* what it knows in an ever-more-comprehensive way. Kant regards this drive as both natural (in the sense of being a part of our psychology) and a good thing, but he devotes a substantial portion of the first *Critique* (the section titled "Transcendental Dialectic") to showing that this drive can also lead the mind into certain cognitive illusions. We might sum up Kant's overall story in the following points, which I have formatted so that some of his more particular hypotheses are represented in the second level of the outline.

1. The mind has a drive to unite the things it knows.
2. The mind has particular types of techniques for uniting knowledge.
   a. These techniques are the three syllogistic forms.
3. The mind pursues its drive to unite knowledge by trying to apply its techniques for uniting knowledge to their ideal limit (a regulative principle).
   a. This takes place through the attempt to ground conditioned premises through construction of prosyllogisms.
4. The mind projects what this ideal limit *would* look like and what would have to be the case if this limit were attained.
   a. There are three of these, the Ideas of Pure Reason: God, soul, and world.
5. The mind can mistake these projected ideal completions for things that it knows in the form of things that must be true unconditionally (dialectical illusion).
   a. The Ideas of Pure Reason cannot be known because the sphere of knowledge is limited to things that could be objects of sensory experience.

I endorse the general points in the first level of the outline, though I do not agree in detail with the more specific Kantian views in the second level. Kant's general insights here are brilliant. But I regard his attempt to characterize understanding (uncapitalized to indicate that I do not mean it in Kant's technical sense) solely in terms of the forms of judgment and argument forms found in deductive logic as one of the main limitations of his account, and perhaps something that he himself would have developed differently had there been more advanced cognitive sciences in his day. (I am likewise at least dubious about his identification of the sphere of possible knowledge with what could be experienced and confirmed through sensory experience.)

One basic difference between my cognitive pluralism and Kant's philosophy is that my theory locates much of understanding as arising from idealized domain-specific models rather than from sentence-sized judgments and arguments composed of such judgments. We have seen that there are ways that the idealization of models can result in a kind of epistemic disunity. I have speculated that sometimes there are incompatibilities between apt models that are principled and enduring, but I have also endorsed attempts to reconcile incompatibilities as a regulative principle. And it is clear that attending to problems that are signaled by having incompatible ways of understanding things often produces intellectual advances, such as corrections to models we have previously used, discoveries of better models (sometimes ones that cover the domains of two preexisting models), and second-order ways of conceiving the relationships between models, such as seeing one as a limiting case of another. Sorting out such problems differs from sorting out inconsistencies between propositions or trying to unify propositions into a more comprehensive system without attending to models. By the same token, the ways we try to unify models can lead to types of illusion that are not apparent if we attend only to propositions and arguments.

The most basic such illusion is perhaps to assume that all the things that we justifiably believe on the basis of apt models should be consistent with one another. Consistency is, after all, a norm of logic. And it is an extremely powerful norm, as it gives rise to a strong intuitive assumption that two inconsistent claims cannot both be true. But we now have several reasons to look past this compelling intuitive assumption.

To begin with, the norm of consistency is a norm *for* sentence-sized units: propositions, judgments, beliefs, claims, assertions, sentences. The introduction of models as a different type of cognitive unit—one that is in some ways *prior to* sentence-sized units, as it is models that define a space of possible propositional representations—complicates matters. What "consistency" means when comparing two models is different from what it means in comparing two propositions. Euclidean and non-Euclidean geometries are inconsistent with each other, but this does not mean that at least one of them is false. Indeed, it makes no sense to speak of a pure mathematical system being "false," or of a pure mathematical claim being either true-full-stop or false-full-stop. Mathematical systems simply are the systems they are, and the truth of a pure mathematical claim like the Pythagorean theorem is truth-in-a-system. When models are applied to the empirical world, there can be formal inconsistencies between their rules and representational systems, but this need not constrain us to endorse contradictory propositions, for reasons explored in chapter 12. Some unificatory projects—such as those involving intertheoretic reduction—have as their aim relations between models. But even to the extent that we endorse the principle of noncontradiction for propositions, this gives us no reason to expect that all apt *models* should be formally consistent.

Of course, the discussion of disunity also included the observation that while there are clear cases in which a proper understanding of how particular models are idealized prevents their committing us to contradictory beliefs (e.g., in pure mathematical models and cases like quantum complementarity), there are probably many more cases where models frame matters in alternative ways that seem to be contradictory without the possibility of any decisive experimental technique for deciding when each is aptly applied. Nonetheless there can be other ways in which they differ with respect to aptness. For example, consequentialist, deontological, and virtue-theoretic models may differ in their assessments of goodness in various cases, with no decisive way to arbitrate between them, and yet still differ with respect to what insights they afford that might prove useful in moral deliberation and assessment. The norm of consistency indeed gives us reason for dialectical engagement in such cases, but it is not the *only* norm to which we are beholden. Even if our concern is only with theoretical understanding, there may be other regulative principles as well—such as having good ways of understanding as many things as possible—which may pull in opposite directions, and the extent to which any regulative principle can be fulfilled completely by a given mind depends on empirical facts about that mind's cognitive architecture. It is an epistemic illusion to assume that a given mind *must* be capable of fulfilling a regulative principle, much less fulfilling all of them simultaneously.

Second, we have various *strategies* for unifying models into a form that is more comprehensive. A number of these strategies have been developed in mathematics and the sciences. One family is called "reductionist" and, in its purest form, involves the axiomatic reconstruction of one domain of understanding in terms of another (e.g., the derivation of arithmetic from set theory, or the attempts to show that the special sciences can be derived from physics). A second is to unify two apparently disparate domains as special cases of something more general (the most historically important example is perhaps Newton's unification of planetary orbits and terrestrial ballistics in a single theory of universal gravitation). A third and distinct strategy is to show that one model is a limiting case of another (a paradigm example would be the claim that the classical universe is a limiting case of the relativistic universe).[1] A fourth, characteristic of Newton's approach to mechanics, is to attempt to view different models as expressing independent *factors* that can be recombined (in Newtonian mechanics, through vector algebra) to get a global picture. There is also a family of strategies, which I hesitate to call "unifying," that we might label "contextual" or "relativistic." Their basic approach is to say that different models allow us to say true things, but in a fashion hedged by *contexts*. I pick these examples out because they are familiar and important, but I do not wish to suggest that this list is exhaustive. Indeed, I regard the list of unifying strategies as open-ended, as there may be unifying strategies that have not yet been recognized.

Each of these strategies has arguably been successfully applied to resolve particular problem cases. What is striking from the standpoint of intellectual history, however, is the confidence with which a particular strategy has often been endorsed, along with a picture of how things would look if it *could* successfully be applied, well in advance of a serious attempt to apply it. This is, to my mind, particularly striking in the case of reductionist strategies. Pre-Socratic philosophers seem to have been remarkably confident that everything *must* be composed of *one* of the "prime elements" of earth, air, fire, water, or aether, though they differed over which one. Pythagoreans shared a similar confidence that everything was ultimately numbers. Early modern materialists and some of the twentieth-century logical empiricists had a similar confidence that the special sciences could be reduced to basic physics, and early proponents who had little scientific evidence to go on (like Hobbes) were no less confident than those who came after. Indeed, the intuitive appeal of reductionism seems to endure even in the face of strong arguments from philosophers of science that true intertheoretic reductions are rare (Horst 2007). Hobbes and Descartes also seemed to have had a further ambition of seeing physical properties as ultimately geometric in nature. And many contemporary theoretical physicists seem to regard the prospect of a unified field theory as something that must be true.

Chapter 12 explored some reasons why such confidence might in fact be ill justified. But we might also ask the equally interesting question: Why would such views recommend themselves to the mind as intuitively appealing, perhaps even as things that must be true? I think that a plausible answer is a variant on Kant's general story about the drive toward unification and its propensity to foster cognitive illusion. The mind has a drive to unify not only its beliefs but the models it uses. There are various strategies for unifying models, which go beyond the syllogism-based strategies identified by Kant. We are capable of projecting what things would have to look like if such a strategy could be carried out—what features the resulting system would have to have. And there is indeed a kind of necessity involved here: *if* such a unification could be carried out, then the resulting system would *have to* have certain features. It seems to be a fact about our psychology that such a vision has a powerful allure, which can produce a peculiar forgetfulness that we are *projecting* an ideal conclusion of the application of a method. We mistake the intuitive appeal of the ideal for a kind of metaphysical necessity, and our confidence is perhaps bolstered by the fact that the drive to unify is a sound regulative ideal. But whether such an ideal can in fact be reached is by no means ensured. It depends not only on empirical facts about the world but also on empirical facts about our cognitive architecture.

14.4 Projective Illusion

I conclude this chapter by discussing a phenomenon that some philosophers have regarded as a type of cognitive illusion, but whose status is more complicated. I introduced the notion of cognitive illusion with reference to Kant, but in fact Hume had already spoken of the mind producing "illusions"—a fact of which Kant was no doubt aware, as at least one of the things Hume described as "illusion" (causation as a real relation outside the mind) had played an important role in waking Kant from his "dogmatic slumbers."

In the *Treatise*, Hume expresses skepticism about several assumptions shared by common sense and previous philosophies. The best known of these is his skepticism about causation as a real relation existing independently of the mind. According to Hume's theory of mind, our only source of knowledge about the world is through sensory impressions, and there is no impression of causation. Rather, what goes on in the mind when we think of an event as involving causation is that we have one impression, followed by another, which the mind associates in a particular way—a kind of associative link to which we give the name "causation"—which disposes us to form expectations of the second type of impression on subsequent occasions when we experience the first. This general view of the psychology of causation attribution is preserved in Hume's later *Enquiry*, albeit in more guarded terms. In the *Treatise*, however, he actually goes so far as to describe the imputation of something called "causation" to the extramental world as an "illusion":

This deficiency in our ideas is not, indeed, perceived in common life, nor are we sensible, that in the most usual conjunctions of cause and effect we are as ignorant of the ultimate principle, which binds them together, as in the most unusual and extraordinary. But this proceeds merely from an *illusion* of the imagination. (Hume 1738, pt. 4, sec. 7; italics mine)

Hume also uses the word 'illusion' in the *Treatise* in conjunction with two additional second skeptical theses. At the outset of part 4, section 2, "Of Skepticism with Regard to the Senses," he writes:

The subject, then, of our present enquiry is concerning the causes which induce us to believe in the existence of body: And my reasonings on this head I shall begin with a distinction, which at first sight may seem superfluous, but which will contribute very much to the perfect understanding of what follows. We ought to examine apart those two questions, which are commonly confounded together, viz. Why we attribute a continued existence to objects, even when they are not present to the senses; and why we suppose them to have an existence DISTINCT from the mind and perception. (Hume 1738, pt. 4, sec. 2)

Here Hume is posing two questions about the psychology of object-oriented cognition: (1) how we come to think of objects as enduring things that exist even when we do not have impressions of them, and (2) how we come to think of objects as *distinct* from the mind and its impressions. Hume speaks of both of these propensities of the mind as "illusions" in the ensuing paragraphs of the section. With respect to continued existence, he writes:

It is a gross *illusion* to suppose, that our resembling perceptions are numerically the same; and it is this illusion, which leads us into the opinion, that these perceptions are uninterrupted, and are still existent, even when they are not present to the senses. (Hume 1738, pt. 4, sec. 2; italics mine)

With respect to external existence:

That our senses offer not their impressions as the images of something distinct, or independent, and external, is evident; because they convey to us nothing but a single perception, and never give us the least intimation of any thing beyond. A single perception can never produce the idea of a double existence, but by some inference either of the reason or imagination. When the mind looks farther than what immediately appears to it, its conclusions can never be put to the account of the senses; and it certainly looks farther, when from a single perception it infers a double existence, and supposes the relations of resemblance and causation betwixt them.

If our senses, therefore, suggest any idea of distinct existences, they must convey the impressions as those very existences, by a kind of fallacy and *illusion*. (Hume 1738, pt. 4, sec. 2; italics mine)

A good deal of scholarly debate has addressed just what *sort* of skepticism Hume wished to endorse on all three of these topics. Hume's views on causation have drawn by far the most scholarly interest, but the general form of the interpretive problem can be generalized to the other topics as well. The *Treatise* was a work of Hume's youth, and he seems to have either backed away from or at least remained silent on some of its more controversial points in his later work, the *Enquiry*. Hume's skepticism about causation can be interpreted in two basic ways. On a milder interpretation, Hume is a skeptic only in the sense that impressions do not give us a basis for *knowledge* of real causal relations in the world, because we have no impressions of causation. This is compatible with the possibility that there *is* such a relation, which our mental habit of imputing such a relation fortuitously gets right, at least in cases where we attribute it *correctly*. According to the stronger interpretation, causation is not a real relation in the world at all but a projection of our own associative interpretation onto the world, painting it from a palette originating in the mind. This stronger interpretation is suggested by Hume's use of the word 'illusion' in his earlier work, the *Treatise*. But as we have seen, Kant came to use the word 'illusion' in a fashion that implied not *falsity of belief* but only a mistaken sense of *knowing* what in fact one cannot know. It is possible to read Hume as presaging Kant's usage, albeit without the careful distinction Kant made to clarify it.

With respect to how we come to think of a single enduring object on the basis of many impressions of its different perceivable qualities, it is clearer that a change has occurred between the two works. In the *Treatise*, Hume endorses a "bundle theory" of perceptions, according to which each "perception" (i.e., impression or idea) is completely distinct and autonomous.

> Since all our perceptions are different from each other, and from everything else in the universe, they are also distinct and separable, and may be considered as separately existent, and may exist separately, and have no need of anything else to support their existence. (Hume 1738, pt. 4, sec. 5)

There is neither something in which impressions reside (a "mind") nor any notion of an "object" or "substance" that is a *bearer* of the properties they represent. We have ideas only as copies of previous impressions, and since we have no impressions of "object" or "substance" in addition to impressions of things like redness and roundness, we can have no idea of "object," either. The common assumption that there are independently existing objects that persist through changing perceptions therefore cannot be grounded in Hume's psychology. So again we are faced with two possible skeptical interpretations: a weaker one, on which we have no *evidence* for a world of independently existing enduring objects (since evidence could come only from impressions), and a stronger one, on which the imputation of their existence is an error of projecting something of the mind's own making onto the world. The problem is that Hume's account does not even tell us how the mind could produce such an illusory idea, so either interpretation is problematic. Hume quietly dropped the bundle theory from his philosophy in his later work, but he never supplied a way of accounting for how we come to think about a world of enduring objects, save for a weak story about how we mistake the *resemblance* of impressions held at different times for *identity* between them.

What I wish to emphasize about these three skeptical theses, however, is this: regardless of whether one favors the milder or the stronger skeptical reading of Hume, he is clearly claiming that our deeply seated tendency to think of a world of independent objects, which endure through time and through changes in how we perceive them and enter into causal relations with one another (and with us as perceivers), is somehow a product of certain peculiarities of the mind that he styles "illusions." He is committed, at the least, to the view that claims about external objects and causation outstrip any kind of evidence we might have, and hence any sense that we have *knowledge* of objects or causation is illusory. He may be committed to the further view that they involve *errors* of imputing to the world features that are really only features of how we think *about* it.

We might characterize the stronger view as positing that the imputation of causation or objects existing independently in the world constitutes a *projective illusion,* in which features that are really only features of the mind are "read onto" the world. Milder versions of such a view had of course already been suggested by philosophers like Descartes and Locke, who claimed that secondary qualities like color are mistakenly assumed to be properties of objects but really are merely ways objects affect us through our senses. But Descartes and Locke had each assumed that there really is a world of substances that stand in causal relations with one another and with human observers. We might be prone to illusions about what properties substances have in their own right, but these illusions could be corrected through philosophical reasoning, and basic metaphysical categories like substance and causation reflect the real metaphysical structure of the world. Hume seems to have rejected this sort of realist assumption, taking a more comprehensive cognitivist turn. Kant, whose views were at least equally cognitivist, made it a major part of his project in the *Critique of Pure Reason* to argue that the categories of substance and causation, while transcendentally ideal, were nonetheless empirically real, and indeed necessary and knowable through synthetic a priori reasoning, because the mind is constrained to apply the categories to all possible objects of experience.

Cognitive pluralism, of course, is also a cognitivist philosophy. Discussion of its possible metaphysical implications will have to wait for another book, but clearly there are important questions to be answered about matters such as its relationship to realism and idealism. I shall confine myself here to a brief discussion of the question of projective illusion. Because aptness is ultimately a pragmatic matter, the fact that object-oriented cognition and schemas for causation are features of our cognitive architecture that provide basic ways of understanding the world does not debar them from counting as apt. Indeed, to the extent that models that allow us to constitute a world of objects standing in causal relations provide good epistemic and practical traction upon the world, this renders them apt. But to this we should add two important caveats. First, contemporary science, unlike the science of Kant's day, includes models that posit phenomena that resist characterization as classical objects. To the extent that the mind strives to force these into the mold of notions of objects or causation, this may well be a process that can produce cognitive illusions. Second, when we move beyond the application of models in ordinary cognition and empirically constrained science and attempt to think about substance and causation as fundamental metaphysical categories, we may well move beyond the realms in which they are aptly applied. In particular, if we adopt the naive realist stance that treats such categories as mind-independent ways the world divides itself up, we indeed fall into something akin to projective illusion.

15 Cognitive Pluralism and Epistemology

The model-based cognitive architecture of cognitive pluralism was developed as an alternative to the three-tiered standard view of the mind, which acknowledges concepts, beliefs, and inferences but accords mental models no fundamental role in understanding. While the standard view is assumed in many accounts of the nature of cognition, it also figures prominently in other areas of philosophy: epistemology, semantics, truth theory, and logic. Indeed, I have suggested that one reason it may recommend itself to philosophers as a view of cognitive architecture stems precisely from the fact that it is assumed in these other areas of philosophy. To the extent that the standard view has served philosophers well by providing a framework for successful theories in their pursuits, that in itself provides some reason to suppose that it supplies an adequate categorization of the basic units of thought. We have now considered the alternative proposal that mental models are fundamental units of understanding, along with several applications of this proposal. In these final chapters, I will consider the implications of cognitive pluralism for theories of knowledge, meaning, and (to a more limited extent) truth.

This chapter will consider the implications of a model-based cognitive pluralism for epistemology. Most contemporary epistemologists treat it as a framing assumption for their discipline that knowledge is warranted true belief. There are some interesting dissenters from this characterization of knowledge (e.g., Sayre 1997), and some philosophers have made a plausible case that it marks more of a departure from classical epistemology than their colleagues suppose, and that, say, Plato's use of the Greek word *epistēmē* might better be translated "understanding" (e.g., Kvanvig 2003; Grimm 2010). There are worthy questions to be pursued here, but for the purposes of this chapter, I will be engaging what is currently the mainstream view, and hence assuming that knowledge is warranted true belief.

The chapter is divided into two parts. The first considers the implications of a model-based view of understanding for epistemology, on the assumption that knowledge is warranted true belief. What is the relationship between models and beliefs? What role do models play in the warrant of beliefs and hence their status as knowledge? Are models themselves things that can be warranted? The second examines the implications of cognitive pluralism for three influential theories of knowledge: foundationalism, coherentism, and reliabilism.

15.1 What Are Beliefs?

Back in chapter 2, I made a terminological decision to speak of occurrent states involving endorsement of propositional content as *judgments*. Such states have sometimes been called "occurrent beliefs" as well, a terminology generally used in contradistinction to "dispositional beliefs." I further decided to use the term 'belief' broadly, to include both judgments and dispositional beliefs, whatever the latter might turn out to be. In subsequent discussions of intuition and cognitive illusion, I concentrated principally on the notion of judgment rather than that of belief, because those chapters dealt with the production of occurrent states from models. We can and will discuss the question of how judgments are warranted; however, epistemologists do not generally confine their discussions of knowledge to occurrent states but employ a notion of belief that encompasses both occurrent and dispositional beliefs. In this chapter we will thus need to discuss the relationship in which models stand to beliefs as well as to judgments. I touched on this topic in chapter 13, in conjunction with a discussion of intuition. Here I take it up again in the context of accounts of beliefs and other intentional states.

As a first step, I would like to explore how mental models might shed further light on the psychological underpinnings of what we describe as beliefs. To begin, let us reconsider several ways of understanding what we are doing when we ascribe beliefs.

1. *Intentional realism*: Belief ascriptions in general involve attributions of mental representations. A statement of the form "S believes P" can be true only if S has a mental representation with the content P. In this account, all beliefs are occurrent states like judgments, though they may not be conscious states and might be represented in ways (e.g., in cognitively impenetrable subsystems) that cannot be brought to conscious awareness. Such an assumption seems necessary for cases in

which beliefs need to figure in causal explanations of mental processes or behavior based on their content (or nonsemantic features that track content). But it does not do justice to the broad range of perfectly ordinary belief attributions where it is implausible to assume that there is an explicit representation of the content—for example, that you believe (and have believed since some indefinite time in your youth) that dogs have kidneys, that 117 + 8 = 125, and so on.

2. *Dispositionalism*: S believes that P just in case S is disposed to readily endorse P when appropriately prompted. The immediacy and confidence with which you assent to "dogs have kidneys" or "117 + 8 = 125" is a sign of, perhaps even constitutive of, the fact that you already *believed* these propositions. This may be true so far as it goes, but it does not tell us much about the psychology of belief. Attributing dispositions does not describe a mechanism so much as it points to a need to specify one.
3. *The intentional stance*: Belief attribution involves adopting a stance of interpreting a system in terms of beliefs and desires—namely, those that the system ought to have if it is to be rational. Sometimes this intentional stance allows us to pick out real patterns in the world; but because the attribution is normatively rational rather than factual, it does not require the attribution of individual occurrent states corresponding to those beliefs and desires. When the stance works for predicting the behavior of a system S, S counts as a genuine believer, regardless of the psychologically real causes of its behavior. This, however, underplays both the fact that at least sometimes the causes are actual conscious judgments and the important possibility that there are also inner states that share much in common with judgments even though they are not conscious. The stance-stance may describe what we are doing in *attributing* intentional states in many cases, but like dispositionalism, it tells us nothing about the underlying psychology. Moreover, assumptions about S's rationality and S's having the information it needs about its environment present problems in using the intentional strategy to explain behavior that is a result of irrational processes (e.g., faulty reasoning) or misinformation.
4. *Normative commitments:* S believes P just in case S is normatively committed to endorsing P or treating P as true. The nature of the normative commitment need not be cashed out in terms of ideal rationality and possession of complete information but can be based on the implications of other commitments S has. If S endorses Q and employs an inference rule such that Q would imply P, S is thereby normatively committed to

P as well. Again, this is a framework for understanding belief attribution (in this case, as holding accountable and scorekeeping [Brandom 1998]) without a psychological story about the basis for the commitments and inference rules.

5. *Sellars's myth of Jones*: In saying "S believes P" we are hypothesizing some kind of inner state that we conceptualize by analogy with S making an assertion that P. The supposition that we conceive of inner states by analogy with assertions involves a kind of agnosticism about the real psychological nature of the states attributed.

We thus have one proposal (intentional realism) that makes a claim about the psychological nature of beliefs (but one that is implausible for many perfectly ordinary belief attributions) and a range of proposals that make claims about what we are doing in attributing beliefs, and set constraints on a psychologically realistic account of beliefs without supplying such an account. Clearly what we should want here is a psychological proposal that can account for our dispositions to form judgments, the success of the intentional strategy, the possession of normative commitments, and the analogy between beliefs and assertions. Mental models help provide such an account.

First, if we are disposed to judge—quickly, automatically, and confidently—that dogs have kidneys and 117 + 8 = 125, on what basis are we so disposed? Our discussion of intuitive judgments suggests that, at least in a wide variety of such cases, the disposition is rooted in the possession of a mental model containing resources that can generate the judgment without explicit reasoning. I have never before considered whether dogs have kidneys; but to do so, all I need to do is consult some kind of inner model of typical animal physiology, and the judgment pops out. I have never added these two numbers before (or if I have, the result is forgotten), but I have mastered a model for adding integers, and when I plug these particular addends into it, the answer pops out. (Note that if we adjusted the example so that the addends exceeded performance limits—say, twenty-digit numbers—your answer to the "do you believe ..." question might well be "I don't know," at least until you worked the problem out with pencil and paper or a calculator.)

It is also natural to speak of facts such as that dogs have kidneys as *commitments* of models if they are somehow contained in, or implied by, the models' rules, representational systems, or data structures. To use language from AI and theoretical cognitive science, if I have a model of animal physiology that contains a node for kidneys linked to the root node as a default feature, or a production rule that generates a representation with the content "… has kidneys" to things to which the model is applied, then by using the model and treating it as apt, I am normatively committed to such inferences, even if I do not actually perform them. If the model *is* apt, there is also a second kind of normativity: it is an inference that one *should* make, because the inferential pattern tracks a real pattern.

We saw in chapter 12, however, that different models involve different, and sometimes conflicting, commitments. If we are committed to the implications of our models—whether their general commitments or the implications of their application to particular cases—it would thus seem that we would turn out to be committed to conflicting beliefs as well. This may initially seem like a problem, but really it is something we should *want* a psychological account of belief to imply or at least accommodate. On the one hand, a wealth of psychological studies show that our judgments are quite sensitive to context and priming. A natural explanation of this is that different contexts and priming conditions activate different models, generating one set of judgments rather than another depending on which model is activated. On the other hand, we often find ourselves inclined, even after much reflection, to endorse two propositions we know to be contradictory when each is intuitively recommended by a model and we are not in a position to see why one model is inapt or otherwise to adjudicate between them. It is not clear that the fact that we have paradoxical sets of beliefs should itself be regarded as paradoxical. But if it is, we seem to gain some resolution by understanding that the situation is not one of simultaneously and assuredly endorsing both P and not-P, but one of feeling the pull of the conflicting implications of two models to which we are committed.

If the phenomenon we call "believing P" amounts to having a model that can readily generate a judgment with the content P in an appropriate evoking context in which the model is treated as apt, and we encode this in the locution "S believes P," clearly information is lost in the translation. Here we see the wisdom behind Dennett's and Sellars's cautions. Belief ascriptions *look like* reports of assertions, and it is tempting to assume that what is reported must, like an assertion, be propositional and episodic. But really it is dispositional and normative. It may be a disposition to produce a certain type of episode—a judgment—but its nature must be understood in terms of the dynamics of models, and models are not judgments or other sentence-sized units, and the processes by which they produce judgments are not logical argument forms operating on sentence-sized units. Moreover, the analogy with assertions may tempt us to suppose that beliefs, or at least judgments, must take place in a language-like medium. But models have their own resources for representing particular states of affairs. Linguistic beings like ourselves can readily transpose many (though not all) of these into language and language-like thought, but this is something *in addition* to a model-based representation. If we assume, on the basis of how we *report* it, that it must be structurally language-like, we are likely to be mistaken. Some, but not all, judgments are made or transposed into language or language-like thought. Some, but not all, beliefs are also judgments.

On my view, the intentional strategy is itself typified by a model—one for interpreting the behavior of agents. Or perhaps better, the intentional strategy may be a feature of any number of more particular models (the core agency system, folk psychology, and more specialized models erected on the scaffolding these provide) for interpreting agents. It may well be that it does so, as Sellars suggests, by characterizing agents *as though* their behavior were driven by language-like inner states. Like any model, this sort of model is idealized and answers to a number of pragmatic factors for its aptness. It is particularly apt in cases in which there are psychological processes involving language-like representations—including, but not limited to, thinking and reasoning in a natural language. It distorts its target more when there are model-based judgments that are not transposed into language or language-like thinking, but it can still latch onto psychologically real episodes even if it (mistakenly) treats them as though they were in a language-like medium. It also provides ways of talking about dispositions and normative commitments based in model-based thinking, though it does so in a fashion that provides no insight into their format or dynamics. The real patterns it tracks are emergent from an underlying model-based dynamics. In my view, it is features of these dynamics that are its real target—a form of psychological realism that is not a realism *about* sentence-like representations.

But the strategy can also be applied, with varying degrees of success, to systems that lack the same model-based dynamics. In the limiting case, we can say things such as "the thermostat believed it was too cool and switched on the furnace." Statements of this kind can allow us to attain some degree of predictive success even while greatly distorting the way we represent the system we are predicting. For Dennett, predictive success would seem to be the only measure of aptness. For me, it is one among several, and I would insist on a richer analysis of how an intentional strategy model is and is not aptly applied to a range of cases instead of a simple prediction-based criterion for what is a "true believer." In my view, the intentional strategy aims at tracking a particular type of target phenomenon—a form of cognition shared by humans and an indefinite number of nonhuman species. It can be apt for predicting the behavior of other types of systems, but when we use it for such purposes, we must remind ourselves that we are in effect taking a model of one target and using it as a model of another.

This approach seems particularly helpful in interpreting ascriptions of beliefs to nonhuman animals. If we suppose that language and language-like thinking are uniquely human, and attributions of beliefs ascribe dispositions to form language-like occurrent states, it would seem that nonhuman animals cannot really have beliefs. On the other hand, ascribing beliefs to a dog does not seem peculiar in the same ways that ascribing beliefs to thermostats does. We can make some sense of this, however, if we recognize that many nonhuman species do have some sorts of mental models, even if they are not able to form models of many of the things we can, and even if their models are surely different in their details from ours. Describing their beliefs in a fashion analogous to reporting human speech involves more of a distortion of the real psychological content of their mental states, because they do not have dispositions to make judgments in linguistic form. But in ascribing beliefs to them, we are nonetheless referring to inner states of the same kind that we refer to when we ascribe dispositional beliefs to one another: to dispositions grounded in the models they possess.

15.2 Models as Epistemic Units

Cashing out the notion of belief in terms of mental models provides a first, but relatively weak, connection to epistemology. In this section, I argue that there are also more robust connections: models are sources of warrant for beliefs and judgments, models are themselves epistemically evaluable, and the fact that models are integral units not reducible to beliefs requires that epistemology accommodate model-sized units.

Models as a Source of Warrant

I have claimed that models are *generative*, in the sense that possessing a model endows a person with a space of possible representations that can be framed in terms of that model. Judgments about particular situations are formed by applying a model that is used to frame the situation, and the resulting judgment is drawn from the space of possible representations generated by the model. (Of course, we can additionally combine representations drawn from multiple framing models using the more ecumenical representational medium of language, but these ultimately derive much of their semantic content from the models employed.) The fact that a judgment is framed within a particular model also connects it to the inferential resources of the model: to constitute something as a chess bishop is to take it to be something that can move diagonally, be captured, capture other pieces, and so on.

We now need to explore the epistemological implications of the relationship between models and judgments. The basic idea that I will develop here is a kind of reliabilist thesis: that models are reliable mechanisms for the production of knowledge in the contexts in which they are aptly applied. If this is the case, then at least part of what warrants many judgments is something about the models used to frame them, in which case mental models are of fundamental importance to epistemology. I will also explore (in the next section) the further question of whether models themselves are subject to epistemological evaluation in a fashion that goes beyond their aptness and reliability—for example, whether we should speak of models, as well as judgments and beliefs, as being warranted.

We should distinguish two types of judgments and their different relationships to models. One type of judgment makes explicit one or more features or commitments of a model. "A body at rest will remain at rest unless acted on by an external force" and "bodies exert an attractive force described by the inverse square law" express basic commitments of Newtonian mechanics. "Bishops may move diagonally" expresses a basic commitment of a model of chess. Insofar as one is using the model, one is normatively committed to such judgments. And to the extent that the model is apt, the judgments are true. In many cases, such judgments also seem intuitively obvious and even necessary to someone conversant with the model. (Indeed, they *are* "necessary" in the sense of being constitutive commitments of the model, much as the parallel postulate is a constitutive commitment of Euclidean geometry and thus necessarily true-within-Euclidean-geometry.) At least when one is using the model naively—that is, not attending to the fact that it is idealized and might not be aptly applied in all cases—such judgments seem *self-evident*. Indeed, I suspect that the vast majority of judgments whose truth seems self-evident do so *because* they express principles of deeply entrenched models that we can access through intuitive processes.[1] Of course, the fact that models are idealized and not apt for all cases implies that we are in danger of epistemic illusion (see chap. 14) if we endorse such propositions as universal and unidealized claims. But they are warranted (a) as claims about the commitments of the model, (b) as claims about what must be the case wherever the model is aptly applied, and often (c) as prima facie but defeasible assumptions if the model is one that is broadly and routinely applicable.

While judgments of this first type are often prominent in philosophy and the sciences, the vast majority of judgments we make are not of this type; they are judgments in which models are employed in how we frame particular states of affairs: "this body is at rest," "this body will take five seconds to reach the ground after it is dropped," "the white bishop is in a position to capture the black queen." Such judgments are neither derivable from nor warranted by models alone, as they also depend on facts in the world. Nevertheless there are important relationships between the framing models and the truth and warrant of the judgments. A judgment can go wrong in one way if it is framed in terms of an inapt model. In the context of cosmology, it makes no sense to speak of absolute rest. The Archbishop of Canterbury and the Queen of England may be standing in a diagonal relation on a checkered dais, and someone glancing at the scene might take it to be part of a game of human chess; but the observer would be mistaken, and the bishop is not poised to capture the queen (because he is not a *chess* bishop or she a *chess* queen at all, and the notion of "capture" employed in chess does not apply to them). If I use an inapt model to frame a situation, any warrant that its inferential rules might provide for a judgment based on them is nullified.

We can also go wrong even when we use apt models, if we do not use their resources to properly represent the particulars of the situation. In traffic court, it does make sense to speak of "being at rest" or "standing still" (with the surface of the Earth or the roads treated as a fixed frame of reference), but people routinely seem to think they have stopped at the intersection at the base of my street when in fact they have never slowed down below five miles per hour. The chess player may not see the board properly or not realize that moving his bishop would expose his king to checkmate. In such cases, the model employed *did* provide ways of representing things in a fashion we would count as true, but the judgment that was actually made represented things in another way, which we count as false. Again, in such cases, any warrant that might have been conferred by the model is nullified. If the board *had* been as I thought it was, my judgment that I was in a position to capture the opposing queen by moving my bishop *would* have been warranted, but the antecedent of the conditional is in fact false.

When we frame a situation using a model, all the inferential power of the model becomes immediately available through intuitive model-based understanding. We are able to attend to further implications through the lens of the model, often in such a fashion that it does not even seem like inference, but merely as "simply seeing" things such as that I can capture the opposing queen. If the framing goes *right*—if we employ an apt model and use the best representation afforded by the model for our judgment—these further judgments are also *warranted*. Models thus provide richly productive mechanisms for generating judgments, and when the model is apt, they are reliable mechanisms for generating true judgments. Moreover, to the extent that the warrant is supplied by the inferential potencies of the model, we may attribute the warrant conferred not only to judgments but also to beliefs in the broader dispositional sense when they are ones we are capable of making into judgments using the resources of model-based understanding.

Are Models Epistemically Evaluable?

Are models themselves epistemically evaluable? Given the standard characterization of knowledge as warranted true belief, models cannot count as knowledge, because models are not beliefs. (And as the terms 'true' and 'false' are applicable only to beliefs and to other units of the same size, such as propositions and sentences, models cannot be true or false, either.) It is true, of course, that we sometimes speak of some types of models, such as scientific theories, as "true" and as "knowledge," but this does not so much present a challenge as it bespeaks an ordinary language usage that is broader than technical usage, or a residual assumption that theories are propositional in character, or perhaps really that we can make statements *about* the aptness of models, and such *statements* can be true or false, warranted or unwarranted.

While I have spoken (albeit briefly) to the reasons for using separate terms to express the alethetic virtues that are appropriately attributed to judgments (and other sentence-sized units) and to models, I have yet to discuss the question of whether models are also units that can be *evaluated* epistemically for warrant or something like it, apart from questions of whether the beliefs constitutively associated with them are individually warranted. We have at least two distinct sorts of questions here. First, are models, as models, subject to any sort of epistemological evaluation at all? Second, if they are, are the features that count as epistemic virtues for models sufficiently like those that count as epistemic virtues for beliefs or judgments that we should use the same terminology for them—for example, should we speak of models as things that can be warranted?

Given the reliabilist characterization I have given earlier, one might be led to suppose that models are not likely to be subject to any sort of epistemological assessment beyond the fact that they can be reliable mechanisms for producing true beliefs and judgments and consequently confer warrant on those beliefs and judgments. Perceptual mechanisms are often used as paradigm cases of reliable judgment-producing mechanisms. But such mechanisms—say, whatever neural processes produce a conceptually laden perceptual gestalt from a pattern of stimulation on sensory transducers—are not themselves the sorts of things that can be warranted or unwarranted, unless perhaps by that we mean simply that they are *reliable*. They deal in concepts and judgments only in their *outputs*, and their operations are subject neither to introspective scrutiny nor to revision. To be sure, we normally proceed on the tacit assumption (a "belief" in a very broad sense) *that* the judgments they produce are trustworthy, and we can judge *that* they are unreliable under particular conditions on the basis of experience. *These* beliefs and judgments can be (or fail to be) warranted, but that does not mean that the mechanisms themselves are warranted or unwarranted. The application of the notion of "warrant" to such perceptual mechanisms seems to be something of a category error.

But this does not imply that belief-producing mechanisms *in general* are not epistemically evaluable beyond the question of whether they reliably produce true beliefs. Special features of the mechanisms involved in perceptual preprocessing may debar them from specifically epistemic evaluation: the mechanisms do not operate on conceptually laden content, operate automatically, and are not subject to scrutiny or revision. Models can also operate automatically; and as some modules may also be models, some models may not be subject to scrutiny or revision, either. But other models have inference patterns that operate on propositional content, can be used deliberately as well as automatically, and are subject to scrutiny and revision in light of *reasons* to do so.

I have used scientific theories as one paradigm example of mental models. And scientific theories are clearly held to standards of epistemic evaluation: they are compared with one another for their adequacy, and we speak of them as being "evidenced" or "confirmed." It is perhaps a matter of semantic choice whether we should speak of theories themselves as being "warranted" or say instead that beliefs *about* their credentials are warranted and use some other term (like "confirmed") for theories themselves. But some form of epistemic evaluation clearly seems appropriate. So at least *some* mental models are subject to epistemic evaluation, whether we call it "warrant" or "confirmation" or by some other name.

But there are other types of mental models with respect to which it would sound decidedly odd to apply the same terms. We would be unlikely to speak of someone's understanding of chess, restaurant dining, or vanity as being justified, warranted, confirmed, or evidenced. If we make normative evaluations of someone's understanding of such matters, it is usually in terms of whether they have gotten the model *right*. Moving a piece in an illegal way, or failing to see an impending checkmate, or going to the restaurant kitchen to prepare one's own meal indicates a limitation or even a lack of understanding. But the deficiencies they show are more akin to falsity (having the wrong model altogether) than to lack of warrant.

On closer scrutiny, however, it does make sense to ask questions about whether one is entitled to all the assumptions implicit in the model one is using. I recall a stage in learning chess as a child when I had observed pawns moving two spaces but had not recognized that this is permitted only on their initial moves. My error was quickly pointed out after I made what I had supposed to be a clever (but in fact illegal) move; and in retrospect I can see that I had incautiously made what turned out to be the wrong generalization based on my observations and whatever instruction I had received in the game at that point. It may well be that this happened so early in my acquaintance with chess that I had not yet fully formed a model (even an incorrect one) of the game and was still operating at the level of memorized rules. But even so, if there had not been early occasion for the error to be corrected, I might very well have formed a deviant model of chess. Forming a deviant model is itself still a matter of getting the model *wrong*. But we might go on to ask questions about the epistemic merits and faults of the process through which I formed the assumption that a pawn can generally move one or two spaces forward. One might debate whether my internalization of a model with that particular set of deviant rules was *justified*—I seem to recall a feeling of outrage at the time that my teacher had not made it clear that pawns could move two spaces *only on the first move*—but the very fact that there is such a question to be asked implies that epistemic evaluation has a place here.

We can frame similar issues about the understanding of when a model can aptly be applied. I recall trying to order wine at a fancy restaurant where such orders are handled not by the waiter or waitress but by a wine specialist called a sommelier. The model of restaurant dining I was accustomed to was one in which a single person takes all types of orders for food and drink, which is indeed an apt model to use in the vast majority of dining establishments I visit. I assumed it would aptly be applied at this particular restaurant as well, but I was mistaken, and the waiter politely replied that he would send over the sommelier. It was an understandable mistake; I had no way of knowing based on a fairly long history of trips to restaurants, no one had bothered to warn me, it was a mistake the restaurant staff probably encounter every day, and so on. Arguably, while I had committed a minor faux pas, I had done so on the basis of assumptions to which I might claim epistemic entitlement. Nonetheless I have subsequently adopted a more cautious attitude toward dining in unfamiliar contexts—particularly restaurants with waitstaff in black tie or where I am faced with an array of more pieces of silverware than I am sure what to do with. *Knowing* that dining contexts are not all alike, I face questions about what I may safely assume. And even before I had this awareness, the assumptions were merely assumptions, and in making them, I deviated from some type of ideal epistemic norm.

Models of chess and restaurant etiquette differ from scientific models in an important way: one has the "right" model of chess or dining simply by dint of having one that corresponds with the models that others are using. There is something of the same in understanding a scientific theory—say, having a correct understanding of classical mechanics—but a further question arises as to whether that model is an adequate model of the phenomena it is applied to, and this latter question is at issue in scientific confirmation. But scientific models are by no means unique in this respect. The models I use to navigate my house or the town I live in have to track features of those environments if they are to enable me to navigate successfully, even though their assumptions and commitments are seldom explicitly expressed. The statistical heuristics we employ in everyday life are apt (or not) for particular contexts, and we can be in better or worse positions to assess their aptness. (Hence the tendency of researchers dealing with heuristics and biases to describe inapt strategies as "irrational," particularly when the person who employs them is in a position to know they are inapt.) Our models of social categories such as race and gender are often woefully inadequate to representing their targets and are produced and reinforced through mechanisms that are notoriously unreliable for tracking the facts. The processes of consciousness-raising that can disabuse us of such social biases are not only morally edifying but also put us in a better epistemic position with respect to our social understanding.

We probably need far more discussion of the epistemic evaluation of models than I am in a position to give here, and different types of models are likely to require different types of evaluation. Epistemic evaluation seems appropriate with respect to ways a model is produced, how (or if) it can be scrutinized and adjusted, and how we can learn to distinguish the cases in which it is apt from those in which it is not and use it only in contexts in which it is apt. Developmentally canalized models like the core systems are produced primarily through evolution and development, but these can perhaps be regarded as prior mechanisms that can be evaluated for their reliability in producing apt models. Learned models can be held accountable to standards of confirmation and can generally also be adjusted and improved on. In both cases, true mastery of the model involves not only *having* the model and being able to use it, but also having an understanding of where it is aptly applied.

Models as Integral Epistemic Units

A model-based view of understanding has implications for epistemology because models turn out to be important epistemic units in their own right, over and above beliefs. If models are themselves epistemically evaluable and furthermore provide warrant for judgments and beliefs, an epistemological theory that deals exclusively with beliefs leaves out something important. But incorporating mental models into epistemology also has a further implication that will prove particularly important in assessing the adequacy of two of the major epistemological theories, foundationalism and holism.

Models are *integral units*. Often the commitments of a model can be made explicit in the form of judgments and implication rules, but a model is not simply an assemblage of beliefs and implication rules that can be independently specified, as these are constitutively interrelated—or, as it is sometimes phrased, "interdefined." This has been a familiar claim about scientific theories for several decades, and that alone should have been enough to cause a rethinking of belief-based epistemology. If it is a general characteristic of mental models, and models are the basis of understanding, the implications are all the more urgent.

In philosophy of science, theories are commonly regarded as principal units for *confirmation*. The credentials of relativistic claims about mass cannot be evaluated apart from relativistic claims about space and time, and vice versa. Likewise, applications of the theory to particular cases are warranted by the aptness of the theory as a whole. An adequate epistemology of science must accommodate the role of models—as integral units—in receiving justification and conferring warrant. If similar considerations apply to models generally, the implications for epistemology are considerably broader. I argue in the next section that this has a serious implication for the field of epistemology: neither foundationalism nor global coherentism can be a fully adequate epistemological theory, because neither accords model-sized units their proper role.

15.3 Cognitive Pluralism and Theories of Knowledge

Now that I have given a rough characterization of the cognitive pluralist view of the interrelationships between models, beliefs, and warrant, let us turn to its relation to familiar theories in epistemology and its implications for those theories. Epistemology is the study of knowledge. Mainstream contemporary epistemology is pursued under the shared assumption that knowledge is warranted true belief, alternative theories differing principally in their accounts of what makes a belief warranted. The traditional alternatives for an account of warrant are foundationalism and coherentism, along with a more recent contender called reliabilism.

Reliabilism

While reliabilism is a relatively new addition to the list of philosophical accounts of epistemic warrant, I will address it first, because the characterization I have given of the warrant conferred by apt models seems in fact to be a reliabilist view. The reliabilist holds that a belief is warranted if and only if it is produced by a reliable mechanism for producing true beliefs in that type of context. Alvin Plantinga, for example, writes that a belief is warranted if and only if it meets the following conditions:

> (1) It has been produced in me by cognitive faculties that are working properly ... in a cognitive environment that is appropriate for my kinds of cognitive faculties, (2) the segment of design plan governing the production of that belief is aimed at the production of true beliefs, and (3) there is a high statistical probability that a belief produced under those conditions will be true. (Plantinga 1993, 46)

Saying that a mechanism for belief production is "reliable" in producing true beliefs does not really tell us much about the mechanism, even if the reliability of the mechanism is cashed out in terms of probability. This is not necessarily a bad characteristic for the reliabilist theory to have, as there may be many reliable mechanisms that are reliable in different ways and for different reasons, and hence a highly flexible account of warrant may actually be required.

One of the core applications of reliabilist accounts has been to the production of beliefs through perception, where the causes of the belief are not themselves beliefs. It seems clear enough that perceptual processes are in some broad sense "mechanisms," and the details of how they work and why they are reliable can be left to the cognitive scientists. But to extend the theory further, we would need some way of clarifying what sort of "faculty" or "mechanism" might be involved in nonperceptual belief formation.

In my account, models provide ways of framing situations and forming judgments about them, as well as licensing judgments about them. Some "beliefs" in the dispositional sense, moreover, can be seen in terms of what inferences are licensed by the inferential rules of the model. Models are thus, among other things, "faculties" or "mechanisms" for judgment production, and when the model is aptly applied, it confers warrant on the judgments thus formed. It is thus natural to view a model-based account as a way of fleshing out a reliabilist epistemology with respect to a large and important class of ways that beliefs are formed through nonperceptual processes.

Foundationalism

Foundationalism is a very old philosophical view. Variations on it were endorsed by Aristotle and his medieval followers, by early modern philosophers in both the rationalist and empiricist camps, and by a significant number of twentieth-century philosophers, including some of the logical positivists, such as Moritz Schlick (1934/1979). Aristotle is often credited with first articulating a foundationalist account of knowledge in his *Posterior Analytics*, in response to a skeptical problem about demonstrative knowledge, the *regress problem*. The skeptic argues that whatever is known must be demonstrated—that is, deduced validly from things that are already known. But for those things to be known, they must be demonstrated from other things that are already known, and so on, in an infinite regress, so that the chain of demonstration cannot get off the ground, and hence knowledge is impossible. Aristotle seems to endorse demonstration as a paradigm for knowledge but rejects both the skeptical regress and the alternative of circular demonstration in favor of the view that there can be another, nondemonstrative form of knowledge of the first and foundational premises. Thus we have two types of knowledge: things that are known directly in their own right, and things validly derived from such foundations. This is the core of a foundationalist view of knowledge: in contemporary terms, a belief is warranted if and only if it is either (1) properly basic or (2) derived in the right sort of way (usually meaning by valid deduction) from properly basic beliefs.

Several brief observations are in order. First, both the core view and the way it is arrived at bear the distinctive marks of the standard view of cognition. Not only are the *units* sentence sized (beliefs, judgments, propositions, or actual sentences) and argument sized (demonstrations), but the starting point of the discussion is a paradigm of knowledge as a product of demonstrative argumentation. Second, while contemporary discussions in epistemology generally employ the word 'belief', it seems clear that the account is really an account of specifically *occurrent* beliefs (i.e., judgments), as it is hard to see how merely dispositional beliefs could plausibly be seen as products of demonstration. And third, I have described what I have presented so far as the "core" of foundationalism because it leaves open the questions of *what sorts* of beliefs can be properly basic, *what kind* of warrant is required for a belief to count as knowledge, and *in what fashion* properly basic beliefs are supposed to be warranted.

Up until fairly recently, the vast majority of foundationalists assumed that the standard for knowledge needed to be *certainty*. As a result, properly basic beliefs were generally assumed to be ones that are in some fashion self-evident, indubitable, or incorrigible. For rationalists, these might be truths of reason that are self-evident because their denial is either self-contradictory or literally inconceivable. For empiricists, they might be Humean impressions or positivist sense data. Foundationalist views that additionally require that properly basic beliefs be in some fashion indubitable are now called *strong foundationalism*.

From a contemporary standpoint, it may seem somewhat remarkable that, for so much of the history of philosophy, the notion of knowledge was restricted to things that could be demonstrated or things so certain that they required no demonstration. I suspect that two factors played a role in this. The first was the specter of skepticism and the perceived need to respond to skeptical challenges. The second was that many generations of philosophers inherited a technical notion of "knowledge" (*epistēmē, scientia*) that in fact had something like Aristotle's demonstrative standards built into it. I see little way of understanding why Locke said that his "sensitive knowledge" might better be called "faith or opinion," or why Hume viewed the nondemonstrative character of induction as leading to a skeptical conclusion, unless they simply used the word 'knowledge' (or the Latin *scientia*) in an already-standardized way to *mean* "things either known directly and indubitably or else derived from these through demonstration."

Locke, of course, also characterized properly basic beliefs as "intuitive knowledge." My discussion of the nature of intuitions and their capacity to generate cognitive illusions provides further reason to doubt that a sense of subjective certainty is any guarantee of truth or warrant. But the market for additional arguments against *strong* foundationalism is slim nowadays, as it fell precipitously from favor in the middle part of the twentieth century, due in large measure to criticisms from Quine (1951) and Sellars (1956). In the 1970s, however, several philosophers proposed versions of a "modest foundationalism," with weaker requirements on the epistemic status of basic beliefs:

> Prior to 1975 foundationalism was largely identified with strong foundationalism. Critics of foundationalism attacked the claims that basic beliefs are infallible,

> incorrigible, or indubitable. However, around this time there was a growing recognition that foundationalism was compatible with basic beliefs that lacked these epistemically exalted properties. William Alston ..., C. F. Delaney ..., and Mark Pastin ... all argued that a foundationalist epistemology merely required that the basic beliefs have a level of positive epistemic status independent of warranting relations from other beliefs. In light of this weaker form of foundationalism the attacks against infallibility, incorrigibility, or indubitability did not touch the core of a foundationalist epistemology. (Poston 2015)

Modest foundationalism avoids one of the problems with strong foundationalism: its overly high epistemic standard for properly basic beliefs or judgments. But it retains the core foundationalist commitment to the assumption that beliefs are warranted individually, some by being properly basic and others by being inferred from properly basic beliefs. And it is this commitment that is most directly at odds with a model-based cognitive pluralism.

From a cognitive pluralist standpoint, the claim that some judgments are warranted because of how they are inferred from other judgments need not itself be regarded as problematic. A cognitive pluralist can certainly recognize that we do sometimes reason using various forms of logical inference ranging over sentence-sized units, and the validity of the inference form and the warrant of the premises are surely relevant to the epistemic status of the conclusions thus derived. The problem is that this is neither the *only* nor even the *most fundamental* source of warrant, and the kind of warrant conferred by models does not simply attach to individual judgments one by one. At the least, *some* of our beliefs that have excellent credentials for counting as knowledge, such as those involved in scientific theories, are neither individually properly basic nor justified by inference from other beliefs that are already warranted as properly basic. Rather, the beliefs and the whole framework of concepts and inferential patterns are warranted together, as a unit. And if the cognitive pluralist is correct that this is not an idiosyncratic feature of scientific theories, but a much more general feature of model-based understanding, foundationalism lacks the resources to explain how an extremely large number of our beliefs are warranted. At the least, foundationalism is incomplete as an account of knowledge and requires resources of a very different sort to supplement it.

From a cognitive pluralist perspective, foundationalism seems best interpreted as an idealized model of a particular dimension of epistemic goodness. It is specifically focused on judgments produced by argumentative inference, and quite reasonably identifies the validity of the inference and the warrant of the premises as factors relevant to the epistemic assessment of judgments thus derived. The legitimate conclusion of the regress argument is that, for such a process of transferring warrant through logical inference to get off the ground, there must be premises that are warranted in some other way. It would perhaps have been best to stop here and regard this as a consequence of how the account is idealized: that it is *just* an account of warrant-via-inference and simply brackets the question of how else judgments can be warranted.

Some versions of modest foundationalism, however, have in fact explored ways that the theory can be supplemented with other sorts of resources, and at least one of these suggests a way that a modest foundationalist might attempt to incorporate insights about mental models. One long-standing problem for epistemology was that of how sensation and perception can warrant beliefs. Alvin Goldman (1979) proposed a reliabilist account of this: perceptual beliefs can be properly basic by dint of being produced by perceptual mechanisms that reliably produce true beliefs. While reliabilism has gone on to be regarded as an epistemological theory supplying its own account of warrant, Goldman's account is often also classified as a foundationalist account, augmented with a reliabilist story about how a certain class of judgments come to be properly basic. This, in turn, suggests a way that a modest foundationalist might embrace models as playing a role in warranting beliefs and judgments. If, like perceptual mechanisms, models reliably generate true judgments when aptly applied, at least some judgments produced from models might count as properly basic.

I prefer to leave it to the foundationalists to decide what should be counted as a variant on foundationalism, or a supplement to it, and what needs to be viewed as a nonfoundationalist account. Clearly the core commitments of foundationalism leave open questions about the noninferential sources of warrant, and I believe that some of these are derived from model-based understanding. Just how model-based understanding can be combined with a foundationalist account is constrained by the fact that foundationalists group all noninferential knowledge together in the category of properly basic beliefs. So whether beliefs that derive their warrant from models can be accommodated within a basically foundationalist framework will depend on how one understands what it means for a belief to be properly basic. Strong foundationalism, of course, has extremely high standards for proper basicality, and judgments produced from model-based mechanisms do not seem to meet those standards. I have explored how models can produce "intuitive" judgments, but we have also seen that such judgments are far from infallible, though they may come with a subjective sense of certainty. But the modest foundationalist is not committed to such high standards, and there seems to be no obvious and uncontentious answer to the question of how the modest foundationalist ought to understand proper basicality.

One possibility is that a belief should be counted as properly basic just in case it is warranted, but not in a fashion derived from logical inference. In this interpretation, judgments produced by the apt application of models might indeed count as properly basic. But if this is all that "properly basic" means, there might be any number of noninferential sources of warrant, and all that groups them into a class is a feature they lack: they are *not* processes of logical inference. This might not be such a problem if they also shared some feature that rendered them in some recognizable sense *foundational*. But it is not clear how the metaphor of a "foundation" should be understood once one has abandoned stronger forms of foundationalism. Perhaps it should be understood epistemically: foundational beliefs are those about which there are no further epistemic questions to ask. Or perhaps it should be understood psychologically, in terms of the origins of beliefs and judgments: the foundational ones are the first epistemically evaluable ones in a process, or perhaps ones without which no other beliefs or judgments could be warranted.

It seems plausible to me that beliefs based in models may fit *some* such *psychological* criteria: without models we would have no beliefs at all, and these are the beliefs on which any other epistemic processes operate. However, most models are learned, and in learning a model, we often engage in some sort of process of confirming that it is a good model, at least for particular epistemic and practical ends. The process of confirmation may be unconscious, implicit, and nonargumentative (though in cases like scientific models, a good deal of explicit reasoning often goes into building and confirming them), but models are not really "epistemic foundations" in the sense of being "givens" whose epistemic credentials can be taken for granted. As a consequence, the production of judgments from models does not seem to be an epistemological foundation, either. There *are* further epistemological questions about judgments produced by models: questions about whether the model is a good model of its domain, whether it is aptly applied in this circumstance, and whether the judgment produced is the best judgment among the options available from the model. There are likewise further epistemological questions about models themselves and the beliefs expressing their systematic commitments: about how they are idealized, and about what sorts of cases the models are apt for and hence about which of the judgments they produce and their implicit constitutive beliefs can be true. (These questions, moreover, can be asked not only about learned models but about developmentally canalized models as well.)

It thus seems decidedly odd to characterize models as epistemic *foundations*. Models may indeed be in some sense prior to beliefs and judgments, and the epistemic credentials of beliefs and judgments ultimately dependent on the qualities of the models in which they are framed. But given that there are multiple learned models, often possessing different constitutive beliefs and licensing contradictory judgments, the resulting picture is closer to Neurath's metaphor of a raft we are always reconfiguring than to Schlick's model of a pyramid of knowledge.

Alternatively, the modest foundationalist might be inclined to view at least some of the processes by which models give rise to judgments as forms of *inference*, distinct from, but not necessarily less respectable than, deductive inference. Some foundationalists have long embraced nondeductive inference forms—induction, retroduction—as conferring warrant, and there seems no principled reason that other inferential processes could not also be included. This strategy, however, strikes me as problematic as well. I have no objection to calling some of the model-based processes that produce judgments "inferences"; indeed, it seems a natural way to characterize them. The potential sticking point is that if we do so, they are inferences *to* things with propositional form but not *from* things with propositional form, as models neither are, nor are made up of, beliefs or judgments. The problem here is that foundationalism characterizes the inferential grounding relation as one between beliefs, which would not be true of an "inferential" process in which model-based processes generate a judgment through their own nonlogical, nonargumentative mechanisms. Broadening the foundationalist notion of "inference" in this way would seem to risk losing a distinction that lies at the heart of foundationalism: the distinction between the kind of warrant conferral that comes through argument and whatever other ways a judgment or belief might come to be warranted.

If models are themselves subject to some form of epistemic evaluation and play a role in conferring warrant on judgments and beliefs, foundationalism is an inadequate account of knowledge. Most fundamentally, it neglects models as a distinct type of epistemic unit, not reducible to beliefs and inferences. Because of this, it also lacks the resources to accommodate the roles models play in warranting beliefs and judgments. Modest foundationalists might seek to treat a model-based account as a supplemental story compatible with foundationalism, whether by treating models as reliable mechanisms for producing properly basic beliefs or by treating model-based processes that produce judgments as a form of inference. But they do so at the risk of abandoning deep foundationalist commitments.

Coherentism

Coherentists locate the warrant for a belief in its being part of a maximally large set of beliefs that are mutually coherent. "Coherence," however, is a rather vague notion. In a minimal construal, coherence requires *consistency*: a coherent set of beliefs must be a set of beliefs that are mutually consistent. But the word 'coherence' also seems to suggest something more than this, ranging from relationships of *mutual support* to *mutual entailment.*

Cognitive pluralism helps to explain the mutually supportive relationships between beliefs associated with a particular model but creates problems for global consistency. The system of meanings, beliefs, rules, and implicational connections involved in a mental model is generally not only mutually consistent but mutually supporting and in some cases mutually implicating. *Within* a model, there should be consistency and fairly strong forms of coherence. But the same need not apply *between* models. Because each model is idealized in a way that is fitted to particular domains, problems, and interests, two models can be mutually contradictory both in their assumptions and in what they imply when applied to particular cases.

As a result, we should distinguish between two kinds of coherentist claims. A claim that requires coherence *within a model* is quite consonant with cognitive pluralism. A model that is internally inconsistent or generates inconsistencies when applied to various particular cases is clearly problematic. But if models are idealized, and two models are idealized in different ways, they can yield implications that are inconsistent. Thus a more *global* coherentist thesis—one that requires consistency between *all* beliefs for any to be warranted or treats inconsistency of a belief with *any* other belief as incompatible with its being warranted—is problematic. Indeed, if the idea, explored in chapter 12, that some such inconsistencies might prove to be principled and abiding is correct, then cognitive pluralism may prove to be incompatible with global consistency. (Of course, there is also a more general concern that, as we all have some number of inconsistent beliefs, a globally coherentist epistemology would imply the strongly skeptical conclusion that none of us knows anything at all.)

As was the case with respect to foundationalism, the cognitive pluralist might see coherentism as expressing a partial and idealized truth. Consistency and stronger forms of coherence are each in themselves *good* things, both within a model and between models. Inconsistency between models is a prima facie indication of a problem and gives us reason for further scrutiny. But it may present an insurmountable problem if it turns out to arise from the ways different models are idealized. On the other hand, incommensurability between models need not always result in inconsistency between beliefs, because models are idealized and not applied aptly to all cases. In some cases, like quantum complementarity, the incommensurability of models may not lead to inconsistency of beliefs because we recognize that one of the models should not be applied in the case at hand, and hence we do not endorse the application of that model to the case by forming a judgment based on that model.

15.4 A View of the Status of Accounts of Epistemology

The basic line of criticism I have leveled at standard epistemological views has been that while each may pick out factors that are relevant to the assessment of warrant, none of them can tell the full story. In the case of foundationalism and coherentism, the problem is that, as general theories, they leave important things out or get some things wrong. In the case of reliabilism, the issue is that a simple appeal to "reliability" does not tell us much about the psychological nature of the reliable mechanisms.

Foundationalism, coherentism, and reliabilism are often treated as rival theories of warrant. In philosophical writing, proponents of one account will often show how their theory handles a particular range of cases well, but argue that the rival accounts yield counterintuitive assessments of which beliefs are warranted. The underlying assumption of such arguments is that each theory is taken to be offered as a comprehensive and unidealized account of the same thing: something called "epistemic warrant" or "knowledge." But cognitive pluralism suggests an alternative way to view them. Each account is an *idealized* account of *one dimension* of epistemic evaluation. Beliefs can be justified by their relationships to other beliefs in various ways, and the ways we reason from some beliefs to others are relevant to the overall status of the derived beliefs, as the foundationalist suggests. The ways various beliefs cohere or conflict with one another are also relevant to their epistemic status, as are the ways they are produced by mechanisms that are or are not reliable in a given type of situation. Arguably, each account provides a useful way of looking at how beliefs can go awry that is not captured by the others. We might thus look at different epistemological theories not as rival accounts but as multiple idealized ways of looking at an issue (epistemic warrant, or how beliefs come to count as knowledge) that is too complex to be handled by a single theory, and as together providing a way of triangulating it for better understanding than we could obtain from any single account.

This is not so uncommon a situation as one might suppose. Consider the concept HEALTH. Medicine employs the concept of health but provides no positive account of it. Instead medicine examines various dimensions along which an organism can approach or deviate from an unarticulated ideal standard: various forms of disease, deformity, injury, and dysfunction. I will call this kind of concept—one of which there is not a single definitive positive account, but various relevant accounts of how an ideal may be approached or deviated from in different ways—a *standard*. HEALTH is a concept of this type, and so, I suggest, are KNOWLEDGE and WARRANT. I suspect that, upon analysis, other important philosophical concepts like GOODNESS might turn out to be standards as well.

This analysis involves putting the interpretive tools of cognitive pluralism to the task of interpreting philosophical theories. Philosophical theories are themselves models, and as such they have particular content domains that they represent in a particular fashion. They are idealized, and unlikely to capture *everything* that might really be important about their target domains, and hence a single theory may not aptly be applied in all cases. Indeed, the kind of situation we find in protracted philosophical debate, in which proponents of "rival" theories each have a set of core cases that seem to be aptly treated by their theory but not by others, can be diagnostic of a situation in which we have multiple models with overlapping domains, none of which is apt for *all* the relevant cases.

16 Cognitive Pluralism and Semantics

Many sorts of things are said to "be meaningful" or "have semantic content." Some of these are mental: concepts, beliefs and other dispositional intentional states, judgments and other occurrent intentional states. Others are linguistic: words, phrases, sentences, speech acts, utterances, inscriptions. Some would say that logic has its own stock of meaningful units, which are neither truly mental nor truly linguistic: bound variables, constants, predicate letters, Fregean propositions, well-formed formulas. And these meaningful entities come in a variety of sizes as well: word-sized units (lexical units in a language, concepts, predicate letters), sentence-sized units (sentences, utterances, speech acts, beliefs, judgments, propositions, well-formed formulas), and larger units such as arguments.

In one sense, we thus have a number of different types of "semantic units," differing in compositional size and in whether they are mental, linguistic, or abstractly logical. But we might also ask whether one of these types of units is *fundamental*. Discussions of this topic generally list either two or three candidate answers. Perhaps most common is a two-way division between the options of atomism and holism:

> Semantic atomism locates meaning at the level of individual words (the meaning of each word is *in* it); semantic holism locates meaning at a broader level, ranging from an indefinite network of linked units to an entire language. (Engler 2011, 266)

> The tradition of formal semantics has been resolutely atomistic, in the sense that the assignment of a semantic interpretant to one element (say, a proper name) is taken to be intelligible independently of the assignment of semantic interpretants to any other elements (for instance, predicates or other proper names). One does not need to know anything about what other dots represent, or what blue wavy lines represent, in order to understand that a particular dot stands for Cleveland on a map. The task of formal semantics is a bottom-up one of explaining how semantically relevant whatsits can systematically be assigned to complex expressions, given that they have already been assigned to simple ones. Atomism adds that the assignments

to the simple ones can be done one by one. By contrast, inferentialist semantics is resolutely *holist.* On an inferentialist account of conceptual content, one cannot have *any* concepts unless one has *many* concepts. For the content of each concept is articulated by its inferential relations to *other* concepts. Concepts, then, must come in packages (though it does not follow that they must come in just one great big one). (Brandom 2000, 15–16)

Brandom's claim that "it does not follow that they must come in just one great big package" invites a further distinction, between globally holist theories and "molecularist" (Block 1998) or "localist" (Weiskopf 2009) theories that define the fundamental semantic unit somewhere between the individual concept and the entire web of concepts, beliefs, and inferential commitments:

Mental (or semantic) holism is the doctrine that the identity of a belief content (or the meaning of a sentence that expresses it) is determined by its place in the web of beliefs or sentences comprising a whole theory or group of theories. It can be contrasted with two other views: atomism and molecularism. Molecularism characterizes meaning and content in terms of relatively *small parts* of the web in a way that allows many different theories to share those parts. For example, the meaning of 'chase' might be said by a molecularist to be *try to catch.* Atomism characterizes meaning and content in terms of *none* of the web; it says that sentences and beliefs have meaning or content independently of their relations to other sentences or beliefs. (Block 1998)

A model-based cognitive pluralism clearly favors a "molecularist" theory of semantics. Most debates in this area, however, treat atomism and holism as the options, and often one theory is touted largely on the grounds that the other is unacceptable. It is widely argued that molecularist accounts are not really a viable third alternative, on the grounds that there is allegedly no way of containing the spread of semantic interrelations short of holism. And this may well be correct, *if* we assume that the only relevant units are concepts, beliefs, and inferences. This, however, is precisely where cognitive pluralism differs from the more familiar theories. A mental model is a distinct type of psychological unit. Moreover, it is not *defined* in terms of interrelationships between concepts, beliefs, and inferences in the sense of being a kind of *construct* out of them. It is conceived to be a real feature of cognitive architecture and as such can provide a principled basis for constitutive relationships *within* a model that can be differentiated from merely epistemic and inferential relationships with things *outside* it. This is just the kind of notion the molecularist needs to escape the charge that there is an unavoidable slippery slope into holism.

16.1 Models and Semantic Value

For several decades, philosophers of science have generally accepted the claim that theoretical terms in a science are constitutively interdefined with other terms in the theory, and their meanings—at least in a broad but plausible sense of 'meaning' that includes the constitutive inferences one is disposed to draw simply by understanding the terms—are derived from the theory as well. But scientific theories are not the only forms of understanding of which this is true. It is part and parcel of having mastered the concept CHESS_BISHOP that one understands that there is such a thing as the game of chess, that bishops are chess pieces, and that they can move diagonally within the game. To grasp the concept WAITER, one must understand that there are dining contexts in which some people have the role of serving food to others and that waiters take customers' orders and bring food and drink. Indeed, it is often through observing such behavior that one decides that a particular concept is applicable in a given case. If I see that two people have drawn an 8 x 8 board in the sand and are moving various rocks between spaces, I might form the judgment that they are playing chess, and decide that one piece is a bishop because it is moved diagonally. In a restaurant, I may conclude that someone is a waiter because I observe him taking orders and delivering food.

The idea that the meaning of a word or concept is partly determined by the inferences in which it is used is central to inferentialist semantics. But not all the inferences we are disposed to draw are plausibly regarded as bound to the semantic values of concepts. On the one hand, there are syntactically based inferences. If I make a syllogistic inference based on the syntactic properties of the propositions, the semantic values of the concepts involved are irrelevant to the "inferential disposition" that is at work, and I can make or be disposed to such inferences without it having any constitutive effect on my concept WAITER. On the other hand, we make inferences that are based in mere association. Someone who lives in New York or Los Angeles may associate waitstaff with out-of-work actors and actresses and perhaps even believe that all waitstaff have dramatic aspirations. This, however, is not part of the *semantics* of WAITER. Rather, it is based in some additional belief, or perhaps merely in a bias that I would not even reflectively endorse as a belief. Confronted with evidence that some or even most waitstaff have no dramatic ambitions would not force me to revise my concept WAITER; it would at most disabuse me of a bias. Connections to *some* other concepts, and to *particular* beliefs and inferences, are part and parcel of the semantic properties of concepts; but not *all* semantic and inferential connections are *constitutive* of a concept's semantic value.

But how are we to distinguish which other concepts and inferences are implicated in the semantics of a concept? In light of the previous chapters, an answer suggests itself: the ones that are involved in the same model are likely to be thus interrelated, the ones outside it are not. Inferential semantics gets something right but does not itself have the resources to differentiate between different types of inferential disposition. We need a principled way of delineating such boundaries. A model-based cognitive pluralism supplies a promising candidate: connections *within* a model are likely to be constitutive of a concept's semantic value, while dispositions grounded in something outside the model are not.

There is, however, an important issue here that I will take up at the end of the chapter: some concepts seem to appear in multiple models. The concept PLANET might occur both in a model of our own solar system (where it applies to a small number of objects) and in a more general model of the mechanics of gravitational systems (where it applies to an open-ended number of objects). Concepts like FRUIT and VEGETABLE seem to appear in biological, culinary, and nutritional models, but their extensions are different in the different models. A tomato is a fruit according to a biological taxonomy, but a vegetable according to a culinary taxonomy. Concepts of mass are found in classical and relativistic theories but have different implications and different meanings within them. This raises problems about the identity conditions for concepts and suggests that we may need to also find a way to say that the life of concepts is not fully contained within individual models, or else say that concepts found in different models are really distinct concepts, though we use the same words to express them.

16.2 Cognitive Pluralism and Other Semantic Theories

Let us now turn to the relationships between a model-based cognitive pluralist semantics and the most familiar alternatives, and the implications of cognitive pluralism for those theories. The principal semantic theories are atomist and holist theories, each of which is problematic from the standpoint of a model-based cognitive pluralism.

Atomism

Discussions of semantic atomism contain several importantly different types of considerations. One strand of atomism is tied to the view that the semantic values of sentence-sized units—natural-language sentences, beliefs, judgments, propositions, and logically well-formed formulas—are a function of (a) the semantic values of their word-sized units and (b) the logical or syntactic structure of the sentence-sized unit. The fundamental locus of semantic value is in word-sized units such as individual concepts or words.

We may distinguish, in terms of analysis, between a weaker and a stronger thesis here:

Compositionality (weak): The semantic value of each sentence-sized unit is a function of (a) the semantic values of its constituent word-sized units plus (b) its combinatorial syntax (or whatever plays a role analogous to syntax within judgments).

Semantic independence (strong): The semantic value of each word-sized unit is independent of the sentence-sized units in which it is used, including the beliefs one has involving the word-sized unit and the judgments using it that one is disposed to make.

Whether these two theses are equivalent depends on the notion of "independence" assumed. There is a clear sense in which the concept BACHELOR cannot be "independent" of a commitment to the judgment or belief that bachelors are unmarried. One cannot be said to possess the concept BACHELOR and not be committed to bachelors being unmarried. But there is also a more directional notion of "dependence" and "independence" that is concerned with what is more fundamental. A theory of semantics that held, for example, that analytic sentences like "all bachelors are unmarried" are true because the concept BACHELOR *contains* the concept UNMARRIED would make the commitment to the judgment "all bachelors are unmarried" a *consequence* of the semantics of the concept BACHELOR, but not the other way around.

But whether or not one holds there to be this unidirectional dependence between semantic value and epistemic commitment, there might nonetheless be a *compositional* dependence: the sentence "all bachelors are unmarried" means what it does as a function of its constituents and its compositional syntactic structure. The potential confusion to be avoided here is one between

1. the relationship between the semantic value M(w) of a word-sized unit w and the semantic value M(s) of a sentence-sized unit s where s contains w; and
2. the relationship between the semantic value M(w) of a word-sized unit w and one's being committed to (a) take a sentence-sized unit s containing w to be true or (b) infer a sentence-sized unit s* from s.

The semantic value of the concept DOG commits us to take "dogs are animals" to be true and licenses inferences from "x is a dog" to "x is an animal." It likewise commits us to take "dogs are vegetables" to be false and incurs no commitment about the truth of "dogs are Pope Francis's favorite type of animal." But the meanings of all three sentences are built up, in the same way, from the semantic values of their constituents and their respective syntactic structures. The difference between atomists, on the one hand, and holists and molecularists, on the other, is not (usually) about whether sentence-sized units get their meanings compositionally, but about whether (or in what cases) epistemic and inferential commitments are a consequence of something that word-sized units have in their own right, or whether word-sized units have such semantic properties because of epistemic and inferential commitments.

A more stringent notion of semantic atomism adds the additional condition that the semantic value of one word-sized unit cannot depend on its relation even to other word-sized units:

> Semantic atomists hold that the meaning of any representation (linguistic, mental, or otherwise) is not determined by the meaning of any other representation. Historically, Anglo-American philosophers in the eighteenth and nineteenth centuries thought that an idea of an X was about X's in virtue of this idea's physically resembling X's. Resemblance theories are no longer thought viable, but a number of contemporary semantic atomists still believe that the basic semantic relation is between a concept and the things to which it applies, and not one among concepts themselves. (Lepore 1999, 829)

By this criterion, if BACHELOR *includes* other concepts (MAN and UNMARRIED), the concept BACHELOR is not a semantic atom. The older definitional approach to semantics claimed that we define many and perhaps most concepts in terms of simpler atomic concepts; and for such a theory to get off the ground, it had to assume that there must be some concepts that are atomic. The question, of course, was what sorts of concepts could really be atomic in this way. One answer, explored by the British empiricists and some of the early logical positivists, was that the best candidates would have to be something like Locke's simple ideas of sensation—or, in more recent formulations, sense data or qualia. Other recent atomisms have attempted to ground semantic atoms in causal and informational relations between a thinker and an object or class of objects. According to one type of causal theory, a concept C means X because Cs are reliably caused by Xs (Fodor 1987). In another type of causal theory, C means X because Cs were caused by Xs in a privileged reference-fixing period (Kripke 1980). In Dretske's teleosemantic account, C means X if and only if C has the function of indicating Xs (Dretske 1986, 1988). Arguably, any such account is applicable only to a subset of the concepts an individual has at his disposal (in one case, sense data; in the other, "natural-kind" concepts).

Let us further distinguish between several claims sometimes associated with the label *atomism*.

A1: *Semantic independence of concepts*: The semantic value of concepts is fixed at the level of concepts and does not depend on *cognitive units other than concepts* (e.g., beliefs, inferential dispositions, mental models).

A2: *Atomic and nonatomic concepts*: At least some concepts are atomic, in the sense that their semantic value does not depend on their relations to other *concepts*.

A3: *No nonatomic concepts*: *All* concepts are atomic in the sense defined in A2.

The first thesis, the semantic independence of concepts, can be held independently of the other two. Someone could hold a network theory of conceptual meaning while denying that a concept's semantic value is determined by any *other* type of cognitive unit like a belief, inferential disposition, or mental model. A1 is a denial of inferentialist semantics and hence is a standard assumption of noninferentialist accounts of semantics, but the distinguishing features of such accounts are found in how they go beyond A1 to make claims about how the semantic values of concepts are fixed. Historically, the predominant account of conceptual semantics was a *definitional* view, which viewed most concepts as constructs out of a relatively small base of atomic concepts that were variously thought to either be innate ideas or acquired through sensory experience. Given that we often *learn* concepts through verbal definitions, this view may strike many as the most intuitively plausible. However, beginning in the 1970s, philosophers began to question the definitional view following its critique by Saul Kripke (1980) and Hilary Putnam (1975). Kripke and Putnam suggested that at least some concepts behave more like proper names, referring to the objects in their extension not by dint of picking out their properties but by a kind of direct reference grounded in causal relationships. Jerry Fodor (1998) and Ruth Millikan (2000) are among the more prominent philosophers who have developed semantic accounts that deny that concepts have an internal structure, and embrace the somewhat counterintuitive view (A3) that all concepts are atomic. There are, of course, potential problems for this theory that do not affect A1 and A2. But as the issues between atomism, holism, and cognitive pluralism can be considered in connection with the first two theses without reference to the further issues posed by A3, I will use the label 'atomism' to refer to views that embrace A1 and A2.

From the cognitive pluralist's standpoint, the problem with semantic atomism is that it has no way of accounting for the constitutive connections within a model between concepts, beliefs, inferential dispositions, and capacities for perception and action. A change in my understanding of how a chess bishop is permitted to move *is* a change in my concept of a chess bishop as well. This is similar to familiar holist arguments against atomism, with the important difference that the cognitive pluralist holds that there is a principled difference between the situations one finds *within* a model and *outside* or *between* models.

Additionally, the cognitive pluralist is *not* required to hold either that the atomist story has nothing right or that there cannot be concepts about which it tells the best story. What the cognitive pluralist needs to insist on is that, *in the case of model-based understanding,* the semantics of a concept must be constitutively linked to those of other concepts, and to the rules, representational system, and inferential commitments of the model. It is compatible with this idea that there are *also* cases where we have concepts that are not grounded in this way in models. Indeed, I think that, in certain learning contexts, concepts can be acquired (at least initially) without being associated with a well-formed model. When we encounter a new kind of thing without yet having any concrete understanding of it, the concept may well take a form very like a conceptual atom. And when we learn a concept through a verbal definition, it may have constitutive dependencies on the other concepts through which it is defined but not yet be embedded in a model. I tend to expect that these are primarily characteristics of learning processes that normally lead toward embeddedness in one or more models, but this is an empirical hypothesis that would need careful exploration.

Holism

Conceptual holism is the thesis that all our concepts are constitutively interdefined, such that a change to any one of them would amount to a change in all of them. Moreover, holist theses are generally not confined to semantics. The holism found in Quine (1951) and Davidson (1967), for example, posits constitutive interconnections between all of a person's concepts, beliefs, and inferential commitments.

A model-based cognitive pluralism also posits constitutive interconnections between concepts, beliefs, and inferential commitments and so has an important resemblance to holism. The difference is in scope. The holist holds that such constitutive interrelations are found between *all* concepts, beliefs, and inferential commitments. The cognitive pluralist finds them only (or *chiefly*) within the concepts, beliefs, and inferential commitments of a model. According to holists, the network relevant to the semantics of a concept is completely *global*. According to cognitive pluralists, it a mental model.

Distinguishing two types of holist claims will help here. The first is that anything, from any domain of understanding (including things we do not yet understand and are completely unaware of), could in principle prove to be relevant to any other concept, belief, or inferential commitment, and hence a change to any unit anywhere could require adjustments anywhere else. The cognitive pluralist can accept this as a "for all we know" kind of claim: I cannot rule out in advance the possibility that some change in my understanding of physics might force me to rethink how I understand some other domain like chess, dining, or ethics. It may seem unlikely, as I cannot see how it could do so. But this is in part because I cannot anticipate what change in my understanding in physics might someday come along that would require such adjustments; and for this reason, I cannot definitively say in advance what other areas of understanding are insulated from it.

The second and stronger type of holist thesis is that each concept, belief, and inferential commitment is *constitutively* bound up with every other, so that any change is *eo ipso* a change to everything else in the system. Cognitive pluralism differs from such a *constitutive* holism in precisely the areas where constitutive holism seems most vulnerable. It seems correct to say that a change in meaning, belief, or inferential commitments has ramifications to other concepts, beliefs, and inferential commitments, but those ramifications have limits. If I change my understanding of, say, how a knight moves in chess, this clearly has constitutive implications for how I need to adjust my understanding of the concept CHESS_KNIGHT and other aspects of my understanding of chess. But it does not follow that I need to similarly adjust my understanding of mechanics or of restaurant etiquette. Similarly, a transition from classical to relativistic mechanics involves a change in concepts of MASS and SPACE and changes in beliefs and inferential commitments *within the domain of mechanics*, but it need not have the slightest impact on my understanding of chess or the semantics of a concept like GRANDMOTHER. What cognitive pluralism adds to this commonsense insight is a theory of why it might be correct: namely, that there is a kind of psychological unit (a mental model) distinct from concepts, beliefs, and inferences, which sets the bounds on how far such changes should ramify. A model-based cognitive pluralism implies that at least some portion of conceptual semantics is best described by a molecularist picture, but cognitive pluralism goes beyond a mere assertion of semantic molecularism in providing a theoretical account of what makes it true.

In short, holism gets some things right: (1) some changes in meaning do or ought to constitute changes in other concepts and in belief and vice versa, and (2) what we believe is in some ways constitutively tied into the semantics of the terms we use. But holism errs in being a completely global and ecumenical thesis. Not all changes in belief and meaning have the same sorts of implications for every other element of a person's epistemic space, and holism lacks the resources to deal with the differences.

16.3 The Multiple Lives of Concepts

Our discussions thus far might suggest a particular view of concepts and their relationship to models. If a concept used within a model—such as CHESS_KNIGHT or the concept MASS used in a relativistic model of gravitation—is constitutively interdefined with the other concepts used in the model and with the model's rules and representational system, this seems to suggest that each concept resides within a single model. It might additionally be used in linguistic inferences that combine concepts drawn from different models, and there may be associations between it and concepts drawn from other models, but these are not *constitutive* of *what concept it is.* I will refer to this thesis that each concept resides in a single model as the *residential* view of concepts.

If we accept the larger thesis that models form tightly integrated units, wherein concepts and inferential commitments are constitutively interconnected, this might recommend the residential view of concepts. But there are also reasons to suspect that the residential view cannot be correct, or at least that it cannot be the full story about concepts. On the one hand, there is reason to think that some concepts—or something closely akin to concepts—do not need to be situated in models at all but play a role more directly connected to grouping stimuli into categories. Much of the literature on concepts in psychology is primarily about categorization, and it is important to examine the relationship between the idea that concepts are tools for categorization and the residential view. On the other hand, there is reason to think that a single concept can play a role in more than one model. At the least, we frequently use the same *word* in expressions of judgments and inferences grounded in multiple models, and at first glance the situation often looks not like a case of simple homonymy but like a single concept straddling several models.

I think that both of these problems are real, and they point to a need to go beyond the residential view of concepts without rejecting the idea that much of the semantic content of concepts is derived from models in which they play a role. There is, of course, a certain amount of terminological choice involved in deciding what to call "a concept," and what to call by some other name—for example, whether the psychological units used for classification in perception are the same sorts of entities as the units used in semantically based reasoning and off-line simulation. But whatever terminological choices we make, we need to respect the same data. And in this case, I believe that the best approach is to view concepts as complicated hybrid entities that must be understood in their relationship to more than one kind of psychological system. A model-based semantics is still a *part* of that story, but it is not the *entire* story.

16.4 Concepts without Models

I will explore three reasons to think that concepts can exist without being embedded in mental models. First, in psychology, concepts are often treated primarily as units of categorization. (And while it is often an assumption left implicit, what is really central to the discussion is often categorization *in perception*.) Categorization *can* take place without a mental model; and so if concepts can be (just) tools for categorization, they need not be embedded in models. Second, in some nonhuman animal species, we might have reason to be skeptical about the presence of the kinds of models that are relevant to inferential semantics, but still have reason to think that the animals have concepts (particularly in the sense of things used in perceptual categorization). And third, concept *learning* does not seem to require the mastery of an associated model. Indeed, the process of learning a model often involves early stages in which we learn to recognize categories of objects that will be relevant to the model without our yet having mastered the model.

In discussions of concepts within psychology, we find a great deal of focus on the role of concepts as tools of categorization, particularly among proponents of prototype and exemplar theories. Eleanor Rosch, the pioneer of the prototype theory of concepts, writes:

> Two general and basic principles are proposed for the formation of categories. The first has to do with the function of category systems and asserts that the task of category systems is to provide maximum information with the least cognitive effort; the second has to do with the structure of the information so provided and asserts that *the perceived world* comes as structured information rather than as arbitrary or

unpredictable attributes. The maximum information with least cognitive effort is achieved if categories map *the perceived world* structure as closely as possible. (Rosch 1978/1999, 190; italics mine)

Rosch's emphasis on "the perceived world" suggests that a core concern of the theory may be more narrowly about categorization *in perception*. This is even more clearly evident with exemplar theories, which treat concepts as being (or involving) stored memory traces of exemplars of a category (Smith and Medin 1981).

Perceptual categorization, or indeed categorization in general, is not the *only* thing a theory of concepts should explain. But it is one important role concepts play in our mental economy. Most theories of categorization in psychology treat the categorization process as one that assigns a category based on *features*. In classical theories, the features might be thought to provide necessary and sufficient conditions for the applicability of the concept. Prototype and exemplar theories, by contrast, are *probabilistic*: a concept is activated by some matching process involving either a set of prototypical features or a set of exemplars stored in memory. The process may involve a comparison of the stimulus to multiple prototypes or exemplars, and its dynamics may be tuned to activate the concept with the best fit with the stimulus. The kinds of inferential relationships that models are supposed to underwrite need not figure in this, and it seems clear that mechanisms for categorization could exist that do not require mental models.

Moreover, it seems reasonable to suppose that there are types of animal minds that possess categorizing mechanisms that stand between perceptual inputs and behavioral control *without* these being embedded in mental models of content domains. There are useful types of cognition that you can obtain only once you have interposed something like a concept (or, if you prefer to restrict that word to humans or other language users or other conscious beings, a *protoconcept* or a *feature aggregator*) between the stimulus end and the motor end of the cognitive architecture. Think of it this way: the success of a motor response often depends little on what sensory channel an affordance or object is detected through, and very much on *what kind* of affordance or object is detected, and the *state* it is in. Moreover, there can be many stimuli that are cues to a single type of affordance (typed relationally) or object (typed "objectively"), and there might be many motor schemas that are useful for responding to the affordance or object. It is thus useful for an animal to have in its cognitive architecture a kind of element that stands in between perception and motor control, as a kind of central clearinghouse for the various sensory and motor schemas that need to go together, yet is not uniquely tied to any one of them. This allows the animal to have a flexible repertoire of ways of detecting and responding to the same class of things—and in a fashion that *treats them as* the same class of things. It seems likely that the interposition of a "conceptual" (or "protoconceptual") layer of this sort between sensory and motor processing is something we should expect to find in a range of species between those in which perceptual inputs are more closely coupled to behavioral response and those capable of acquiring learned models (fig. 16.1).

In addition, there are times when we humans learn a concept *without* having the sort of understanding that allows us to reason with it. We may have concepts we can use only to identify things as members of a category. And even in the process of acquiring a model—say, when we learn to play chess—there is an early stage at which we are able to do things like identify the various classes of pieces but do not yet know how to use or think about them *as chess pieces*. At this stage, I have a concept that I use the word 'knight' to express. It is not yet the concept CHESS_KNIGHT, because I do not understand the game or the roles knights play in it. (Nor is it the same concept KNIGHT used to understand a particular feudal role.) But it is still *some kind* of concept. (Indeed, I will argue later in the chapter that there is a psychological sense in which it may even be the selfsame concept—that is, the same psychological continuant—which gets edited over time.)

There are, of course, issues of terminological choice here: Should we talk about cognitive units that are used *only* for categorization as "concepts" or find some other term for them? And, particularly in the case of much simpler animal minds, should we use the word 'concept' at all, or some other term, such as 'protoconcept'? The choices we make here might make a big difference in avoiding misunderstandings in the context of a comprehensive account of concepts, but my ambitions here are far more modest. And because I will be arguing that there is a sense in which "the same concept" can endure as a psychological continuant over changes in its properties, I will not make any such terminological distinctions yet.

16.5 Concepts with Multiple Models

There is also a second and opposite issue, for it also appears, at first glance, that a single concept can appear in multiple models. At the least, we often use the same *words* to express insights grounded in different models. Words can, of course, be homonymous and express more than one concept. The word 'bank' can mean a kind of place where money is kept and the side of a riverbed, and in those two uses it expresses distinct concepts. But the

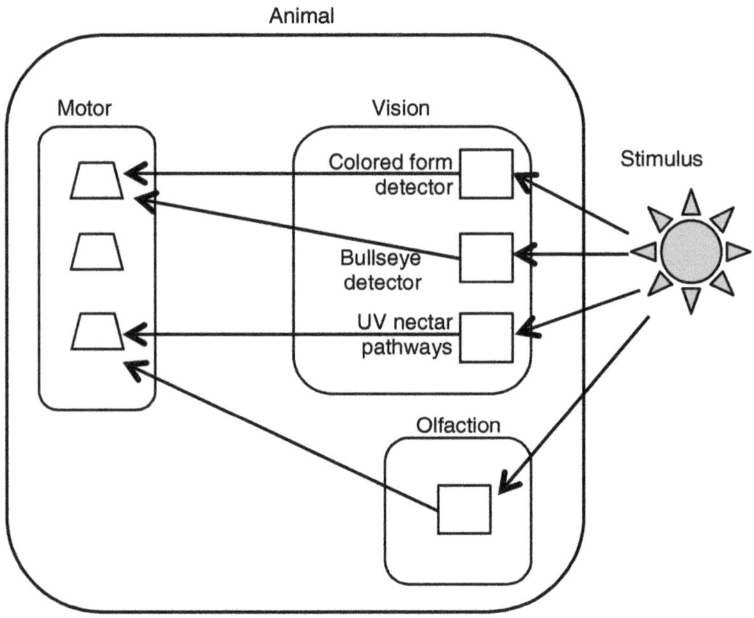
Animal
Motor
Vision
Colored form detector
Bullseye detector
UV nectar pathways
Olfaction
Stimulus

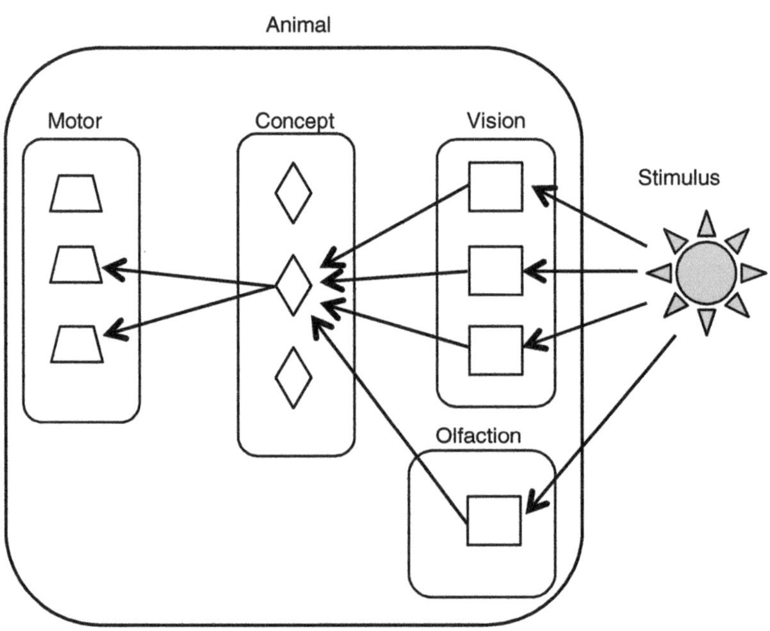
Animal
Motor
Concept
Vision
Olfaction
Stimulus

Figure 16.1
Concepts as feature aggregators. Diagrams of two types of simple cognitive architectures. In the first figure, perceptual feature detectors (for features like colored forms and nectar pathways visible on flowers to birds or insects) directly drive motor control. In the second figure, a middle layer is interposed, which responds to many features, either individually or in combination, and can drive multiple motor responses. This architecture allows an organism to detect the same objects, properties, or affordances from multiple distinct cues, and to treat them as equivalent for purposes of behavioral response.

residential view would have far more radical consequences. If there are multiple models of banking institutions—say, architectural models, models used by patrons of how to make various kinds of financial transactions, economic models used in finance—they would all have to employ distinct concepts, all expressed by the English word 'bank'. This might be attractive from the standpoint of a molecularist inferential semantics; the uses associated with different models yield different types of inferences. But it also seems rather counterintuitive, and it presents a problem of how we know (or at least why we have the sense that) we are talking about the same institutions when we are thinking about them through different models. It cannot *merely* be that we use the same *word*, because a single word can be used to talk about distinct classes of things, like a financial institution and the side of a riverbed, and in those sorts of cases we have no similar temptation to suppose that there is only a single concept.

Consider some other examples. There are no doubt a number of distinct models that deal with cows. People who herd cattle probably have mental models of how cattle roam about singly and in groups and how to interact with them to cajole them into going where the herder wishes them to go. Cattle breeders have mental models of different heritable traits of cows and how they can be combined through selective breeding. Wildlife ecologists dealing with open ranges have ecological models involving cattle and their impact on particular environments, and so on. They all use the word 'cow' to talk about the same class of things, and it takes no special mental gymnastics to recognize that it is the same class of things that is understood, in different ways, through the different models. The natural conclusion to draw would seem to be that there is a single concept, COW, that appears in many models.

A slightly more difficult case involves theory difference and theory change. Classical and relativistic mechanics both employ the word 'mass', though the constitutive implications of the word differ importantly in the two theories. In one sense, we might be inclined to say that the concept used in classical mechanics M_c is a distinct concept from the concept M_r used in relativistic mechanics. There are probably quite a number of publications that contrast "the classical concept of mass" with "the relativistic concept of mass." Yet it also seems natural to say that they are "concepts of the same thing," and that M_r is a refinement of M_c, albeit one that could be obtained only by a more wholesale change at the level of an entire theory.

How are we able to make sense of conceptual change? Here we seem to be presented with a dilemma. On the one hand, a *semantic* approach to concepts inclines us to the view that a concept is constituted by its semantic properties, broadly construed to include constitutive inferences. If concepts C_1 and C_2 have different semantic properties, they are different concepts. On the other hand, we seem to be able to make at least minor adjustments *to* a concept. A child may initially suppose that dogs are domestic animals and that each dog is someone's pet. This might even be so deeply bound up with his understanding of dogs as to be a constitutive implication of his concept DOG. If this is the case, when he is disabused of this assumption, his concept DOG undergoes a certain amount of revision. It does not seem natural to say that he lost (or stopped using) *one* concept and acquired (and began using) *another*.

The dilemma can also be posed as a problem about whether two people share the same concept. Four-year-old Alex and his older sister Betty talk about dogs frequently. Alex's concept of dogs involves an implicit assumption that dogs are people's pets. He would be quite baffled the first time someone told him that a certain dog was *nobody's* dog. Betty understands that there are wild dogs, feral dogs, stray dogs, and the like. Do Alex and Betty have "the same concept"? On the one hand, we might be inclined to say *no*, because the concepts they express using the word 'dog' have different semantic properties. On the other hand, they are able to talk with each other about the same class of animals without difficulty, and if Betty tries to disabuse Alex of his assumption that dogs must be pets, Alex may at first react with confusion or even disbelief, but he knows well enough what class of things she is talking about: the same class of things picked out by his concept DOG. (Indeed, it is precisely because he is able to do this that he is perplexed by her claim, which violates the constitutive rules of *his* concept DOG.)

Resolving the dilemma requires two sorts of additional resources. The first involves separating two *aspects* (perhaps even *components*) of concepts: an "intensional" component involving constitutive semantic connections, and an "extensional" or "ostensive" component with the function of tracking particular individuals, kinds, properties, relations, and processes. The second involves distinguishing two ways of typing concepts: by their semantic properties and as psychological continuants in the mind of an individual.

Intension and Ostension

First, the distinction between intensional and ostensive components. It is easier to find a point of entry by considering our concepts of individuals rather than our concepts of kinds, properties, relations, or processes. We think of individuals as having enduring properties and properties that change over time. Being able to think about an individual person or object at all requires the ability to think of it as one and the same thing despite the constant changes in variable properties: Socrates is first sitting then standing, asleep then chatting in the Agora, bearded and then newly shaven. The boundary between "enduring" and "changing" properties is a bit hazy, though: we may think of a person's physiognomy and personality as relatively stable things, but both in fact change substantially throughout the life cycle and can undergo rapid and dramatic changes as well. Just *what* can change without forcing us to conclude that the individual we knew is no longer present seems somewhat open-ended, a fact reflected in rival theories of personal identity and of the identity conditions of objects, as well as in the problems involved in deciding when an embryo becomes a human person.

The simplest and perhaps the only explanation for this would seem to be that concepts of individuals (whether persons or objects) involve multiple components. Some of these involve attribution of properties, and it may be that there is an important psychological difference in the way we represent properties we regard as enduring (perhaps even assumed to be part of an essence) and those that are evanescent. But in addition to these there seems to have to be a component that is more *ostensive*—one that has the function of tracking a particular individual through changes. And the "changes" we need to track through are not only real changes in the individual but changes in our beliefs *about* that individual. I see Socrates and take *him* to be a Spartan but then change my mind and decide *he* is an Athenian. I see something on the path and take *it* to be a rock until it moves and I see it from a different angle and conclude that *it* is some kind of animal. I do not simply first think, "There is a Spartan," and then deny that and think, "There is an Athenian," or first think, "There is a rock" and then "There is an animal." I am not simply registering (and changing my mind about) the presence of property-instances; I am tracking *one individual* throughout, to which I am imputing different attributes. Indeed, in some cases, I may have so little information that I simply conclude that there is *something* (some particular individual thing) there, without *any* concrete hypotheses about what it is. Camping in the dark of the night, I hear a noise in the woods and think, "There's *something* out there," and then perhaps feverishly imagine various hypotheses about *what* it might be.

For readers who take an interest in philosophical history, I take it that I am developing an insight that has appeared several times before in the history of philosophy. Aristotle claimed that a thing is always a "this-such" (*tode ti*). I interpret this in cognitive terms: to *think* of a thing (*as* a thing, i.e., as a particular individual) is to have a concept of it with components that (1) aim to track it as a particular individual (the "this-ness") and (2) ascribe particular attributes to it (the "such-ness"). My claim also seems closely akin to Husserl's statement that the core of the sense of a thought about a thing (what he calls the "noematic Sinn") includes not only descriptive components but also what he calls the "pure x," which I take to mean an ostensive namelike element that has the function of tracking an individual (Husserl 1913/1989, §131).

I think that there is a similar story to be told about other kinds of concepts: concepts of kinds, properties, relations, and processes. Or, more carefully, that at least *some* concepts of kinds, properties, relations, and processes have an analogous feature. The situation seems clearest with kind concepts. Our minds are disposed not only to think in sortal kinds but also to form concepts for new sortal kinds quickly when confronted with something that does not seem to fit into the kinds we already know about. When we encounter an unfamiliar phenomenon—particularly a new kind of animal or plant, though those are not the only cases—we quickly form a sort of stub of a kind concept. This involves an implicit assumption that there *is* a kind of thing there to be tracked, though as yet we do not understand much about it. There is thus a kind of ostensive act in which we cast our semantic hook, hoping it latches onto a kind, and then over time reel it in, gaining more concrete understanding of it, which we incorporate into our concept of it. The initial tacit assumption may, of course, prove mistaken: what we observed may be a member of a known kind in an unfamiliar state, or we may have mistaken what are really two kinds for one, or there may not be a phenomenon robust enough to count as a kind at all. But the ability to acquire greater understanding about a kind—or more broadly about any sort of phenomenon—depends on our first being able to fix our ostensive hook into it as a target to be tracked by a concept, and to keep tracking the same target in spite of semantic and epistemic changes.[1]

Semantic and Psychological Typing

This leads quite naturally to the second observation: a distinction between semantic and psychological typing. There is a perfectly reasonable sense in which, for purposes of semantics, one might wish to type concepts according to their semantic properties, even at a maximally fine-grained level. Many philosophers have taken this approach, some concluding that it is unlikely that two people share many concepts, because there are likely to be many fine-grained differences in the semantic properties of your concept of something versus mine. But the foregoing discussion suggests that we need to think about the identity of concepts as *psychological* entities in a different way. I coin a new concept on the basis of thinking I have encountered a new phenomenon that is not captured by my existing concepts. I do not yet understand it well, but a new psychological structure is created—a kind of stub—that has the function of tracking *that* sort of phenomenon. I then explore and learn, and different hypotheses about *it* are encoded in my concept of it. The "concept" in this psychological sense is a *psychological continuant,* whose semantic properties (particularly its sense) can change over time. Alex formed a concept DOG at some point, probably through early encounters with dogs. To this were attached various semantic properties, such as that dogs are pets. When he learns the disconcerting truth that not all dogs are *someone's* dog, it is *that* concept (in the psychologically individuated sense) that will undergo revision.

Stepping back a bit, we can see that this is really an application of the basic insights of cognitive pluralism to theoretical models of concepts—say, those that individuate concepts semantically and those that do so psychologically. We speak of these as accounts, views, models, or theories *of* concepts out of an assumption that there is something real that they are all seeking to characterize. But they characterize "it" in different ways, emphasizing different things that provide distinct insights and forms of reasoning, but idealized in ways that can generate puzzles when we try to put them back together.[2] For example, viewing concepts as being individuated by their semantic properties may provide a way of seeing how they express things (properties) that could potentially be shared because they are abstract entities, though there are also problems for *actual* property sharing if the defining semantic properties are sufficiently fine grained. A psychological notion of a concept, by contrast, seems ill suited to this, as it is a notion of a continuant within a single mind. On the other hand, the psychological notion allows us to account for the continuity in our mental lives: we do not simply delete an element in a mental lexicon and coin another but make revisions to a preexisting structure that persists through the change, albeit in an altered form.

Concepts within and outside Models

What we need now is a way of combining this picture with a model-based account of understanding, semantics, and reasoning. To do this, it seems to me that we need to posit more than one kind of psychological entity. On the one hand, we need models and things that are components of models, including the representational types that are related to model-based judgments and reasoning. The latter are what we might have taken to be "concepts" in a purely "residential" model-based account. (Call them CMs, for the notion of *concepts* involved in *models*.) On the other, we need a kind of entity that stands outside models—a kind of quasi-lexical entity. (Call these CLs, for a *concept* in a *lexical* sense.) Because CLs need to be able to change their semantic properties, their identity is not fixed by such properties. This leaves the question of what psychological story we should tell about their semantic properties, and one natural hypothesis is that they have semantic properties by dint of being linked in the right ways to CMs inside models. But because CLs stand outside models, a single CL can have links to multiple CMs in different models. I can have "a single concept" cow in the sense of having a single CL that is "involved in" or "linked to" various models of herding, species biology, prairie ecology, and so on. Such a concept can also change its semantic properties over time, either by changes in which models the CL is linked to, or by changes within the models to which it is linked.

It also seems plausible that it is not CMs but their associated CLs that are used in language-like thought. CLs may literally be the quasi-lexical units employed in language-like thinking. Ordinary language may very well, by default, assign words at the level of CLs, and garden-variety homonymy (*bank*/*bank*) involves an accidental assignment of the same phonetically typed units of a public language to two different CLs. If we count such homonyms as different words, a word is composed of an associated pair consisting of a phonetically typed unit and a CL.

16.6 Toward a Schematic Multifactor Account of Concepts

We now have several constraints on an account of concepts:

1. We need a way of situating concepts as things that are involved in semantically based inferences. In a pluralistic model-based account, the

natural way of doing this is to look to some element within a model that plays this role. In the residential view, these units would count as concepts. In a more complex view, they would need to be some sort of units (CMs) that can be constituents of concepts.

2. We need a kind of psychological unit that can serve as an aggregator of features used in classification. It should be possible for such a unit to exist independently of a mental model, but also for it to be incorporated into a model—at the least as part of a mechanism used for selecting what concept to apply to a stimulus.
3. We need a way of accommodating the fact that we can use what seem to be the same concepts in multiple models.
4. We need to be able to treat concepts as psychological continuants that can be altered in the process of learning.
5. In addition to being used for classification and inference, concepts have a function of *tracking* objects, kinds, properties, relations, and processes. Moreover, this is not a fact that is apparent only from the outside, but is part of the phenomenology and logical structure of conceptual thinking.
6. We need to account for the relationships between concepts and lexical units in a language.
7. We need to account for how concepts can play a role in language-like thinking.

A number of possible theoretical models could serve as candidates for meeting all these constraints. Fully enumerating them and arbitrating between them is a large project in its own right, which would require a good deal of both philosophical analysis and empirical testing. This task would arguably call for a book of its own. What I will attempt to do here is therefore more modest: to present elements that I think such a model *needs* to have, and to suggest, for further discussion and development, several different ways it might be completed. I will do this with reference to diagrams, which I will build up from simpler to more complex.

Let individual models be indicated by boxes with double lines and elements corresponding to concepts (CMs) within those models by boxes with single lines. The constitutive relations between CMs that play a role in the inferential patterns employed within the model are represented as links between these (fig. 16.2).

In addition to models, we need units that stand outside models, which can serve as quasi-lexical units (CLs) in language-like thinking and can be specially linked with CMs and thereby draw on model-based inference. I will indicate these with shaded rounded rectangles. The linkage to

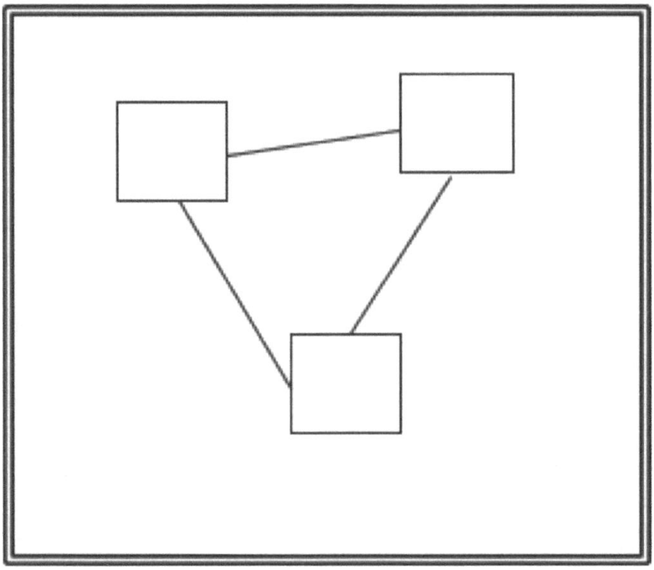

Figure 16.2
Models and their elements. A model is indicated by a double box. Elements within a model corresponding to concepts are indicated by single boxes, with constitutive links between them indicated schematically by lines.

particular CMs will be indicated with dotted lines, and the linked CMs shaded (fig. 16.3). A CL can be linked to CMs in multiple models—say, a CL for cow with CMs contained within models used in herding, selective breeding, and wildlife ecology.

As a first approximation, a *concept* involves the linked network of a CL and the various CMs it is linked to (the shaded parts of the figure and the links between them). The inferential power associated with concepts is derived in large measure from its CMs and the way they are embedded in models. But the CL gives the concept an independence from particular models that allows for application of multiple models, and for cross-model reasoning.

Second, we need a way of accommodating the relationships between concepts and features. There are possible relationships in two directions here. On the one hand, concepts can be activated by feature detectors, perhaps through the mediation of a mechanism that consolidates features into categories. On the other hand, the features associated with a concept project a classification scheme onto the world: groupings of features define a

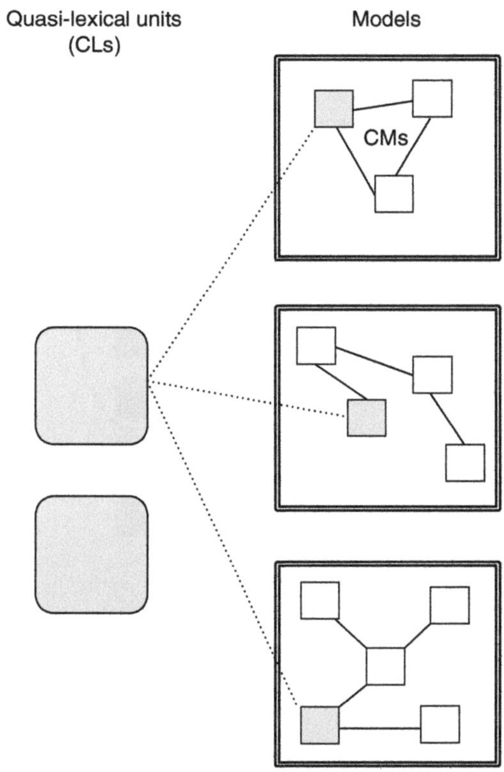

Figure 16.3
Concepts, CLs, and CMs. The shaded rounded rectangles indicate CLs standing outside models. These are linked to elements within models. A concept includes both a CL and its linked CMs. Model-based reasoning can be applied to a concept on the basis of its CMs.

subjective ontology of objects. Likewise, the constitutive rules of a model project a class of objects to which its CMs apply (and these may be different for two CMs that appear in different models linked to the same CL) (fig. 16.4).

Here there are several possible hypotheses about *where* in our schema feature aggregators should lie. And the answer may be different for different types of cognitive architecture. Some animals may have *only* such aggregators and would be more properly depicted by figure 16.1. In *human* cognition, there is an obvious question about whether feature aggregators (classifiers, categorizers) directly activate CLs, CMs, or both (fig. 16.5).

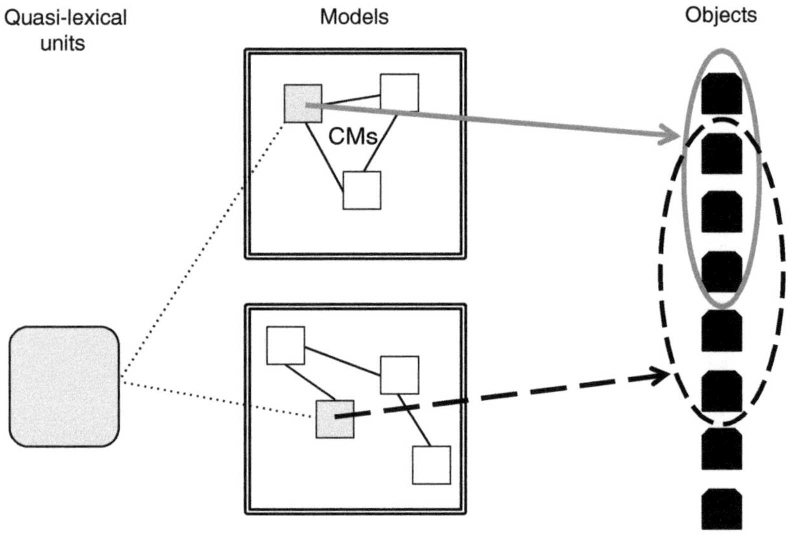

Figure 16.4
Projective classification. The constitutive rules of models can determine projective classes of objects or properties for their CMs—i.e., the set of things characterized by the rules of the model applying to the CMs.

The "outward" relationships between CMs and features, however, are potentially of special interest, as they might provide a way of explaining what is right about theories of semantics that treat concepts as *determining* the conditions for membership in a class of objects. There is a sense in which this is correct: models do project a subjective ontology, and CMs are the natural locus for the units that determine a projective class of objects.

On the other hand, concepts also have an ostensive element. We can in some sense have a concept of a thing—in the sense of a concept that has the function of tracking that sort of thing—without having a concrete understanding of its nature. And we can still have a concept of a kind of thing even if we misunderstand its nature and the projective ontological classification arising from a CM does not in fact pick out something as we suppose. This requires that we add another element to our schema: a component of a concept like Husserl's "pure x" that has the function of tracking an object, kind, property, relation, or process. Again, there are variants on this theme where the normative ostensive relationship is tied to a CM, to a CL, to both individually, or to the whole network that constitutes the concept. The diagram (fig. 16.6) here presents only one variant but highlights

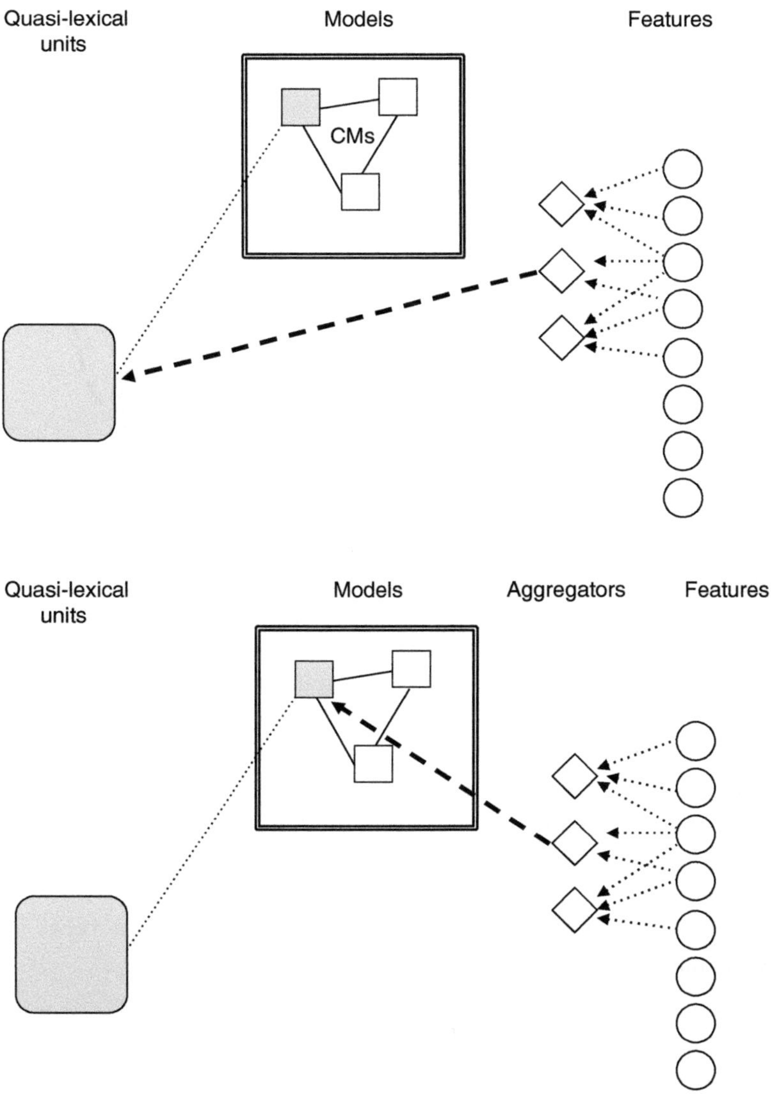

Figure 16.5
Two possible locations of aggregators. Feature aggregators (diamond boxes) consolidate detectable features and activate elements of a concept. In the upper diagram, they do so by activating the CL; in the lower, by activating a CM contained in a model.

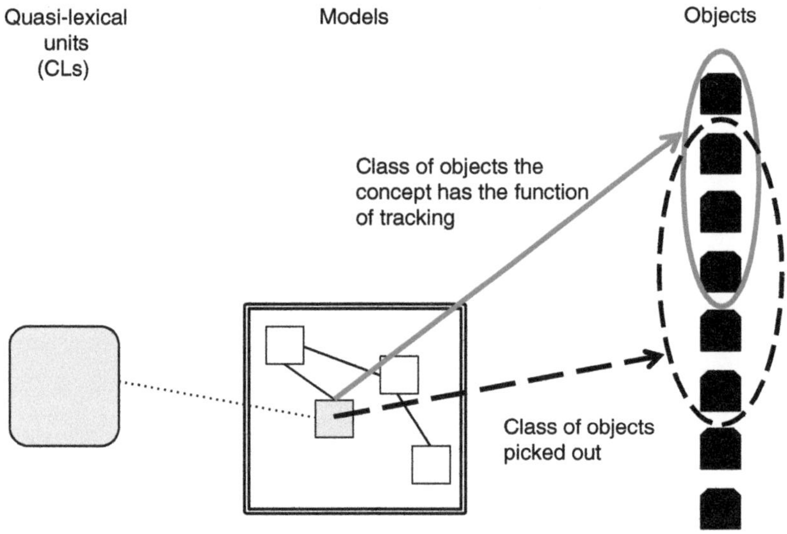

Figure 16.6
Two projective classes. The constitutive rules of a model determine one extension for a CM, consisting of the things that really conform to the way the model frames them (dark dashed lines). But a concept may also aim at tracking some class of things that may be mischaracterized by the model, which has a different extension (solid gray lines).

the fact that the tracking target of a concept may have a different extension from the set of things meeting the conditions of whatever rules are implicit within the model.

This overall schema also suggests several ways that different types of conceptual change might occur.

1. The links between a CL and models change:
   a. A new model is associated with a CL.
   b. A model ceases to be associated with a CL.
2. There are adjustments to a model resulting in the constitutive features of a CM within it being linked with a given CL.
3. There are changes to feature aggregators used for categorization that are associated with a CL, CM, or model.

The independent status of CLs also allows them to serve as units in non-model-based cognition, particularly in language-like thinking. In them we find discrete lexical or quasi-lexical units that can be fundamental units in language-like cognition. Their semantic properties are determined not by their nature as language-like units but by the fact that they are also constituents of concepts. Language and language-like thinking have CLs as their constituents, but because these are bound to CMs as conceptual structures, language incorporates or ties into model-based semantics as well. This distinction between what is a *constituent* of language and what is incorporated into it is quite useful, as it provides some insight into how there can be formal languages without semantic properties (CLs that are not tied to models or feature aggregators), and how we are capable of making syntactically based inferences even without fully grasping the relevant concepts.

16.7 Possible Implications for Disputes about Concepts and Semantics

What I have proposed in the previous section should be viewed as something on the order of a schematic outline of a theory of conceptual semantics. It is not quite so concrete as a theory, but it enumerates some of the components that might need to go into a theory, and explores some basic suggestions for how they might be related.

The proposal for looking forward involves a reframing of how we think about philosophical debates in semantics. Debates about semantic theories are too often framed as either-or debates between accounts that each focus on different, but arguably equally legitimate, features and concerns: sense, reference, truth, or inferential commitments; internalism versus externalism; intension versus extension; atomism versus holism. There have been some attempts to move beyond this, such as the two-factor accounts proposed by Field (1977) and Block (1986). My suggestion is to broaden this approach to recognize the multiple factors that not only *are* a part of semantic competence but would *have to be* present in any organism that possesses the kinds of cognitive capacities that human beings possess, even if that means that they cannot all be accommodated in a compact and tidy account.

Conceptual atomists and direct-reference theorists have something right. There is an element of a concept whose function is to track individuals, kinds, properties, relations, or processes (depending on what kind of concept it is): an ostensive element akin to what Husserl called the "pure x." Just as was the case with the triangulation of individuals through different ways of thinking about them, so it is crucially important for us to be able to have a way of thinking about "the same thing" through changes in our understanding of it, and thus it seems quite necessary that concepts have such an element. In my view, however, the ostensive tracking relationship here is not a purely naturalistic relationship, such as causal covariation, but a *normative* relationship. And it is not the *only* important semantic feature of concepts.

Inferentialists also have something right. The semantic nature of a concept is in large measure determined by its relationships to other concepts and its constitutive implications. In my view, these are to be found *not* in the language-like CLs but in the CMs embedded in models. Feature-based accounts also have something right. There must be something like feature-aggregating mechanisms that constitute classes as categories for perception, and these are part of the network of systems that is a concept. It may be that different feature-based accounts—say, those involving descriptions, prototypes, and exemplars—may each have something right at a more fine-grained level as well. There is no reason to assume that there is *just one way* that the mind detects and groups features and uses them as a basis for categorization. Intensionalist approaches have something right: CMs define a projective class of things constituted by the properties *imputed* by a model. But extensionalists also have something right: concepts, or elements thereof, aim at tracking some class of things that may not be characterized, or may be mischaracterized, by the CMs.

On this view, particular semantic theories are best viewed as *idealized models* of *particular factors* of semantic understanding. It may be dialectically useful for a time to set them up as direct competitors, each of which attempts to explain *everything* about semantics, because this is an efficient way of revealing the limitations of each. The moral to draw when such limitations are revealed, however, is not always that a model gets *nothing* right, but that there are things the model is not suited to addressing at all, which might be illuminated by other models. The approach I suggest is to look for the kinds of psychological mechanisms that might be needed to underwrite different aspects of concepts and inference, and for how these might interact with one another, even when their interaction does not appear to be logically tidy. Logical tidiness is a commendable ideal; however, our minds were designed not to be logically tidy but to do things like track objects and kinds, make inferences and predictions about them, and learn from our mistakes. The result of such an investigation may well be that "concepts" and "semantics" turn out to be more complicated affairs than we might have supposed. But that is the road forward.

Notes

2 A Standard Philosophical View of Cognitive Architecture

1. Definitions of "cognitive architecture" (and, more broadly, "cognitive science") vary on the question of just what mental, psychological, and neural phenomena fall within its scope. Some (e.g., Pylyshyn 1991) limit its scope to phenomena involving semantic content, while others use it as a general term for the basic functional architecture of mind and brain. These uses differ, for example, on the question of whether connectionist architectures should be counted as *cognitive* architectures.

2. Dennett, of course, is aware of all of this and addresses such topics in his 1977 review of Fodor's *The Language of Thought*.

3. Classic precedents for this approach include Heidegger 1927/1996; Merleau-Ponty 1945/1962; Gibson 1979. Some more recent important publications include M. Anderson 1997; Chemero 2009; Clark 1997, 2008; Gallagher 2005; Gallagher and Zahavi 2008; Lakoff and Johnson 1999; Noë 2004; Shapiro 2004; Silberstein and Chemero 2011; Thompson 2007; Thompson and Stapleton 2008; Varela, Thompson, and Rosch 1991; Zahavi 2005.

5 Cognitive Pluralism

1. *Alethetic* meaning truth related. *Aletheia* is the Greek word for truth.

7 Mental Models

1. My use of the term 'affordance' here may be closer to Donald Norman's (1988) usage than to Gibson's. For Gibson, an affordance was a perfectly objective feature of the environment that also happened to be relevant to the organism. Norman's characterization locates affordances more at the interface of organism and environment.

2. This is sometimes obscured in the psychological literature, where one finds claims that these systems are "domain general." This seems to be a carryover from Fodor's characterization of domain specificity in terms of responsiveness only to inputs from a particular perceptual modality. This has always struck me as an odd thing to call "domain specificity," but if clarification is needed, *my* claims about domain specificity are about *content* domains, not about input modality.

8 Relations between Models

1. The problem Dreyfus identifies by this name is related to, but distinct from, an earlier and broader usage in AI, identified by McCarthy and Hayes (1969).

2. We might also note that early perceptual processing is itself less encapsulated than Fodor's work might suggest. Retinal processing is a fairly simple feed-forward structure. But even the primary visual areas are highly interconnected with feedback relationships both to one another and to the thalamus.

11 The Complementarity of Models and Language

1. For critiques of this assumption, see Horst 1996, 1999.

12 Disunities of Knowledge, Science, and Understanding

1. More specifically, he claimed this in the A edition of the *Critique of Pure Reason*. In the B edition, he emended this to the more limited view that we must regard all *effects* as determined by a cause, leaving open the possibility that there are events that are not also effects. Even in the B edition, however, he retained a deterministic thesis about the phenomenal world as one half of one of the antinomies of reason.

14 Cognitive Illusion

1. If A is a limiting case of B, A is not reducible to B, and A and B are not both special cases of a third system C, at least in cases where the limiting case is in fact inconsistent with the rules of B. A flat Euclidean space is a limiting case of relativistic space, in the sense that relativistic space asymptotically approaches Euclidean space as the quantity of matter decreases and is spread homogeneously. However, there cannot be a relativistic space altogether devoid of matter because matter and space are interdefined, and a space with matter would not be perfectly Euclidean.

15 Cognitive Pluralism and Epistemology

1. This is potentially the basis for a critique of many intuition-based views in metaphysics. Chapters addressing these views were originally part of this project; but in

the process of producing the book, in consultation with editor and referees, I decided that such a critique was better suited to a book of its own.

16 Cognitive Pluralism and Semantics

1. Robert Cummins (1996) explores a related notion of the target of an intentional attitude.

2. Machery (2009) makes a similar point, arguing at greater length that the various accounts of concepts in philosophy and psychology do not play well together. I believe I am more inclined than Machery to see this as a case of multiple models of the same thing, idealized in different ways, though as the sections that follow indicate, I also think we a need to decompose the notion of a concept into finer-grained elements.

References

Agnoli, F. 1991. Development of judgmental heuristics and logical reasoning: Training counteracts the representativeness heuristic. *Cognitive Development* 6:195–217.

Agnoli, F., and D. H. Krantz. 1989. Suppressing natural heuristics by formal instruction: The case of the conjunction fallacy. *Cognitive Psychology* 21:515–550.

Allen, Colin, and Marc Bekoff. 2005. Animal play and the evolution of morality: An ethological approach. *Topoi* 24 (2): 125–135.

Anderson, John R. 1983. *The Architecture of Cognition*. Cambridge, MA: Harvard University Press.

Anderson, Michael L. 1997. *Content and Comportment: On Embodiment and the Epistemic Availability of the World*. Lanham, MD: Rowman & Littlefield.

Anderson, Michael L. 2006. *The Incorporated Self: Interdisciplinary Perspectives on Embodiment*. Lanham, MD: Rowman & Littlefield.

Anderson, Michael L. 2007. Massive redeployment, exaptation, and the functional integration of cognitive operations. *Synthese* 159 (3): 329–345.

Anderson, Michael L. 2010. Neural reuse: A fundamental organizational principle of the brain. *Behavioral and Brain Sciences* 33:245–313.

Anderson, Michael L. 2014. *After Phrenology: Neural Reuse and the Interactive Brain*. Cambridge, MA: MIT Press.

Armstrong, David. 1968. *A Materialist Theory of the Mind*. London: Routledge.

Armstrong, David. 1984. Consciousness and causality. In *Consciousness and Causality*, ed. D. Armstrong and N. Malcolm. Oxford: Blackwell.

Austin, J. L. 1962. *How to Do Things with Words*. Oxford: Clarendon Press.

Bailer-Jones, Daniela M., and C. A. L. Bailer-Jones. 2002. Modeling data: Analogies in neural networks, simulated annealling, and genetic algorithms. In *Model-Based*

Reasoning: Science, Technology, Values, ed. L. Magnani and N. Nersessian. New York: Kluwer Academic/Plenum Publishers.

Barad, Karen. 2007. *Meeting the Universe Halfway: Quantum Physics and the Entanglement of Matter and Meaning*. Durham, NC: Duke University Press.

Barrett, Justin. 2004 a. The naturalness of religious concepts: An emerging cognitive science of religion. In *New Approaches to the Study of Religion*, ed. P. Antes, A. Geertz, and A. A. Wayne. Berlin: Walter de Gruyter.

Barrett, Justin. 2004 b. *Why Would Anyone Believe in God?* Walnut Grove, CA: Alta Vista Press.

Barrett, Justin. 2009. Coding and quantifying counterintuitiveness in religious concepts: Theoretical and methodological reflections. *Method and Theory in the Study of Religion* 20:308–338.

Bartlett, Frederick. 1932. *Remembering: A Study in Experimental and Social Psychology*. Cambridge: Cambridge University Press.

Bell, C. Gordon, and Allen Newell. 1971. *Computer Structures: Readings and Examples*. McGraw-Hill Computer Science Series. New York: McGraw-Hill.

Berglund, B., U. Berglund, T. Engen, and G. Ekman. 1973. Multidimensional analysis of twenty-one odors. *Scandinavian Journal of Psychology* 14:131–137.

Bering, Jesse. 2002. Intuitive conceptions of dead agents' minds: The natural foundations of afterlife beliefs as phenomenal boundary. *Journal of Cognition and Culture* 2:263–308.

Bering, Jesse. 2006. The folk psychology of souls. *Behavioral and Brain Sciences* 20:1–46.

Bering, Jesse, and D. F. Bjorklund. 2004. The natural emergence of reasoning about the afterlife as a developmental regularity. *Developmental Psychology* 40:217–233.

Bless, H., G. L. Clore, N. Schwarz, V. Golisano, C. Rabe, and M. Wolk. 1996. Mood and the use of scripts: Does a happy mood really lead to mindlessness? *Journal of Personality and Social Psychology* 71 (4): 665–679.

Block, Ned. 1986. Advertisement for a semantics for psychology. *Midwest Studies in Philosophy* 10:615–678.

Block, Ned. 1998. Holism, mental and semantic. In *The Routledge Encyclopedia of Philosophy*, ed. E. Craig. London: Routledge.

Bloom, Paul. 2004. *Descartes' Baby: How the Science of Child Development Explains What Makes Us Human*. New York: Basic Books.

Bloom, Paul. 2007. Religion is natural. *Developmental Science* 10:147–151.

Bohr, Niels. 1949. Discussion with Einstein on epistemological problems in atomic physics. In *Albert Einstein: Philosopher-Scientist*, ed. P. Schilpp. Peru, IL: Open Court.

Boyer, Pascal. 2001. *Religion Explained*. New York: Basic Books.

Brandom, Robert B. 1998. *Making It Explicit: Reasoning, Representing, and Discursive Commitment*. Cambridge, MA: Harvard University Press.

Brandom, Robert B. 2000. *Articulating Reasons: An Introduction to Inferentialism*. Cambridge, MA: Harvard University Press.

Brentano, Franz. 1874. *Psychologie vom empirischen Standpunkt*. Berlin: Duncker & Humblot.

Byrne, Ruth, and Philip N. Johnson-Laird. 1989. Spatial Reasoning. *Journal of Memory and Language* 28 (5): 564–575.

Carey, Susan. 2011. *The Origin of Concepts*. New York: Oxford University Press.

Carruthers, Peter. 2006 a. *The Architecture of the Mind*. Oxford: Clarendon.

Carruthers, Peter. 2006 b. The case for massively modular models of mind. In *Contemporary Debates in Cognitive Science*, ed. R. J. Standton. Malden, MA: Wiley-Blackwell.

Cartwright, Nancy. 1999. *The Dappled World: A Study in the Boundaries of Science*. New York: Cambridge University Press.

Cat, Jordi, Nancy Cartwright, and Hasok Chang. 1996. Otto Neurath: Politics and the unity of science. In *The Disunity of Science: Boundaries, Contexts, and Power*, ed. P. Galison and D. Stump. Stanford: Stanford University Press.

Chaiken, S., and Y. Trope. 1999. *Dual-Process Theories in Social Psychology*. New York: Guilford Press.

Chalmers, David. 1996. *The Conscious Mind: In Search of a Fundamental Theory*. Oxford: Oxford University Press.

Chemero, Anthony. 2009. *Radical Embodied Cognitive Science*. Cambridge, MA: MIT Press.

Chomsky, Noam. 1965. *Aspects of the Theory of Syntax*. Cambridge, MA: MIT Press.

Chomsky, Noam. 1966. *Cartesian Linguistics*. New York: Harper & Row.

Churchland, Patricia S. 1986. *Neurophilosophy*. Cambridge, MA: MIT Press.

Churchland, Patricia S., and Terrence J. Sejnowski. 1989. Neural representation and neural computation. In *From Reading to Neurons*, ed. A. M. Galaburda. Cambridge, MA: MIT Press.

Churchland, Paul M. 1981. Eliminative materialism and the propositional attitudes. *Journal of Philosophy* 78 (2): 67–90.

Clark, Andy. 1997. *Being There: Putting Mind, Body, and World Together Again*. Cambridge, MA: MIT Press.

Clark, Andy. 2008. *Supersizing the Mind: Embodiment, Action, and Cognitive Extension*. New York: Oxford University Press.

Clark, Andy, and David J. Chalmers. 1998. The extended mind. *Analysis* 58:7–19.

Cohen, Paul R., and Edward A. Feigenbaum. 1982. *The Handbook of Artificial Intelligence*, vol. 3. Los Altos, CA: William Kaufmann.

Collins, Allan M., and Dedre Gentner. 1983. Multiple models of evaporation processes. Paper presented at the Fifth Annual Conference of the Cognitive Science Society, Rochester, New York.

Collins, Allan M., and M. Ross Quillian. 1969. Retrieval time from semantic memory. *Journal of Verbal Learning and Verbal Behavior* 8 (2): 240–247.

Craik, Kenneth. 1943. *The Nature of Exploration*. Cambridge: Cambridge University Press.

Craik, Kenneth. 1947. Theory of the human operator in control systems. I: The operator as an engineering system. *British Journal of Psychology: General Section* 38 (2): 56–61.

Craik, Kenneth. 1948. Theory of the human operator in control systems. II: Man as an element in a control system. *British Journal of Psychology: General Section* 38 (3): 142–148.

Crick, Francis, and Christof Koch. 1993. A framework for consciousness. *Nature Neuroscience* 6 (2): 119–126.

Cummins, Robert. 1996. *Representations, Targets, and Attitudes*. Cambridge, MA: MIT Press.

Da Costa, Newton. 2003. *Science and Partial Truth: A Unitary Approach to Models and Scientific Reasoning*. Oxford: Oxford University Press.

Damasio, Antonio. 2010. *The Self Comes to Mind: Constructing the Conscious Brain*. New York: Pantheon (Random House).

Danks, David. 2014. *Unifying the Mind: Cognitive Representations as Graphical Models*. Cambridge, MA: MIT Press.

Davidson, Donald. 1967. Truth and meaning. *Synthese* 17:304–323.

Dennett, Daniel C. 1971. Intentional systems. *Journal of Philosophy* 67 (4): 87–106.

Dennett, Daniel C. 1977. Critical notice. *Mind* 86 (342): 265–280.

Dennett, Daniel C. 1981/1997. True believers: The intentional strategy and why it works. In *Mind Design II*, ed. J. Haugeland. Cambridge, MA: MIT Press.

Dennett, Daniel C. 1987. *The Intentional Stance*. Cambridge, MA: MIT Press.

Dennett, Daniel C. 1991 a. *Consciousness Explained*. Boston: Little, Brown.

Dennett, Daniel C. 1991 b. Real patterns. *Journal of Philosophy* 88 (1): 27–51.

Descartes, René. 1988. *The Philosophical Writings of Descartes*. 3 vols. Trans. J. Cottingham, R. Stoothoff, D. Murdoch, and A. Kenny. Cambridge: Cambridge University Press.

DeWitt, B. 1970. Quantum mechanics and reality. *Physics Today* 23 (9): 155–165.

Dobzhansky, Theodosius. 1937. *Genetics and the Origin of Species*. New York: Columbia University Press.

Donald, Merlin. 1991. *Origins of the Modern Mind: Three Stages in the Evolution of Culture and Cognition*. Cambridge, MA: Harvard University Press.

Dretske, Fred. 1981. *Knowledge and the Flow of Information*. Cambridge, MA: MIT Press.

Dretske, Fred. 1986. Misrepresentation. In *Belief*, ed. R. Bogdan. Oxford: Oxford University Press.

Dretske, Fred. 1988. *Explaining Behavior*. Cambridge, MA: MIT Press.

Dreyfus, Hubert. 1979. *What Computers Can't Do: A Critique of Artificial Reason*. New York: Harper & Row.

Dreyfus, Hubert L., and Stuart E. Dreyfus. 1992. What is moral maturity? Towards a phenomenology of ethical expertise. In *Revisioning Philosophy*, ed. J. Ogilvy. Albany, NY: SUNY Press.

Dupré, John. 1993. *The Disorder of Things: Metaphysical Foundations of the Disunity of Science*. Cambridge, MA: Harvard University Press.

Dupré, John. 2001. *Human Nature and the Limits of Science*. Oxford University Press.

Elgin, Catherine. 1996. *Considered Judgment*. Princeton: Princeton University Press.

Engler, Steven. 2011. Grounded theory. In *The Routledge Handbook of the Research Methods in the Study of Religion*, ed. M. Stausberg and S. Engler. New York: Routledge.

Field, Hartry. 1977. Logic, meaning, and conceptual role. *Journal of Philosophy* 74:379–409.

Finucane, M. L., A. Alhakami, P. Slovic, and S. M. Johnson. 2000. The affect heuristic in judgments of risks and benefits. *Journal of Behavioral Decision Making* 13:1–17.

Fodor, Jerry. 1975. *The Language of Thought*. New York: Thomas Crowell.

Fodor, Jerry. 1978. Propositional attitudes. *Monist* 61 (4): 501–523.

Fodor, Jerry A. 1983. *Modularity of Mind: An Essay on Faculty Psychology*. Cambridge, MA: MIT Press.

Fodor, Jerry. 1987. *Psychosemantics*. Cambridge, MA: MIT Press.

Fodor, Jerry. 1990. *A Theory of Content and Other Essays*. Cambridge, MA: MIT Press.

Fodor, Jerry. 1998. *Concepts: Where Cognitive Science Went Wrong*. New York: Oxford University Press.

Fodor, Jerry A. 2001. *The Mind Doesn't Work That Way: The Scope and Limits of Computational Psychology*. Cambridge, MA: MIT Press.

Forbus, Kenneth D., and Dedre Gentner. 1997. Qualitative mental models: Simulations of memories? Paper presented at the Eleventh International Workshop on Qualitative Reasoning, Cortona, Italy.

Gallagher, Shaun. 2005. *How the Body Shapes the Mind*. Oxford: Oxford University Press.

Gallagher, Shaun, and Dan Zahavi. 2008. *The Phenomenological Mind: An Introduction to Philosophy of Mind and Cognitive Science*. New York: Routledge.

Gelman, Susan. 2004. Psychological essentialism in children. *Trends in Cognitive Sciences* 8 (9): 404–409.

Gelman, S., J. Coley, and G. Gottfried. 1994. Essentialist beliefs in children: The acquisition of concepts and theories. In *Mapping the Mind*, ed. L. Hirschfield and S. Gelman. Cambridge: Cambridge University Press.

Gentner, Dedre, and Donald R. Gentner. 1983. Flowing water or teeming crowds: Mental models of electricity. In *Mental Models*, ed. D. Gentner and A. L. Stevens. Hillsdale, NJ: Erlbaum.

Gentner, Dedre, and Albert L. Stevens, eds. 1983. *Mental Models*. Hillsdale, NJ: Erlbaum.

Gibson, James J. 1966. *The Senses Considered as Perceptual Systems*. Boston: Houghton Mifflin.

Gibson, James J. 1977. The theory of affordances. In *Perceiving, Acting, and Knowing: Toward an Ecological Psychology*, ed. R. Shaw and J. Bransford. Hillsdale, NJ: Erlbaum.

Gibson, James J. 1979. *The Ecological Approach to Visual Perception*. Boston: Houghton Mifflin.

Giere, Ronald. 1988. *Explaining Science: A Cognitive Approach*. Chicago: University of Chicago Press.

Giere, Ronald. 1992. *Cognitive Models of Science*. Minnesota Studies in the Philosophy of Science, vol. 15. Minneapolis: University of Minnesota Press.

Giere, Ronald. 2004. How models are used to represent reality. *Philosophy of Science* 71 (5): 742–752.

Gigerenzer, Gerd. 1991. How to make cognitive illusions disappear: Beyond "heuristics and biases." *European Review of Social Psychology* 2:83–115.

Gigerenzer, Gerd, Ulrich Hoffrage, and Heinz Kleinbölting. 1999. Probabilistic mental models: A Brunswikian theory of confidence. *Psychological Review* 4:506–528.

Gigerenzer, Gerd, P. M Todd, and the ABC Research Group. 1999. *Simple Heuristics That Make Us Smart*. New York: Oxford University Press.

Gilbert, D. 1999. What the mind's not. In *Dual-Process Theories in Social Psychology*, ed. S. Chaiken and Y. Trope. New York: Guilford Press.

Gilovich, Thomas, Dale Griffin, and Daniel Kahneman. 2002. *Heuristics and Biases: The Psychology of Intuitive Judgment*. New York: Cambridge University Press.

Goldman, Alvin. 1979. What is justified belief? In *Justification and Knowledge*, ed. G. Pappas. Dordrecht: D. Reidel.

Gopnik, Alison. 1996. The child as scientist. *Philosophy of Science* 63 (4): 485–514.

Gopnik, A., and A. N. Meltzoff. 1997. *Words, Thoughts, and Theories*. Cambridge, MA: MIT Press.

Gopnik, A., and H. Wellman. 1994. The theory theory. In *Mapping the Mind: Domain Specificity in Cognition and Culture*, ed. L. Hirschfield and S. Gelman. New York: Cambridge University Press.

Gould, Stephen Jay. 1997. Nonoverlapping magisteria. *Natural History* 106:16–22.

Grimm, Stephen. 2010. Understanding. In *The Routledge Companion to Epistemology*, ed. S. Bernecker and D. Pritchard. New York: Routledge.

Guthrie, Stewart. 2002. Animal animism: Evolutionary roots of religious cognition. In *Current Approaches in the Cognitive Science of Religion*, ed. I. Pyysiänen and V. Anttonen. London: Continuum.

Hacking, Ian. 1996. The disunities of the sciences. In *The Disunity of Science: Boundaries, Contexts, and Power*, ed. P. Galison and D. Stump. Stanford: Stanford University Press.

Hacking, Ian. 2007. Putnam's theory of natural kinds and their names is not the same as Kripke's. *Principia* 11 (1): 1–24.

Hawking, Stephen. 2002. *Gödel and the End of Physics*. http://www.hawking.org.uk/godel-and-the-end-of-physics.html (accessed September 20, 2014).

Hawking, Stephen, and Leonard Mlodinow. 2010. *The Grand Design*. New York: Bantam.

Heidegger, Martin. 1927/1996. *Being and Time*. Trans. J. Macquarrie and E. Robinson. New York: Harper & Row.

Hellman, D. H., ed. 1988. *Analogical Reasoning*. Dordrecht: Kluwer.

Hesse, Mary. 1963. *Models and Analogies in Science*. Notre Dame, IN: University of Notre Dame Press.

Hesse, Mary. 1974. *The Structure of Scientific Inference*. London: Macmillan.

Hintikka, Jakko. 1999. The emperor's new intuitions. *Journal of Philosophy* 96 (3): 127–147.

Hodge, K. Mitch. 2011. On imagining the afterlife. *Journal of Cognition and Culture* 11:367–389.

Holyoak, Keith, and Paul Thagard. 1995. *Mental Leaps: Analogy in Creative Thought*. Cambridge, MA: MIT Press.

Hooker, Clifford. 1972. The nature of quantum mechanical reality. In *Paradigms and Paradoxes*, ed. R. G. Colodny. Pittsburgh: University of Pittsburgh Press.

Horst, Steven. 1995. Eliminativism and the ambiguity of "belief." *Synthese* 104: 123–145.

Horst, Steven. 1996. *Symbols, Computation, and Intentionality: A Critique of the Computational Theory of Mind*. Berkeley: University of California Press.

Horst, Steven. 1999. Symbols and computation. *Minds and Machines* 9 (3): 347–381.

Horst, Steven. 2007. *Beyond Reduction: Philosophy of Mind and Post-reductionist Philosophy of Science*. New York: Oxford University Press.

Horst, Steven. 2011. *Laws, Mind, and Free Will*. Cambridge, MA: MIT Press.

Horst, Steven. 2014. Beyond reduction: From naturalism to cognitive pluralism. *Mind and Matter* 12 (2): 197–244.

Hume, David. 1738. *A Treatise of Human Nature*. London: John Noon.

Husserl, Edmund. 1900/1973. *Logical Investigations*. Trans. J. N. Findlay. London: Routledge.

Husserl, Edmund. 1913/1989. *Ideas Pertaining to a Pure Phenomenology and to a Phenomenological Philosophy—First Book: General Introduction to Pure Phenomenology*. Trans. F. Kersten. The Hague: Nijhoff.

Isen, A. M., T. E. Nygren, and F. G. Ashby. 1988. Influence of positive affect on the subjective utility of gains and losses—it is just not worth the risk. *Journal of Personality and Social Psychology* 55 (5): 710–717.

Jackson, Frank. 1982. Epiphenomenal qualia. *Philosophical Quarterly* 32:127–166.

James, William. 1890. *The Principles of Psychology*. Boston: Henry Holt.

Johnson-Laird, Philip N. 1983. *Mental Models*. Cambridge: Harvard University Press.

Jordan, Michael. 2004. Graphical models. *Statistical Science* 19 (1): 140–155.

Kahneman, Daniel, and Shane Frederick. 2002. Representativeness revisited: Attribute substitution in intuitive judgment. In *Heuristics of Intuitive Judgment: Extensions and Applications*, ed. T. Gilovich, D. Griffin, and D. Kahneman. New York: Cambridge University Press.

Kahneman, Daniel, and Amos Tversky. 1973. On the psychology of prediction. *Psychological Review* 80:237–251.

Kahneman, Daniel, and Amos Tversky. 1982. On the study of statistical intuitions. *Cognition* 11:123–141.

Kandel, Eric R., James H. Schwartz, and Eric R. Jessell. 2000. *Principles of Neural Science*, 4th ed. New York: McGraw-Hill.

Keil, Frank C. 1992. The origins of an autonomous biology. In *Modularity and Constraints in Language and Cognition: Minnesota Symposium on Child Psychology*, ed. M. R. Gunnar and M. Maratsos. Hillsdale, NJ: Erlbaum.

Kelemen, Deborah. 1999. Function, goals, and intention: Children's teleological reasoning about objects. *Trends in Cognitive Sciences* 3 (12): 461–468.

Kitcher, Philip. 1984. 1953 and all that: A tale of two sciences. *Philosophical Review* 93:335–373.

Kitcher, Philip. 1985. *Vaulting Ambition: Sociobiology and the Quest for Human Nature*. Cambridge, MA: MIT Press.

Kripke, Saul. 1980. *Naming and Necessity*. Cambridge, MA: Harvard University Press.

Kroes, Peter. 1989. Structural analogies between physical systems. *British Journal for the Philosophy of Science* 40:145–154.

Kuhn, Thomas S. 1962. *The Structure of Scientific Revolutions*. Chicago: University of Chicago Press.

Kuhn, Thomas S. 1983/2000. Rationality and theory choice. *Journal of Philosophy* 80 (10): 563–570.

Kuhn, Thomas S. 1989/2000. Possible worlds in the history of science. In *The Road since Structure*, ed. J. Conant and J. Haugeland. Chicago: University of Chicago Press.

Kuhn, Thomas S. 1993/2000. Afterwords. In *The Road since Structure*, ed. J. Conant and J. Haugeland. Chicago: University of Chicago Press.

Kvanvig, Jonathan. 2003. *The Value of Knowledge and the Pursuit of Understanding*. Cambridge: Cambridge University Press.

Lakoff, George, and Mark Johnson. 1999. *Philosophy in the Flesh: The Embodied Mind and Its Challenge to Western Thought*. New York: Basic Books.

Lakoff, George, and Mark Johnson. 2003. *Metaphors We Live By*. Chicago: University of Chicago Press.

Lepore, Ernest. 1999. Semantic holism. In *The Cambridge Dictionary of Philosophy*, ed. R. Audi. Cambridge: Cambridge University Press.

Leslie, Sarah-Jane. 2013. Essence and natural kinds: When science meets preschooler intuition. *Oxford Studies in Epistemology* 4:108–166.

Levine, Joseph. 1983. Materialism and qualia: The explanatory gap. *Pacific Philosophical Quarterly* 64:354–361.

Lewontin, Richard. 1983. Biological determinism. *Tanner Lectures on Human Values*, vol. 4, ed. Sterling M. McMurrin, 147–183. Cambridge: Cambridge University Press.

Locke, John. 1690/1995. *An Essay concerning Human Understanding*. New York: Prometheus Books.

Lycan, William. 1996. *Consciousness and Experience*. Cambridge, MA: MIT Press.

Machery, Edouard. 2009. *Doing Without Concepts*. New York: Oxford University Press.

Mackie, Penelope. 2006. *How Things Might Have Been: Individuals, Kinds, and Essential Properties*. New York: Oxford University Press.

Magnani, Lorenzo, and Nancy J. Nersessian, eds. 2002. *Model-Based Reasoning: Scientific Discovery, Technological Innovation, Values*. New York: Kluwer Academic/Plenum.

Magnani, Lorenzo, Nancy Nersessian, and Paul Thagard, eds. 1999. *Model-Based Reasoning in Scientific Discovery*. Dordrecht: Kluwer.

Marr, David. 1982. *Vision: A Computational Investigation into the Human Representation and Processing of Visual Information*. New York: Freeman.

Mayden, Richard L. 1997. A hierarchy of species concepts: the denouement in the saga of the species problem. In *Species: The Units of Biodiversity*, ed. M. F. Claridge, H. A. Dawah and M. R. Wilson. London: Chapman & Hall.

Mayr, Ernst. 1942. *Systematics and the Origin of Species, from the Viewpoint of a Zoologist*. Cambridge, MA: Harvard University Press.

McCarthy, John, and P. J. Hayes. 1969. Some philosophical problems from the standpoint of artificial intelligence. In *Machine Intelligence 4*, ed. D. Michie and B. Meltzer. Edinburgh: Edinburgh University Press.

McGinn, Colin. 1991. *The Problem of Consciousness: Essays towards a Resolution*. Oxford: Blackwell.

Medin, Douglas L., and A. Ortony. 1989. Psychological essentialism. In *Similarity and Analogical Reasoning*, ed. S. Vosniadou and A. Ortony. Cambridge: Cambridge University Press.

Mellor, D. H. 1977. Natural kinds. *British Journal for the Philosophy of Science* 28 (4): 299–312.

Merleau-Ponty, Maurice. 1945/1962. *Phenomenology of Perception*. Trans. C. Smith. London: Routledge & Kegan Paul.

Meyer, M., Sarah-Jane Leslie, S. Gelman, and S. M. Stilwell. 2013. Essentialist beliefs about bodily transplants in the United States and India. *Cognitive Science* 37 (1): 668–710.

Millikan, Ruth. 2000. *On Clear and Confused Ideas*. Cambridge: Cambridge University Press.

Minsky, Marvin. 1974. A framework for representing knowledge. http://web.media.mit.edu/%7Eminsky/papers/Frames/frames.html.

Minsky, Marvin. 1985. *The Society of Mind*. New York: Simon & Schuster.

Minsky, Marvin, and Simon Papert. 1972. Progress report on artificial intelligence. AI Memo. MIT, Cambridge, Massachusetts.

Moreau, D., A. Mansy-Dannay, J. Clerc, and A. Juerrién. 2011. Spatial ability and motor performance: Assessing mental rotation processes in elite and novice athletes. *International Journal of Sport Psychology* 42 (6): 525–547.

Morgan, Mary S., and Margaret Morrison, eds. 1999. *Models as Mediators*. Cambridge: Cambridge University Press.

Nagel, Thomas. 1974. What is it like to be a bat? *Philosophical Review* 4:435–450.

Nagel, Thomas. 1979. The limits of objectivity. In *Tanner Lectures on Human Values*, vol. 1, ed. Sterling M. McMurrin, 75–139. Salt Lake City: University of Utah Press; Cambridge: Cambridge University Press.

Nersessian, Nancy J. 1999. Model-based reasoning in conceptual change. In *Model-Based Reasoning in Scientific Discovery*, ed. L. Magnani, N. J. Nersessian, and P. Thagard. New York: Kluwer Academic/Plenum.

Newell, Allen. 1990. *Unified Theories of Cognition*. Cambridge, MA: Harvard University Press.

Newell, Allen, and Herbert Simon. 1972. *Human Problem Solving*. Englewood Cliffs, NJ: Prentice Hall.

Newell, Allen, John C. Shaw, and Herbert Simon. 1959. Report on a general problem-solving program. Paper presented at International Conference on Information Processing.

Newell, Allen, and Herbert A. Simon. 1956. The logic theory machine: A complex information processing system. *IRE Transactions on Information Theory* IT-2 (3): 61–79.

Nisbett, R. E., D. H. Krantz, C. Jepson, and Z. Kunda. 1983. The use of statistical heuristics in everyday inductive reasoning. *Psychological Review* 90 (4): 339–363.

Nisbett, Richard, and Lee Ross. 1980. *Human Inference: Strategies and Shortcomings of Social Judgment.* Englewood Cliffs, NJ: Prentice Hall.

Noë, Alva. 2004. *Action in Perception.* Cambridge, MA: MIT Press.

Norman, Donald. 1988. *The Design of Everyday Things.* New York: Basic Books.

O'Hear, A. 1997. *Beyond Evolution.* Oxford: Oxford University Press.

Oppenheim, Paul, and Hilary Putnam. 1958. Unity of science as a working hypothesis. In *Concepts, Theories, and the Mind-Body Problem,* ed. H. Feigl, M. Scriven and G. Maxwell. Minneapolis: University of Minnesota Press.

Papineau, David. 2002. *Thinking about Consciousness.* Oxford: Oxford University Press.

Pearl, Judea. 2000. *Causality: Models, Reasoning, and Inference.* New York: Cambridge University Press.

Pietsch, S., and P. Jansen. 2012. Different mental rotation performance in students of music, sport, and education. *Learning and Individual Differences* 22 (1): 159–163.

Pinker, Steven. 1997. *How the Mind Works.* New York: W. W. Norton.

Plantinga, Alvin. 1993. *Warrant and Proper Function.* New York: Oxford University Press.

Poston, Ted. 2015. Foundationalism. In *Internet Encyclopedia of Philosophy,* http://www.iep.utm.edu/found-ep/#SSH4aii.

Prinz, Jesse J. 2006. Is the mind really modular? In *Contemporary Debates in Cognitive Science,* ed. R. J. Standton. Malden, MA: Blackwell.

Psillos, Stathis. 1995. The cognitive interplay between theories and models: The case of 19th century optics. In *Theories and Models in Scientific Process,* ed. W. Herfel, W. Krajewski, I. Niiniluoto, and R. Wojcicki. Amsterdam: Rodopi.

Putnam, Hilary. 1975. The meaning of "meaning." *Minnesota Studies in the Philosophy of Science* 7:131–193.

Pylyshyn, Zenon W. 1991. The role of cognitive architecture in theories of cognition. In *Architectures for Intelligence,* ed. K. VanLehn. Hillsdale, NJ: Erlbaum.

Quine, W. O. 1951. Two dogmas of empiricism. *Philosophical Review* 60:20–43.

Ramsey, William, Stephen Stich, and David Rumelhart. 1991. *Philosophy and Connectionist Theory*. Hillsdale, NJ: Erlbaum.

Redhead, Michael. 1980. Models in physics. *British Journal for the Philosophy of Science* 31:145–163.

Rosch, Eleanor. 1978/1999. Principles of categorization. In *Concepts: Core Readings*, ed. E. Margolis and S. Laurence. Cambridge, MA: MIT Press. Original edition, *Cognition and Categorization*, ed. E. Rosch and B. Lloyd (Erlbaum, 1978).

Rosen, Gideon. 1994. Objectivity and modern idealism: What is the question? In *Philosophy in Mind*, ed. M. Michael and J. O'Leary-Hawthorne. Dodrecht: Kluwer.

Rosenthal, David. 1986. Two concepts of consciousness. *Philosophical Studies* 49:329–359.

Rosenthal, David. 1993. Thinking that one thinks. In *Consciousness: Psychological and Philosophical Essays*, ed. M. Davies and G. Humphreys. Oxford: Blackwell.

Rosenthal, David. 2005. *Consciousness and Mind*. Oxford: Oxford University Press.

Rumelhart, David, and James McClelland. 1986. *Parallel Distributed Processing*, vol. 1. Cambridge, MA: MIT Press.

Russell, Bertrand. 1912/1997. *The Problems of Philosophy*. New York: Oxford University Press.

Sachs, Joe. Aristotle: Metaphysics. In *Internet Encyclopedia of Philosophy*, ed. J. Fieser and B. Dowden. http://www.iep.utm.edu/aris-met/.

Salmon, Nathan. 1979. How not to derive essentialism from the theory of reference. *Journal of Philosophy* 76 (12): 703–725.

Sayre, Kenneth. 1997. *Belief and Knowledge: Mapping the Cognitive Landscape*. New York: Rowman & Littlefield.

Schank, Roger C., and R. Abelson. 1977. *Scripts, Plans, Goals, and Understanding*. Hillsdale, NJ: Erlbaum.

Schlick, Moritz. 1934/1979. The foundations of knowledge. In *Philosophical Papers*, ed. H. L. Mulder and B. F. B. van de Velde-Schlick. Dordrecht: D. Reidel.

Selfridge, Oliver G. 1959. Pandemonium: A paradigm for learning. Paper presented at the Symposium on Mechanisation of Thought Processes, London.

Sellars, Wilfrid. 1956. Empiricism and the philosophy of mind. In *Minnesota Studies in the Philosophy of Science*, ed. I. H. Feigl and M. Scriven. Minneapolis: University of Minnesota Press.

Shapiro, Lawrence A. 2004. *The Mind Incarnate*. Cambridge, MA: MIT Press.

Shepard, R. N., and L. A. Cooper 1982. *Mental Images and Their Transformations*. Cambridge, MA: MIT Press.

Shepard, R. N., and J. Meltzer. 1971. Mental rotation of three-dimensional objects. *Science* 171:701–703.

Silberstein, Michael. 2002. Reduction, emergence, and explanation. In *The Blackwell Guide to the Philosophy of Science*, ed. P. Machamer and M. Silberstein. Malden, MA: Blackwell.

Silberstein, Michael, and Anthony Chemero. 2011. Complexity and extended phenomenological-cognitive systems. *Topics in Cognitive Science* 2011:1–16.

Simon, Herbert. 1977. *Models of Discovery, and Other Topics in the Methods of Science*. Dordrecht: Reidel.

Sloman, Aaron. 1971. Interactions between philosophy and artificial intelligence: The role of intuition and non-logical reasoning in intelligence. *Artificial Intelligence* 2:209–225.

Smith, Edward E., and Douglas L. Medin. 1981. *Categories and Concepts*. Cambridge, MA: Harvard University Press.

Spelke, Elizabeth S. 2000. Core knowledge. *American Psychologist* 55 (11): 1233–1243.

Spelke, Elizabeth S., and Katherine D. Kinzler. 2007. Core knowledge. *Developmental Science* 10 (1): 89–96.

Springer, Elise. 2013. *Communicating Moral Concern: An Ethics of Critical Responsiveness*. Cambridge, MA: MIT Press.

Stanovich, K. E., and R. West. 2002. Individual differences in reasoning: Implications for the rationality debate. In *Heuristics and Biases: The Psychology of Judgment*, ed. T. Gilovich, D. Griffin, and D. Kahneman. New York: Cambridge University Press.

Stich, Stephen. 1983. *From Folk Psychology to Cognitive Science*. Cambridge, MA: MIT Press.

Strevens, Michael. 2000. The essentialist aspect of naive theories. *Cognition* 74:149–175.

Suppe, Frederick. 1960. A comparison of the meaning and uses of models in mathematics and the empirical sciences. *Synthese* 12:287–301.

Suppe, Frederick. 1989. *The Semantic Conception of Theories and Scientific Realism*. Urbana: University of Illinois Press.

Suppes, Patrick. 1974. The structure of theories and the analysis of data. In *The Structure of Scientific Theories*, ed. F. Suppe. Urbana: University of Illinois Press.

Suppes, Patrick. 2002. *Representation and the Invariance of Scientific Structures*. Stanford, CA: CSLI Publications.

Thompson, Evan. 2007. *Mind and Life*. Cambridge, MA: Harvard University Press.

Thompson, Evan, and Mog Stapleton. 2008. Making sense of sense-making: Reflections on enactive and extended mind theories. *Topoi* 28 (1): 23–30.

Tooby, John, and Leda Cosmides. 2005. Conceptual foundations of evolutionary psychology. In *The Handbook of Evolutionary Psychology*, ed. D. M. Buss. Hoboken, NJ: Wiley.

Toulmin, Stephen. 1974. The structure of scientific theories. In *The Structure of Scientific Theories*, ed. F. Suppe. Urbana: University of Illinois Press.

van Fraassen, Bas. 1980. *The Scientific Image*. Oxford: Oxford University Press.

van Fraassen, Bas. 1989. *Laws and Symmetry*. Oxford: Oxford University Press.

Varela, Francesco, Evan Thompson, and Eleanor Rosch. 1991. *The Embodied Mind: Cognitive Science and Human Experience*. Cambridge, MA: MIT Press.

Waskan, Jonathan A. 2003. Intrinsic cognitive models. *Cognitive Science* 27:259–283.

Waskan, Jonathan A. 2006. *Models and Cognition*. Cambridge, MA: MIT Press.

Weiskopf, Daniel A. 2009. Atomism, pluralism, and conceptual content. *Philosophy and Phenomenological Research* 79 (1): 131–163.

Wilson, Mark. 2006. *Wandering Significance*. Oxford: Oxford University Press.

Wittgenstein, Ludwig. 1922. *Tractatus Logico-Philosophicus*. Trans. C. K. Ogden. London: Kegan Paul, Trench, Trubner.

Zahavi, Dan. 2005. *Subjectivity and Selfhood: Investigating the First-Person Perspective*. Cambridge, MA: MIT Press.

Index

9 780262 552615

Printed by Libri Plureos GmbH in Hamburg,
Germany